THE THIRTEEN BOOKS

OF

EUCLID'S ELEMENTS

THE THIRTEEN BOOKS OF EUCLID'S ELEMENTS

TRANSLATED FROM THE TEXT OF HEIBERG

WITH INTRODUCTION AND COMMENTARY

BY

Sir THOMAS L. HEATH,

K.C.B., K.C.V.O., F.R.S.,

SC.D. CAMB., HON. D.SC. OXFORD

HONORARY FELLOW (SOMETIME FELLOW) OF TRINITY COLLEGE CAMBRIDGE

SECOND EDITION

REVISED WITH ADDITIONS

VOLUME II

BOOKS III—IX

CAMBRIDGE

AT THE UNIVERSITY PRESS

1926

CAMBRIDGE
UNIVERSITY PRESS

University Printing House, Cambridge CB2 8BS, United Kingdom

Cambridge University Press is part of the University of Cambridge.

It furthers the University's mission by disseminating knowledge in the pursuit of
education, learning and research at the highest international levels of excellence.

www.cambridge.org
Information on this title: www.cambridge.org/9781107480469

© Cambridge University Press 1926

First published 1926
First paperback edition 2014

A catalogue record for this publication is available from the British Library

ISBN 978-1-107-48046-9 Paperback

CONTENTS OF VOLUME II.

BOOK III.

DEFINITIONS.

1. **Equal circles** are those the diameters of which are equal, or the radii of which are equal.

2. A straight line is said to **touch a circle** which, meeting the circle and being produced, does not cut the circle.

3. **Circles** are said to **touch one another** which, meeting one another, do not cut one another.

4. In a circle straight lines are said **to be equally distant from the centre** when the perpendiculars drawn to them from the centre are equal.

5. And that straight line is said to be **at a greater distance** on which the greater perpendicular falls.

6. A **segment of a circle** is the figure contained by a straight line and a circumference of a circle.

7. An **angle of a segment** is that contained by a straight line and a circumference of a circle.

8. An **angle in a segment** is the angle which, when a point is taken on the circumference of the segment and straight lines are joined from it to the extremities of the straight line which is the **base of the segment,** is contained by the straight lines so joined.

9. And, when the straight lines containing the angle cut off a circumference, the angle is said to **stand upon** that circumference.

10. **A sector of a circle** is the figure which, when an angle is constructed at the centre of the circle, is contained by the straight lines containing the angle and the circumference cut off by them.

11. **Similar segments of circles** are those which admit equal angles, or in which the angles are equal to one another.

DEFINITION 1.

Ἴσοι κύκλοι εἰσίν, ὧν αἱ διάμετροι ἴσαι εἰσίν, ἢ ὧν αἱ ἐκ τῶν κέντρων ἴσαι εἰσίν.

Many editors have held that this should not have been included among definitions. Some, e.g. Tartaglia, would call it a *postulate*; others, e.g. Borelli and Playfair, would call it an *axiom*; others again, as Billingsley and Clavius, while admitting it as a *definition*, add explanations based on the mode of constructing a circle; Simson and Pfleiderer hold that it is a *theorem*. I think however that Euclid would have maintained that it is a definition in the proper sense of the term; and certainly it satisfies Aristotle's requirement that a "definitional statement" (ὁριστικὸς λόγος) should not only state the *fact* (τὸ ὅτι) but should indicate the *cause* as well (*De anima* II. 2, 413 a 13). The equality of circles with equal radii can of course be proved by superposition, but, as we have seen, Euclid avoided this method wherever he could, and there is nothing technically wrong in saying "By *equal circles* I mean circles with equal radii." No flaw is thereby introduced into the system of the *Elements*; for the definition could only be objected to if it could be proved that the equality predicated of the two circles in the definition was not the same thing as the equality predicated of other equal figures in the *Elements* on the basis of the Congruence-Axiom, and, needless to say, this cannot be proved because it is not true. The *existence* of equal circles (in the sense of the definition) follows from the existence of equal straight lines and I. Post. 3.

The Greeks had no distinct word for *radius*, which is with them, as here, *the (straight line drawn) from the centre* ἡ ἐκ τοῦ κέντρου (εὐθεῖα); and so definitely was the expression appropriated to the radius that ἐκ τοῦ κέντρου was used without the article as a predicate, just as if it were one word. Thus, e.g., in III. 1 ἐκ κέντρου γάρ means "for they are radii": cf. Archimedes, *On the Sphere and Cylinder* II. 2, ἡ BE ἐκ τοῦ κέντρου ἐστὶ τοῦ...κύκλου, *BE is a radius of the circle*.

DEFINITION 2.

Εὐθεῖα κύκλου ἐφάπτεσθαι λέγεται, ἥτις ἁπτομένη τοῦ κύκλου καὶ ἐκβαλλομένη οὐ τέμνει τὸν κύκλον.

Euclid's phraseology here shows the regular distinction between ἅπτεσθαι and its compound ἐφάπτεσθαι, the former meaning "to *meet*" and the latter "to *touch*." The distinction was generally observed by Greek geometers from Euclid onwards. There are however exceptions so far as ἅπτεσθαι is concerned; thus it means "to *touch*" in Eucl. IV. Def. 5 and sometimes in Archimedes. On the other hand, ἐφάπτεσθαι is used by Aristotle in certain cases where the orthodox geometrical term would be ἅπτεσθαι. Thus in *Meteorologica* III. 5 (376 b 9) he says a certain circle *will pass through all the angles* (ἁπασῶν ἐφάψεται τῶν γωνιῶν), and (376 a 6) M *will lie on a given* (*circular*) *circumference* (δεδομένης περιφερείας ἐφάψεται τὸ M). We shall find ἅπτεσθαι used in these senses in Book IV. Deff. 2, 6 and Deff. 1, 3 respectively. The latter of the two expressions quoted from Aristotle means that *the locus of M is a given circle*, just as in Pappus ἅψεται τὸ σημεῖον θέσει δεδομένης εὐθείας means that *the locus of* the point is a straight line given in position.

DEFINITION 3.

Κύκλοι ἐφάπτεσθαι ἀλλήλων λέγονται οἵτινες ἁπτόμενοι ἀλλήλων οὐ τέμνουσιν ἀλλήλους.

Todhunter remarks that different opinions have been held as to what is, or should be, included in this definition, one opinion being that it only means that the circles do not cut in the neighbourhood of the point of contact, and that it must be shown that they do not cut elsewhere, while another opinion is that the definition means that the circles do not cut at all. Todhunter thinks the latter opinion correct. I do not think this is proved ; and I prefer to read the definition as meaning simply that the circles meet at a point but do not cut *at that point*. I think this interpretation preferable for the reason that, although Euclid does practically assume in III. 11—13, without stating, the theorem that circles touching at one point do not intersect anywhere else, he has given us, before reaching that point in the Book, means for proving for ourselves the truth of that statement. In particular, he has given us the propositions III. 7, 8 which, taken as a whole, give us more information as to the general nature of a circle than any other propositions that have preceded, and which can be used, as will be seen in the sequel, to solve any doubts arising out of Euclid's unproved assumptions. Now, as a matter of fact, the propositions are not used in any of the genuine proofs of the theorems in Book III. ; III. 8 is required for the second proof of III. 9 which Simson selected in preference to the first proof, but the first proof only is regarded by Heiberg as genuine. Hence it would not be easy to account for the appearance of III. 7, 8 at all unless as affording means of answering possible *objections* (cf. Proclus' explanation of Euclid's reason for inserting the second part of I. 5).

External and *internal* contact are not distinguished in Euclid until III. 11, 12, though the *figure* of III. 6 (not the *enunciation* in the original text) represents the case of internal contact only. But the definition of touching circles here given must be taken to imply so much about *internal* and *external* contact respectively as that (*a*) a circle touching another internally must, immediately before "meeting" it, have passed through points *within* the circle that it touches, and (*b*) a circle touching another externally must, immediately before meeting it, have passed through points *outside* the circle which it touches. These facts must indeed be admitted if *internal* and *external* are to have any meaning at all in this connexion, and they constitute a minimum admission necessary to the proof of III. 6.

DEFINITION 4.

Ἐν κύκλῳ ἴσον ἀπέχειν ἀπὸ τοῦ κέντρου εὐθεῖαι λέγονται, ὅταν αἱ ἀπὸ τοῦ κέντρου ἐπ' αὐτὰς κάθετοι ἀγόμεναι ἴσαι ὦσιν.

DEFINITION 5.

Μεῖζον δὲ ἀπέχειν λέγεται, ἐφ' ἣν ἡ μείζων κάθετος πίπτει.

DEFINITION 6.

Τμῆμα κύκλου ἐστὶ τὸ περιεχόμενον σχῆμα ὑπό τε εὐθείας καὶ κύκλου περιφερείας.

DEFINITION 7.

Τμήματος δὲ γωνία ἐστὶν ἡ περιεχομένη ὑπό τε εὐθείας καὶ κύκλου περιφερείας.

This definition is only interesting historically. The *angle of a segment*, being the "angle" formed by a straight line and a "circumference," is of the kind described by Proclus as "mixed." A particular "angle" of this sort is the "*angle of a semicircle*," which we meet with again in III. 16, along with the so-called "horn-like angle" (κερατοειδής), the supposed "angle" between a tangent to a circle and the circle itself. The "angle of a semicircle" occurs once in Pappus (VII. p. 670, 19), but it there means scarcely more than the *corner* of a semicircle regarded as a point to which a straight line is directed. Heron does not give the definition of the *angle of a segment*, and we may conclude that the mention of it and of the *angle of a semicircle* in Euclid is a survival from earlier text-books rather than an indication that Euclid considered either to be of importance in elementary geometry (cf. the note on III. 16 below).

We have however, in the note on I. 5 above (Vol. I. pp. 252—3), seen evidence that the *angle of a segment* had played some part in geometrical proofs up to Euclid's time. It would appear from the passage of Aristotle there quoted (*Anal. prior.* I. 24, 41 b 13 sqq.) that the theorem of I. 5 was, in the text-books immediately preceding Euclid, proved by means of the equality of the two "*angles of*" any one segment. This latter property must therefore have been regarded as more elementary (for whatever reason) than the theorem of I. 5; indeed the definition as given by Euclid practically implies the same thing, since it speaks of only *one* "angle of a segment," namely "*the* angle contained by a straight line and a circumference of a circle." Euclid abandoned the actual use of the "angle" in question, but no doubt thought it unnecessary to break with tradition so far as to strike the definition out also.

DEFINITION 8.

Ἐν τμήματι δὲ γωνία ἐστίν, ὅταν ἐπὶ τῆς περιφερείας τοῦ τμήματος ληφθῇ τι σημεῖον καὶ ἀπ' αὐτοῦ ἐπὶ τὰ πέρατα τῆς εὐθείας, ἥ ἐστι βάσις τοῦ τμήματος, ἐπιζευχθῶσιν εὐθεῖαι, ἡ περιεχομένη γωνία ὑπὸ τῶν ἐπιζευχθεισῶν εὐθειῶν.

DEFINITION 9.

Ὅταν δὲ αἱ περιέχουσαι τὴν γωνίαν εὐθεῖαι ἀπολαμβάνωσί τινα περιφέρειαν, ἐπ' ἐκείνης λέγεται βεβηκέναι ἡ γωνία.

DEFINITION 10.

Τομεὺς δὲ κύκλου ἐστίν, ὅταν πρὸς τῷ κέντρῳ τοῦ κύκλου συσταθῇ γωνία, τὸ περιεχόμενον σχῆμα ὑπό τε τῶν τὴν γωνίαν περιεχουσῶν εὐθειῶν καὶ τῆς ἀπολαμβανομένης ὑπ' αὐτῶν περιφερείας.

A scholiast says that it was the *shoemaker's knife*, σκυτοτομικὸς τομεύς, which suggested the name τομεύς for a sector of a circle. The derivation of the name from a resemblance of shape is parallel to the use of ἄρβηλος (also a *shoemaker's knife*) to denote the well known figure of the Book of Lemmas partly attributed to Archimedes.

A wider definition of a sector than that given by Euclid is found in a Greek scholiast (Heiberg's Euclid, Vol. v. p. 260) and in an-Nairīzī (ed. Curtze, p. 112). "There are two varieties of sectors; the one kind have the angular vertices at the centres, the other at the circumferences. Those others which have their vertices neither at the circumferences nor at the centres, but at some other points, are for that reason not called sectors but sector-like figures (τομοειδῆ σχήματα)." The exact agreement between the scholiast and an-Nairīzī suggests that Heron was the authority for this explanation.

The *sector-like figure* bounded by an arc of a circle and two lines drawn from its extremities to meet at any point actually appears in Euclid's book *On divisions* (περὶ διαιρέσεων) discovered in an Arabic MS. and edited by Woepcke (cf. Vol. I. pp. 8.—10 above). This treatise, alluded to by Proclus, had for its object the division of figures such as triangles, trapezia, quadrilaterals and circles, by means of straight lines, into parts equal or in given ratios. One proposition e.g. is, *To divide a triangle into two equal parts by a straight line passing through a given point on one side*. The proposition (28) in which the *quasi-sector* occurs is, *To divide such a figure by a straight line into two equal parts*. The solution in this case is given by Cantor (*Gesch. d. Math.* I₃, pp. 287—8).

If *ABCD* be the given figure, *E* the middle point of *BD* and *EC* at right angles to *BD*, the broken line *AEC* clearly divides the figure into two equal parts.

Join *AC*, and draw *EF* parallel to it meeting *AB* in *F*.

Join *CF*, when it is seen that *CF* divides the figure into two equal parts.

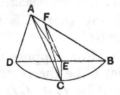

DEFINITION 11.

Ὅμοια τμήματα κύκλων ἐστὶ τὰ δεχόμενα γωνίας ἴσας, ἢ ἐν οἷς αἱ γωνίαι ἴσαι ἀλλήλαις εἰσίν.

De Morgan remarks that the use of the word *similar* in "similar segments" is an anticipation, and that similarity *of form* is meant. He adds that the definition is a theorem, or would be if "similar" had taken its final meaning.

BOOK III. PROPOSITIONS.

PROPOSITION 1.

To find the centre of a given circle.

Let ABC be the given circle;

thus it is required to find the centre of the circle ABC.

Let a straight line AB be drawn
5 through it at random, and let it be bisected
at the point D;

from D let DC be drawn at right angles
to AB and let it be drawn through to E;
let CE be bisected at F;

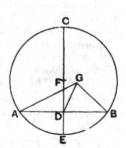

10 I say that F is the centre of the circle
ABC.

For suppose it is not, but, if possible,
let G be the centre,

and let GA, GD, GB be joined.

15   Then, since AD is equal to DB,

and DG is common,

the two sides AD, DG are equal to the two sides
BD, DG respectively;

and the base GA is equal to the base GB, for they are
20 radii;

therefore the angle ADG is equal to the angle GDB. [I. 8]

But, when a straight line set up on a straight line makes
the adjacent angles equal to one another, each of the equal
angles is right;   [I. Def. 10]

25   therefore the angle GDB is right.

But the angle FDB is also right;
therefore the angle FDB is equal to the angle GDB, the greater to the less : which is impossible.

Therefore G is not the centre of the circle ABC.

30   Similarly we can prove that neither is any other point except F.

Therefore the point F is the centre of the circle ABC.

PORISM. From this it is manifest that, if in a circle a straight line cut a straight line into two equal parts and at
35 right angles, the centre of the circle is on the cutting straight line.

Q. E. F.

12. **For suppose it is not.** This is expressed in the Greek by the two words Μὴ γάρ, but such an elliptical phrase is impossible in English.

17. **the two sides AD, DG are equal to the two sides BD, DG respectively.** As before observed, Euclid is not always careful to put the equals in corresponding order. The text here has "*GD, DB.*"

Todhunter observes that, when, in the construction, DC is said to be *produced* to E, it is assumed that D is within the circle, a fact which Euclid first demonstrates in III. 2. This is no doubt true, although the word διήχθω, "let it be *drawn through*," is used instead of ἐκβεβλήσθω, "let it be *produced*." And, although it is not necessary to assume that D is within the circle, it is necessary for the success of the construction that the straight line drawn through D at right angles to AB shall meet the circle in two points (and no more): an assumption which we are not entitled to make on the basis of what has gone before only.

Hence there is much to be said for the alternative procedure recommended by De Morgan as preferable to that of Euclid. De Morgan would first prove the fundamental theorem that "the line which bisects a chord perpendicularly must contain the centre," and then make III. 1, III. 25 and IV. 5 immediate corollaries of it. The fundamental theorem is a direct consequence of the theorem that, if P is *any* point equidistant from A and B, then P lies on the straight line bisecting AB perpendicularly. We then take any two chords AB, AC of the given circle and draw DO, EO bisecting them perpendicularly. Unless BA, AC are in one straight line, the straight lines DO, EO must meet in some point O (see note on IV. 5 for possible methods of proving this). And, since both DO, EO must contain the centre, O must be the centre.

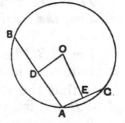

This method, which seems now to be generally preferred to Euclid's, has the advantage of showing that, in order to find the centre of a circle, it is sufficient to know three points on the circumference. If therefore two circles have three points in common, they must have the same centre and radius, so that two circles cannot have three points in common without coinciding entirely. Also, as indicated by De Morgan, the same construction enables us (1) to draw the complete circle of which a segment or arc only is given (III. 25), and (2) to circumscribe a circle to any triangle (IV. 5).

But, if the Greeks had used this construction for finding the centre of a circle, they would have considered it necessary to add a proof that no other point than that obtained by the construction can be the centre, as is clear both from the similar *reductio ad absurdum* in III 1 and also from the fact that Euclid thinks it necessary to prove as a separate theorem (III. 9) that, if a point within a circle be such that three straight lines (at least) drawn from it to the circumference are equal, that point must be the centre. In fact, however, the proof amounts to no more than the remark that the two perpendicular bisectors can have no more than one point common.

And even in De Morgan's method there is a yet unproved assumption. In order that DO, EO may meet, it is necessary that AB, AC should not be in one straight line or, in other words, that BC should not pass through A. This results from III. 2, which therefore, strictly speaking, should precede.

To return to Euclid's own proposition III. 1, it will be observed that the demonstration only shows that the centre of the circle cannot lie on either side of CD, so that it must lie on CD or CD produced. It is however taken for granted rather than proved that the centre must be the middle point of CE. The proof of this by *reductio ad absurdum* is however so obvious as to be scarcely worth giving. The same consideration which would prove it may be used to show that *a circle cannot have more than one centre*, a proposition which, if thought necessary, may be added to III. 1 as a corollary.

Simson observed that the proof of III. 1 could not but be by *reductio ad absurdum*. At the beginning of Book III. we have nothing more to base the proof upon than the definition of a circle, and this cannot be made use of unless we assume some point to be the centre. We cannot however assume that the point found by the construction is the centre, because that is the thing to be proved. Nothing is therefore left to us but to assume that some other point is the centre and then to prove that, whatever other point is taken, an absurdity results; whence we can infer that the point found is the centre.

The Porism to III. 1 is inserted, as usual, parenthetically before the words ὅπερ ἔδει ποιῆσαι, which of course refer to the problem itself.

PROPOSITION 2.

If on the circumference of a circle two points be taken at random, the straight line joining the points will fall within the circle.

Let ABC be a circle, and let two points A, B be taken at random on its circumference;

I say that the straight line joined from A to B will fall within the circle.

For suppose it does not, but, if possible, let it fall outside, as AEB; let the centre of the circle ABC be taken [III. 1], and let it be D; let DA, DB be joined, and let DFE be drawn through.

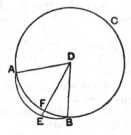

Then, since DA is equal to DB,

the angle DAE is also equal to the angle DBE.　[I. 5]

And, since one side AEB of the triangle DAE is produced,

the angle DEB is greater than the angle DAE.　[I. 16]

But the angle DAE is equal to the angle DBE;

therefore the angle DEB is greater than the angle DBE.

And the greater angle is subtended by the greater side; [I. 19]

therefore DB is greater than DE.

But DB is equal to DF;

therefore DF is greater than DE,

the less than the greater: which is impossible.

Therefore the straight line joined from A to B will not fall outside the circle.

Similarly we can prove that neither will it fall on the circumference itself;

therefore it will fall within.

Therefore etc.

Q. E. D.

The *reductio ad absurdum* form of proof is not really necessary in this case, and it has the additional disadvantage that it requires the destruction of two hypotheses, namely that the chord is (1) outside, (2) on the circle. To prove the proposition directly, we have only to show that, if E be any point on the straight line AB between A and B, DE is less than the radius of the circle. This may be done by the method shown above, under I. 24, for proving what is assumed in that proposition, namely that, in the figure of the proposition, F falls below EG if DE is not greater than DF. The assumption amounts to the following proposition, which De Morgan would make to precede I. 24: "Every straight line drawn from the vertex of a triangle to the base is less than the greater of the two sides, or than either if they be equal." The case here is that in which the two sides are equal; and, since the angle DAB is equal to the angle DBA, while the exterior angle DEA is greater than the interior and opposite angle DBA, it follows that the angle DEA is greater than the angle DAE, whence DE must be less than DA or DB.

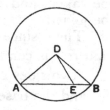

Camerer points out that we may add to this proposition the further statement that all points on AB *produced* in either direction are outside the circle. This follows from the proposition (also proved by means of the theorems that the exterior angle of a triangle is greater than either of the interior and opposite angles and that the greater angle is subtended by the greater side) which De Morgan proposes to introduce after I. 21, namely,

"The perpendicular is the shortest straight line that can be drawn from a given point to a given straight line, and of others that which is nearer to the perpendicular is less than the more remote, and the converse; also not more than two equal straight lines can be drawn from the point to the line, one on each side of the perpendicular."

The fact that not more than two equal straight lines can be drawn from a given point to a given straight line not passing through it is proved by Proclus on I. 16 (see the note to that proposition) and can alternatively be proved by means of I. 7, as shown above in the note on I. 12. It follows that

A straight line cannot cut a circle in more than two points.

a proposition which De Morgan would introduce here after III. 2. The proof given does not apply to a straight line *passing through the centre*; but that such a line only cuts the circle in two points is self-evident.

Proposition 3.

If in a circle a straight line through the centre bisect a straight line not through the centre, it also cuts it at right angles; and if it cut it at right angles, it also bisects it.

Let ABC be a circle, and in it let a straight line CD
5 through the centre bisect a straight line
AB not through the centre at the point
F;

I say that it also cuts it at right angles.

For let the centre of the circle ABC
10 be taken, and let it be E; let EA, EB
be joined.

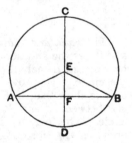

Then, since AF is equal to FB,
and FE is common,

two sides are equal to two sides;

15   and the base EA is equal to the base EB;

therefore the angle AFE is equal to the angle BFE. [I. 8]

But, when a straight line set up on a straight line makes the adjacent angles equal to one another, each of the equal angles is right;   [I. Def. 10]

20   therefore each of the angles AFE, BFE is right.

Therefore CD, which is through the centre, and bisects AB which is not through the centre, also cuts it at right angles.

Again, let CD cut AB at right angles;
25 I say that it also bisects it. that is, that AF is equal to FB.

For, with the same construction,
　　　　　　　since *EA* is equal to *EB*,
the angle *EAF* is also equal to the angle *EBF*.　　　　　[I. 5]
　　But the right angle *AFE* is equal to the right angle *BFE*,
30 therefore *EAF*, *EBF* are two triangles having two angles
equal to two angles and one side equal to one side, namely
EF, which is common to them, and subtends one of the equal
angles;
　　therefore they will also have the remaining sides equal to
35 the remaining sides;　　　　　　　　　　　　　[I. 26]
　　　　　　　therefore *AF* is equal to *FB*.
　　Therefore etc.

　　　　　　　　　　　　　　　　　　　Q. E. D.

26. **with the same construction,** τῶν αὐτῶν κατασκευασθέντων.

　　This proposition asserts the two *partial* converses (cf. note on I. 6) of the
Porism to III. 1.　De Morgan would place it next to III. 1.

PROPOSITION 4.

*If in a circle two straight lines cut one another which are
not through the centre, they do not bisect one another.*

　　Let *ABCD* be a circle, and in it let the two straight lines
AC, *BD*, which are not through the
centre, cut one another at *E*;
I say that they do not bisect one
another.

　　For, if possible, let them bisect one
another, so that *AE* is equal to *EC*,
and *BE* to *ED*;
let the centre of the circle *ABCD* be
taken [III. 1], and let it be *F*; let *FE* be
joined.

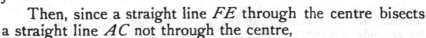

　　Then, since a straight line *FE* through the centre bisects
a straight line *AC* not through the centre,
　　　　　　　it also cuts it at right angles;　　　　　[III. 3]
　　therefore the angle *FEA* is right.
　　Again, since a straight line *FE* bisects a straight line *BD*,
　　　　　　　it also cuts it at right angles;　　　　　[III. 3]
　　therefore the angle *FEB* is right.

But the angle *FEA* was also proved right;

therefore the angle *FEA* is equal to the angle *FEB*, the less to the greater : which is impossible.

Therefore *AC*, *BD* do not bisect one another.
Therefore etc.

Q. E. D.

PROPOSITION 5.

If two circles cut one another, they will not have the same centre.

For let the circles *ABC*, *CDG* cut one another at the points *B, C*;

I say that they will not have the same centre.

For, if possible, let it be *E*; let *EC* be joined, and let *EFG* be drawn through at random.

Then, since the point *E* is the centre of the circle *ABC*,

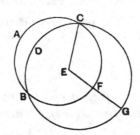

EC is equal to *EF*. [I. Def. 15]

Again, since the point *E* is the centre of the circle *CDG*,

EC is equal to *EG*.

But *EC* was proved equal to *EF* also;

therefore *EF* is also equal to *EG*, the less to the greater : which is impossible.

Therefore the point *E* is not the centre of the circles *ABC*, *CDG*.

Therefore etc.

Q. E. D.

The propositions III. 5, 6 could be combined in one. It makes no difference whether the circles cut, or meet without cutting, so long as they do not coincide altogether; in either case they cannot have the same centre. The two cases are covered by the enunciation : *If the circumferences of two circles meet at a point they cannot have the same centre.* On the other hand, *If two circles have the same centre and one point in their circumferences common, they must coincide altogether.*

PROPOSITION 6.

If two circles touch one another, they will not have the same centre.

For let the two circles *ABC*, *CDE* touch one another at the point *C*;

I say that they will not have the same centre.

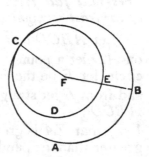

For, if possible, let it be *F*; let *FC* be joined, and let *FEB* be drawn through at random.

Then, since the point *F* is the centre of the circle *ABC*,

FC is equal to *FB*.

Again, since the point *F* is the centre of the circle *CDE*,

FC is equal to *FE*.

But *FC* was proved equal to *FB*;

therefore *FE* is also equal to *FB*, the less to the greater: which is impossible.

Therefore *F* is not the centre of the circles *ABC*, *CDE*.

Therefore etc.

Q. E. D.

The English editions enunciate this proposition of circles touching *internally*, but the word (ἐντός) is a mere interpolation, which was no doubt made because Euclid's figure showed only the case of *internal* contact. The fact is that, in his usual manner, he chose for demonstration the more difficult case, and left the other case (that of *external* contact) to the intelligence of the reader. It is indeed sufficiently self-evident that circles touching externally cannot have the same centre; but Euclid's proof can really be used for this case too.

Camerer remarks that the proof of III. 6 seems to assume tacitly that the points *E* and *B* cannot coincide, or that circles which touch internally at *C* cannot meet in any other point, whereas this fact is not proved by Euclid till III. 13. But no such general assumption is necessary here; it is only necessary that *one* line drawn from the assumed common centre should meet the circles in different points; and the very notion of internal contact requires that, before one circle *meets* the other on its inner side, it must have passed through points *within* the latter circle.

PROPOSITION 7.

If on the diameter of a circle a point be taken which is not
the centre of the circle, and from the point straight lines fall
upon the circle, that will be greatest on which the centre is, the
remainder of the same diameter will be least, and of the rest
5 *the nearer to the straight line through the centre is always*
greater than the more remote, and only two equal straight
lines will fall from the point on the circle, one on each side
of the least straight line.

Let *ABCD* be a circle, and let *AD* be a diameter of it;
10 on *AD* let a point *F* be taken which is not the centre of the
circle, let *E* be the centre of the circle,

and from *F* let straight lines *FB, FC, FG* fall upon the circle
ABCD;

I say that *FA* is greatest, *FD* is least, and of the rest *FB* is
15 greater than *FC*, and *FC* than *FG*.

For let *BE, CE, GE* be joined.

Then, since in any triangle two
sides are greater than the remaining
one,   [I. 20]

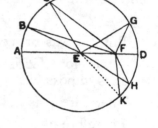

20   *EB, EF* are greater than *BF*.

But *AE* is equal to *BE*;

therefore *AF* is greater than *BF*.

Again, since *BE* is equal to *CE*,
and *FE* is common,
25 the two sides *BE, EF* are equal to the two sides *CE, EF*.

But the angle *BEF* is also greater than the angle *CEF*;

therefore the base *BF* is greater than the base *CF*.   [I. 24]

For the same reason

       CF is also greater than *FG*.

30   Again, since *GF, FE* are greater than *EG*,
and *EG* is equal to *ED*,

        GF, FE are greater than *ED*.

Let *EF* be subtracted from each;

therefore the remainder *GF* is greater than the remainder
35 *FD*.

Therefore *FA* is greatest, *FD* is least, and *FB* is greater
than *FC*, and *FC* than *FG*.

I say also that from the point F only two equal straight lines will fall on the circle $ABCD$, one on each side of the least FD.

For on the straight line EF, and at the point E on it, let the angle FEH be constructed equal to the angle GEF [1. 23], and let FH be joined.

Then, since GE is equal to EH,

and EF is common,

the two sides GE, EF are equal to the two sides HE, EF;

and the angle GEF is equal to the angle HEF;

therefore the base FG is equal to the base FH. 　　[1. 4]

I say again that another straight line equal to FG will not fall on the circle from the point F.

For, if possible, let FK so fall.

Then, since FK is equal to FG, and FH to FG,

FK is also equal to FH,

the nearer to the straight line through the centre being thus equal to the more remote: which is impossible.

Therefore another straight line equal to GF will not fall from the point F upon the circle;

therefore only one straight line will so fall.

Therefore etc.

　　　　　　　　　　　　　　　　　　　　　　Q. E. D.

4. **of the same diameter.** I have inserted these words for clearness' sake. The text has simply ἐλαχίστη δὲ ἡ λοιπή, "and the remaining (straight line) least."

7, 39. **one on each side.** The word "one" is not in the Greek, but is necessary to give the force of ἐφ' ἐκάτερα τῆς ἐλαχίστης, literally "on both sides," or "on each of the two sides, of the least."

De Morgan points out that there is an unproved assumption in this demonstration. We draw straight lines *from F*, as *FB, FC*, such that the angle DFB is greater than the angle DFC and then assume, with respect to the straight lines drawn *from the centre E to B, C*, that the angle DEB is greater than the angle DEC. This is most easily proved, I think, by means of the converse of part of the theorem about the lengths of different straight lines drawn to a given straight line from an external point which was mentioned above in the note on III. 2. This converse would be to the effect that, *If two unequal straight lines be drawn from a point to a given straight line which are not perpendicular to the straight line, the greater of the two is the further from the perpendicular from the point to the given straight line.* This can either be proved from its converse by *reductio ad absurdum*, or established directly by means of I. 47. Thus, in the accompanying figure, FB must cut EC in some point M, since the angle BFE is less than the angle CFE.

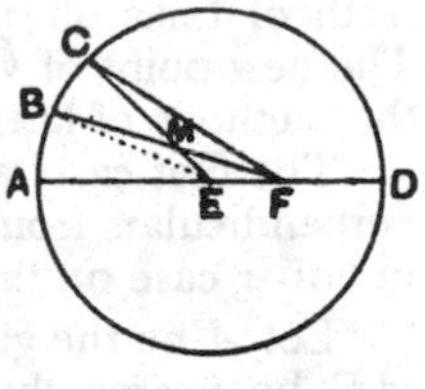

Therefore EM is less than EC, and therefore than EB.

Hence the point *B* in which *FB* meets the circle is further from the foot of the perpendicular from *E* on *FB* than *M* is;

therefore the angle *BEF* is greater than the angle *CEF*.

Another way of enunciating the first part of the proposition is that of Mr H. M. Taylor, viz. "Of all straight lines drawn to a circle from an internal point not the centre, the one which passes through the centre is the greatest, and the one which when produced passes through the centre is the least; and of any two others the one which *subtends the greater angle at the centre* is the greater." The substitution of the *angle subtended at the centre* as the criterion no doubt has the effect of avoiding the necessity of dealing with the unproved assumption in Euclid's proof referred to above, and the similar substitution in the enunciation of the first part of III. 8 has the effect of avoiding the necessity for dealing with like unproved assumptions in Euclid's proof, as well as the complication caused by the distinction in Euclid's enunciation between lines falling from an external point on the *convex circumference* and on the *concave circumference* of a circle respectively, terms which are not defined but taken as understood.

Mr Nixon (*Euclid Revised*) similarly substitutes as the criterion the angle subtended at the centre, but gives as his reason that the words "nearer" and "more remote" in Euclid's enunciation are scarcely clear enough without some definition of the sense in which they are used. Smith and Bryant make the substitution in III. 8, but follow Euclid in III. 7.

On the whole, I think that Euclid's plan of taking straight lines drawn from the point which is not the centre direct to the circumference and making greater or less angles *at that point* with the straight line containing it and the centre is the more instructive and useful of the two, since it is such lines drawn in any manner to the circle from the point which are immediately useful in the proofs of later propositions or in resolving difficulties connected with those proofs.

Heron again (an-Nairīzī, ed. Curtze, pp. 114—5) has a note on this proposition which is curious. He first of all says that Euclid proves that lines *nearer the centre* are greater than those more remote *from it*. This is a different view of the question from that taken in Euclid's proposition as we have it, in which the lines are not nearer to and more remote from the *centre* but from *the line through the centre*. Euclid takes lines inclined to the latter line at a greater or less angle; Heron introduces distance *from the centre* in the sense of Deff. 4, 5, i.e. in the sense of *the length of the perpendicular* drawn to the line from the centre, which Euclid does not use till III. 14, 15. Heron then observes that in Euclid's proposition the lines compared are all drawn on one side of the line through the centre, and sets himself to prove the same truth of lines on *opposite* sides which are more or less distant *from the centre*. The new point of view necessitates a quite different line of proof, anticipating the methods of later propositions.

The first case taken by Heron is that of two straight lines such that the perpendiculars from the centre on them fall on the lines themselves and not in either case on the line produced.

Let *A* be the given point, *D* the centre, and let *AE* be nearer the centre than *AF*, so that the perpendicular *DG* on *AE* is less than the perpendicular *DH* on *AF*.

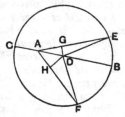

Then　sqs. on *DG, GE* = sqs. on *DH, HF*,

and　sqs. on *DG, GA* = sqs. on *DH, HA*.

But　　　　sq. on *DG* < sq. on *DH*.

Therefore   sq. on *GE* > sq. on *HF*,

and   sq. on *GA* > sq. on *HA*,

whence   *GE* > *HF*,

                    GA > *HA*.

Therefore, by addition, *AE* > *AF*.

The other case taken by Heron is that where one perpendicular falls on the line produced, as in the annexed figure. In this case we prove in like manner that   *GE* > *HF*,

and   *GA* > *AH*.

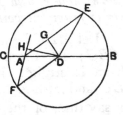

Thus *AE* is greater than the sum of *HF*, *AH*, whence, *a fortiori*, *AE* is greater than the difference of *HF*, *AH*, i.e. than *AF*.

Heron does not give the third possible case, that, namely, where *both* perpendiculars fall on the lines produced, The fact is that, in this case, the foregoing method breaks down. Though *AE* be nearer to the centre than *AF* in the sense that *DG* is less than *DH*,

        AE is not greater but *less* than *AF*.

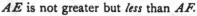

Moreover this cannot be proved by the same method as before.

For, while we can prove that

        GE > *HF*,

        GA > *AH*,

we cannot make any inference as to the comparative length of *AE*, *AF*.

To judge by Heron's corresponding note to III. 8, he would, to prove this case, practically prove III. 35 first, i.e. prove that, if *EA* be produced to *K* and *FA* to *L*,

        rect. *FA*, *AL* = rect. *EA*, *AK*,

from which he would infer that, since *AK* > *AL* by the first case,

        AE < *AF*.

An excellent moral can, I think, be drawn from the note of Heron. Having the appearance of supplementing, or giving an alternative for, Euclid's proposition, it cannot be said to do more than confuse the subject. Nor was it necessary to find a new proof for the case where the two lines which are compared are on *opposite* sides of the diameter, since Euclid shows that for each line from the point to the circumference on one side of the diameter there is another of the same length equally inclined to it on the other side.

PROPOSITION 8.

If a point be taken outside a circle and from the point straight lines be drawn through to the circle, one of which is through the centre and the others are drawn at random, then, of the straight lines which fall on the concave circumference, that through the centre is greatest, while of the rest

the nearer to that through the centre is always greater than the more remote, but, of the straight lines falling on the convex circumference, that between the point and the diameter is least, while of the rest the nearer to the least is always less than the more remote, and only two equal straight lines will fall on the circle from the point, one on each side of the least.

Let ABC be a circle, and let a point D be taken outside ABC; let there be drawn through from it straight lines DA, DE, DF, DC, and let DA be through the centre; I say that, of the straight lines falling on the concave circumference $AEFC$, the straight line DA through the centre is greatest,

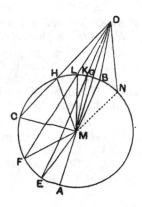

while DE is greater than DF and DF than DC;

but, of the straight lines falling on the convex circumference $HLKG$, the straight line DG between the point and the diameter AG is least; and the nearer to the least DG is always less than the more remote, namely DK than DL, and DL than DH.

For let the centre of the circle ABC be taken [III. 1], and let it be M; let ME, MF, MC, MK, ML, MH be joined.

Then, since AM is equal to EM, let MD be added to each;

therefore AD is equal to EM, MD.

But EM, MD are greater than ED;   [I. 20]

therefore AD is also greater than ED.

Again, since ME is equal to MF,

and MD is common,

therefore EM, MD are equal to FM, MD;

and the angle EMD is greater than the angle FMD;

therefore the base ED is greater than the base FD.

         [I. 24]

Similarly we can prove that FD is greater than CD; therefore DA is greatest, while DE is greater than DF, and DF than DC.

Next, since MK, KD are greater than MD,   [I. 20]
and MG is equal to MK,
therefore the remainder KD is greater than the remainder GD,

so that GD is less than KD.

And, since on MD, one of the sides of the triangle MLD, two straight lines MK, KD were constructed meeting within the triangle,
therefore MK, KD are less than ML, LD;   [I. 21]
and MK is equal to ML;

therefore the remainder DK is less than the remainder DL.

Similarly we can prove that DL is also less than DH;

therefore DG is least, while DK is less than DL, and DL than DH.

I say also that only two equal straight lines will fall from the point D on the circle, one on each side of the least DG.

On the straight line MD, and at the point M on it, let the angle DMB be constructed equal to the angle KMD, and let DB be joined.

Then, since MK is equal to MB,
and MD is common,
the two sides KM, MD are equal to the two sides BM, MD respectively;
and the angle KMD is equal to the angle BMD;

therefore the base DK is equal to the base DB.   [I. 4]

I say that no other straight line equal to the straight line DK will fall on the circle from the point D.

For, if possible, let a straight line so fall, and let it be DN.

Then, since DK is equal to DN,
while DK is equal to DB,

DB is also equal to DN,

that is, the nearer to the least DG equal to the more remote: which was proved impossible.

Therefore no more than two equal straight lines will fall on the circle ABC from the point D, one on each side of DG the least.

Therefore etc.   Q. E. D.

As De Morgan points out, there are here two assumptions similar to that tacitly made in the proof of III. 7, namely that K falls within the triangle DLM and E outside the triangle DFM. These facts can be proved in the same way as the assumption in III. 7. Let DE meet FM in Y and LM in Z. Then, as before, MZ is less than ML and therefore than MK. Therefore K lies further than Z from the foot of the perpendicular from M on DE. Similarly E lies further than Y from the foot of the same perpendicular.

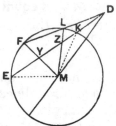

Heron deals with lines on *opposite* sides of the diameter through the external point in a manner similar to that adopted in his previous note.

For the case where E, F are the *second* points in which AE, AF meet the circle the method answers well enough.

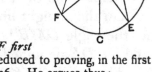

If AE is *nearer the centre D* than AF is,

$$\text{sqs. on } DG, GE = \text{sqs. on } DH, HF$$

and   sqs. on DG, GA = sqs. on DH, HA,

whence, since $DG < DH$,

it follows that $GE > HF$,

and $AG > AH$,

so that, by addition, $AE > AF$.

But, if K, L be the points in which AE, AF *first* meet the circle, the method fails, and Heron is reduced to proving, in the first instance, the property usually deduced from III. 36. He argues thus:

AKD being an obtuse angle,

sq. on AD = sum of sqs. on AK, KD and twice rect. AK, KG. [II. 12]

ALD is also an obtuse angle, and it follows that

sum of sqs. on AK, KD and twice rect. AK, KG is equal to

sum of sqs. on AL, LD and twice rect. AL, LH.

Therefore, the squares on KD, LD being equal,

sq on AK and twice rect. AK, KG = sq. on AL and twice rect. AL, LH,

or   sq on AK and rect. AK, KE = sq. on AL and rect. AL, LF,

i.e.   rect. AK, AE = rect. AL, AF.

But, by the first part, $AE > AF$.

Therefore $AK < AL$.

III. 7, 8 deal with the lengths of the several lines drawn to the circumference of a circle (1) from a point within it, (2) from a point outside it; but a similar proposition is true of straight lines drawn from a point on the circumference itself: *If any point be taken on the circumference of a circle, then, of all the straight lines which can be drawn from it to the circumference, the greatest is that in which the centre is; of any others that which is nearer to the straight line which passes through the centre is greater than one more remote; and from the same point there can be drawn to the circumference two straight lines, and only two, which are equal to one another, one on each side of the greatest line.*

The converses of III. 7, 8 and of the proposition just given are also true and can easily be proved by *reductio ad absurdum*. They could be employed to throw light on such questions as that of internal contact, and the relative position of the centres of circles so touching. This is clear when part of the converses is stated : thus (1) if from any point in the plane of a circle a number of straight lines be drawn to the circumference of the circle, and one of these is greater than any other, the centre of the circle must lie on that one, (2) if one of them is less than any other, then, (*a*) if the point is within the circle, the centre is on the minimum straight line produced *beyond the point*, (*b*) if the point is outside the circle, the centre is on the minimum straight line produced *beyond the point in which it meets the circle.*

PROPOSITION 9.

If a point be taken within a circle, and more than two equal straight lines fall from the point on the circle, the point taken is the centre of the circle.

Let *ABC* be a circle and *D* a point within it, and from *D* let more than two equal straight lines, namely *DA*, *DB*, *DC*, fall on the circle *ABC*;
I say that the point *D* is the centre of the circle *ABC*.

For let *AB*, *BC* be joined and bisected at the points *E*, *F*, and let *ED*, *FD* be joined and drawn through to the points *G*, *K*, *H*, *L*.

Then, since *AE* is equal to *EB*, and *ED* is common,

the two sides *AE*, *ED* are equal to the two sides *BE*, *ED* ; and the base *DA* is equal to the base *DB* ;

therefore the angle *AED* is equal to the angle *BED*.

[I. 8]

Therefore each of the angles *AED*, *BED* is right ;

[I. Def. 10]

therefore *GK* cuts *AB* into two equal parts and at right angles.

And since, if in a circle a straight line cut a straight line into two equal parts and at right angles, the centre of the circle is on the cutting straight line,   [III. 1, Por.]

the centre of the circle is on *GK*.

For the same reason

the centre of the circle ABC is also on HL.

And the straight lines GK, HL have no other point common but the point D;

therefore the point D is the centre of the circle ABC.

Therefore etc.   Q. E. D.

The result of this proposition is quoted by Aristotle, *Meteorologica* III. 3, 373 a 13—16 (cf. note on I. 8).

III. 9 is, as De Morgan remarks, a *logical* equivalent of part of III. 7, where it is proved that every *non*-central point is *not* a point from which three equal straight lines can be drawn to the circle. Thus III. 7 says that every *not-A* is *not-B*, and III. 9 states the equivalent fact that every *B* is *A*. Mr H. M. Taylor does in effect make a *logical* inference of the theorem that, *If from a point three equal straight lines can be drawn to a circle, that point is the centre*, by making it a corollary to his proposition which includes the part of III. 7 referred to. Euclid does not allow himself these logical inferences, as we shall have occasion to observe elsewhere also.

Of the two proofs of this proposition given in earlier texts of Euclid, August and Heiberg regard that translated above as genuine, relegating the other, which Simson gave alone, to a place in an Appendix. Camerer remarks that the genuine proof should also have contemplated the case in which one or other of the straight lines AB, BC passes through D. This would however have been a departure from Euclid's manner of taking the most obscure case for proof and leaving others to the reader.

The other proof, that selected by Simson, is as follows:

"For let a point D be taken within the circle ABC, and from D let more than two equal straight lines, namely AD, DB, DC, fall on the circle ABC;

I say that the point D so taken is the centre of the circle ABC.

For suppose it is not; but, if possible, let it be E, and let DE be joined and carried through to the points F, G.

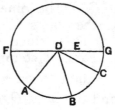

Therefore FG is a diameter of the circle ABC.

Since, then, on the diameter FG of the circle ABC a point has been taken which is not the centre of the circle, namely D,

DG is greatest, and DC is greater than DB, and DB than DA.

But the latter are also equal: which is impossible

Therefore E is not the centre of the circle.

Similarly we can prove that neither is any other point except D;

therefore the point D is the centre of the circle ABC.

Q. E. D."

On this Todhunter correctly points out that the point E might be supposed to fall *within* the angle $A\hat{D}C$. It cannot then be shown that DC is greater than DB and DB than DA, but only that either DC or DA is less than DB; this however is sufficient for establishing the proposition.

PROPOSITION 10.

A circle does not cut a circle at more points than two.

For, if possible, let the circle *ABC* cut the circle *DEF* at more points than two, namely *B*, *C*, *F*, *H*;

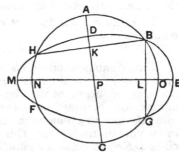

let *BH*, *BG* be joined and bisected at the points *K*, *L*, and from *K*, *L* let *KC*, *LM* be drawn at right angles to *BH*, *BG* and carried through to the points *A*, *E*.

Then, since in the circle *ABC* a straight line *AC* cuts a straight line *BH* into two equal parts and at right angles,

the centre of the circle *ABC* is on *AC*.   [III. 1, Por.]

Again, since in the same circle *ABC* a straight line *NO* cuts a straight line *BG* into two equal parts and at right angles,

the centre of the circle *ABC* is on *NO*.

But it was also proved to be on *AC*, and the straight lines *AC*, *NO* meet at no point except at *P*;

therefore the point *P* is the centre of the circle *ABC*.

Similarly we can prove that *P* is also the centre of the circle *DEF*;

therefore the two circles *ABC*, *DEF* which cut one another have the same centre *P*: which is impossible.   [III. 5]

Therefore etc.   Q. E. D.

1. The word circle (κύκλος) is here employed in the unusual sense of the *circumference* (περιφέρεια) of a circle. Cf. note on I. Def. 15.

There is nothing in the demonstration of this proposition which assumes that the circles *cut* one another; it proves that two circles cannot *meet* at more than two points, whether they cut or meet without cutting, i.e. *touch* one another.

Here again, of two demonstrations given in the earlier texts, Simson chose the second, which August and Heiberg relegate to an Appendix and which is as follows:

"For again let the circle *ABC* cut the circle *DEF* at more points than two, namely *B*, *G*, *H*, *F*;

let the centre *K* of the circle *ABC* be taken, and let *KB*, *KG*, *KF* be joined.

Since then a point K has been taken within the circle DEF, and from K more than two straight lines, namely KB, KF, KG, have fallen on the circle DEF, the point K is the centre of the circle DEF. [III. 9]

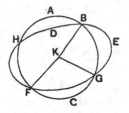

But K is also the centre of the circle ABC.

Therefore two circles cutting one another have the same centre K: which is impossible.   [III. 5]

Therefore a circle does not cut a circle at more points than two.

Q. E. D."

This demonstration is claimed by Heron (see an-Nairīzī, ed. Curtze, pp. 120—1). It is incomplete because it assumes that the point K which is taken as the centre of the circle ABC is *within* the circle DEF. It can however be completed by means of III. 8 and the corresponding proposition with reference to a point *on* the circumference of a circle which was enunciated in the note on III. 8. For (1) if the point K is *on* the circumference of the circle DEF, we obtain a contradiction of the latter proposition which asserts that only *two* equal straight lines can be drawn from K to the circumference of the circle DEF; (2) if the point K is *outside* the circle DEF, we obtain a contradiction of the corresponding part of III. 8.

Euclid's proof contains an unproved assumption, namely that the lines bisecting BG, BH at right angles *will* meet in a point P. For a discussion of this assumption see note on IV. 5.

PROPOSITION 11.

If two circles touch one another internally, and their centres be taken, the straight line joining their centres, if it be also produced, will fall on the point of contact of the circles.

For let the two circles ABC, ADE touch one another internally at the point A, and let the centre F of the circle ABC, and the centre G of ADE, be taken;

I say that the straight line joined from G to F and produced will fall on A.

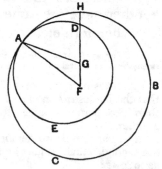

For suppose it does not, but, if possible, let it fall as FGH, and let AF, AG be joined.

Then, since AG, GF are greater than FA, that is, than FH,

let FG be subtracted from each;

therefore the remainder AG is greater than the remainder GH.

But AG is equal to GD;

therefore GD is also greater than GH,

the less than the greater : which is impossible.

Therefore the straight line joined from F to G will not fall outside ;

therefore it will fall at A on the point of contact.

Therefore etc.

Q. E. D.

2. **the straight line joining their centres,** literally "the straight line joined to their centres" (ἡ ἐπὶ τὰ κέντρα αὐτῶν ἐπιζευγνυμένη εὐθεῖα).

3. **point of contact** is here συναφή, and in the enunciation of the next proposition ἐπαφή.

Again August and Heiberg give in an Appendix the additional or alternative proof, which however shows little or no variation from the genuine proof and can therefore well be dispensed with.

The genuine proof is beset with difficulties in consequence of what it tacitly assumes in the figure, on the ground, probably, of its being obvious to the eye. Camerer has set out these difficulties in a most careful note, the heads of which may be given as follows :

He observes, first, that the straight line joining the centres, when produced, must necessarily (though this is not stated by Euclid) be produced *in the direction of the centre of the circle which touches the other internally.* (For brevity, I shall call this circle the " inner circle," though I shall imply nothing by that term except that it is the circle which touches the other on the inner side of the latter, and therefore that, in accordance with the definition of *touching,* points on it in the immediate neighbourhood of the point of contact are necessarily *within* the circle which it touches.) Camerer then proceeds by the following steps.

1. The two circles, touching at the given point, cannot *intersect* at any point. For, since points on the "inner" in the immediate neighbourhood of the point of contact are within the "outer" circle, the inner circle, if it intersects the other anywhere, must pass outside it and then return. This is only possible (*a*) if it passes out at one point and returns at another point, or (*b*) if it passes out and returns through one and the same point. (*a*) is impossible because it would require two circles to have *three* common points ; (*b*) would require that the inner circle should have a *node* at the point where it passes outside the other, and this is proved to be impossible by drawing any radius cutting both loops.

2. Since the circles cannot intersect, one must be *entirely* within the other.

3. Therefore the outer circle must be greater than the inner, and the radius of the outer greater than that of the inner.

4. Now, if F be the centre of the greater and G of the inner circle, and if FG produced beyond G does *not* pass through A, the given point of contact, then there are three possible hypotheses.

(*a*) A may lie on GF produced beyond F.

(*b*)   *A* may lie outside the line *FG* altogether, in which case *FG* produced beyond *G* must, in consequence of result 2 above, either

(i) meet the circles in a point common to both, or

(ii) meet the circles in two points, of which that which is on the inner circle is nearer to *G* than the other is.

(*a*) is then proved to be impossible by means of the fact that the radius of the inner circle is less than the radius of the outer.

(*b*) (ii) is Euclid's case; and his proof holds equally of (*b*) (i), the hypothesis, namely, that *D* and *H* in the figure coincide.

Thus all alternative hypotheses are successively shown to be impossible, and the proposition is completely established.

I think, however, that this procedure may be somewhat shortened in the following manner.

In order to make Euclid's proof absolutely conclusive we have only (1) to take care to produce *FG* beyond *G*, the centre of the "inner" circle, and then (2) to prove that the point in which *FG* so produced meets the "inner" circle is *not further* from *G* than is the point in which it meets the other circle. Euclid's proof is equally valid whether the first point is nearer to *G* than the second or the first point and the second coincide.

If *FG* produced beyond *G* does not pass through *A*, there are two

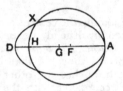

conceivable hypotheses: (*a*) *A* may lie on *GF* produced beyond *F*, or (*b*) *A* may be outside *FG* produced either way. In either case, if *FG* produced meets the "inner" circle in *D* and the other in *H*, and if *GD* is greater than *GH*, then the "inner" circle must cut the "outer" circle at some point between *A* and *D*, say *X*.

But, if two circles have a common point *X* lying on one side of the line of centres, they must have another corresponding point on the other side of the line of centres. This is clear from III. 7, 8; for the point is determined by drawing from *F* and *G*, on the opposite side to that where *X* is, straight lines *FY*, *GY* making with *FD* angles equal to the angles *DFX*, *DGX* respectively.

Hence the two circles will have at least three points common: which is impossible.

Therefore *GD* cannot be greater than *GH*; accordingly *GD* must be either equal to, or less than, *GH*, and Euclid's proof is valid.

The particular hypothesis in which *FG* is supposed to be in the same straight line with *A* but *G* is on the side of *F* away from *A* is easily disposed of, and would in any case have been left to the reader by Euclid.

For *GD* is either equal to or less than *GH*.

Therefore *GD* is less than *FH*, and therefore less than *FA*.

But *GD* is equal to *GA*, and therefore greater than *FA*: which is impossible.

Subject to the same preliminary investigation as that required by Euclid's proof, the proposition can also be proved directly from III. 7.

For, by III. 7, GH is the shortest straight line that can be drawn from G to the circle with centre F;

therefore GH is less than GA,

and therefore less than GD: which is absurd.

This proposition is the crucial one as regards circles which touch internally; and, when it is once established, the relative position of the circles can be completely elucidated by means of it and the propositions which have preceded it. Thus, in the annexed figure, if F be the centre of the outer circle and G the centre of the inner, and if any radius FQ of the outer circle meet the two circles in Q, P respectively, it follows, from III. 7, III. 8, or the corresponding theorem with reference to a point *on* the circumference, that FA is the maximum straight line from F to the circumference of the inner circle, FP is less than FA, and FP diminishes in length as FQ moves round from FA until FP reaches its minimum length FB. Hence the circles do not meet at any other point than A, and the distance PQ cut off between them on any radius FQ of the outer circle becomes greater and greater as FQ moves round from FA to FC and is a maximum when FQ coincides with FC, after which it diminishes again on the other side of FC.

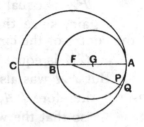

The same consideration gives the partial converse of III. 11 which forms the 6th lemma of Pappus to the first book of the *Tactiones* of Apollonius (Pappus, VII. p. 826). This is to the effect that, *if* AB, AC *are in one straight line, and on one side of* A, *the circles described on* AB, AC *as diameters touch (internally at the point* A). Pappus concludes this from the fact that the circles have a common tangent at A; but the truth of it is clear from the fact that FP diminishes as FQ moves away from FA on either side; whence the circles meet at A but do not cut one another.

Pappus' 5th lemma (VII. p. 824) is another partial converse, namely that, *given two circles touching internally at* A, *and a line* ABC *drawn from* A *cutting both, then, if the centre of the outer circle lies on* ABC, *so does the centre of the inner*. Pappus himself proves this, by means of the common tangent to the circles at A, in two ways. (1) The tangent is at right angles to AC and therefore to AB: therefore the centre of the inner circle lies on AB. (2) By III. 32, the angles in the alternate segments of both circles are right angles, so that ABC is a diameter of both.

[PROPOSITION 12.

If two circles touch one another externally, the straight line joining their centres will pass through the point of contact.

For let the two circles ABC, ADE touch one another externally at the point A, and let the centre F of ABC, and the centre G of ADE, be taken;

I say that the straight line joined from F to G will pass through the point of contact at A.

10 For suppose it does not, but, if possible, let it pass as $FCDG$, and let AF, AG be joined.

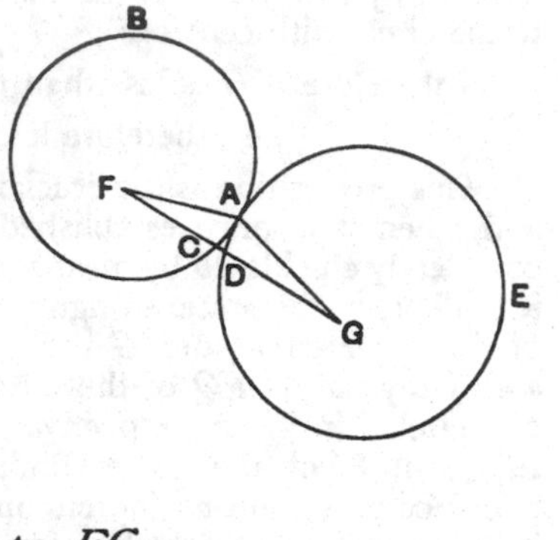

Then, since the point F is the centre of the circle ABC,

15   FA is equal to FC.

Again, since the point G is the centre of the circle ADE,

GA is equal to GD.

But FA was also proved equal to FC;

20   therefore FA, AG are equal to FC, GD,

so that the whole FG is greater than FA, AG;

but it is also less [I. 20]: which is impossible.

Therefore the straight line joined from F to G will not fail to pass through the point of contact at A;

25   therefore it will pass through it.

Therefore etc.   Q. E. D.]

23. **will not fail to pass.** The Greek has the double negative, οὐκ ἄρα ἡ...εὐθεῖα... οὐκ ἐλεύσεται, literally "the straight line...will not *not*-pass...."

Heron says on III. 11: "Euclid in proposition 11 has supposed the two circles to touch internally, made his proposition deal with this case and proved what was sought in it. *But I will show how it is to be proved if the contact is external.*" He then gives substantially the proof and figure of III. 12. It seems clear that neither Heron nor an-Nairīzī had III. 12 in this place.

Campanus and the Arabic edition of Naṣīraddīn aṭ-Ṭūsī have nothing more of III. 12 than the following addition to III. 11. "In the case of external contact the two lines *ae* and *eb* will be greater than *ab*, whence *ad* and *cb* will be greater than the whole *ab*, which is false." (The points *a*, *b*, *c*, *d*, *e* correspond respectively to G, F, C, D, A in the above figure.) It is most probable that Theon or some other editor added Heron's proof in his edition and made Prop. 12 out of it (an-Nairīzī, ed. Curtze, pp. 121—2). An-Nairīzī and Campanus, conformably with what has been said, number Prop. 13 of Heiberg's text Prop. 12, and so on through the Book.

What was said in the note on the last proposition applies, *mutatis mutandis*, to this. Camerer proceeds in the same manner as before; and we may use the same alternative argument in this case also.

Euclid's proof is valid provided only that, if FG, joining the assumed centres, meets the circle with centre F in C and the other circle in D, C is not within the circle ADE and D is not within the circle ABC. (The proof is equally valid whether C, D coincide or the successive points are, as drawn in the figure, in the order F, C, D, G.) Now, if C is within the circle ADE

and *D* within the circle *ABC*, the circles must have cut between *A* and *C* and between *A* and *D*. Hence, as before, they must also have another corresponding point common on the other side of *CD*. That is, the circles must have *three* common points : which is impossible.

Hence Euclid's proof is valid if *F, A, G* form a triangle, and the only hypothesis which has still to be disproved is the hypothesis which he would in any case have left to the reader, namely that *A* does not lie on *FG* but on *FG produced* in either direction. In this case, as before, either *C, D* must coincide or *C* is nearer *F* than *D* is. Then the radius *FC* must be equal to *FA* : which is impossible, since *FC* cannot be greater than *FD*, and must therefore be *less* than *FA*.

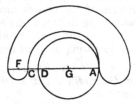

Given the same preliminaries, III. 12 can be proved by means of III. 8.

Again, when the proposition III. 12 is once proved, III. 8 helps us to prove at once that the circles lie entirely outside each other and have no other common point than the point of contact.

Among Pappus' lemmas to Apollonius' *Tactiones* are the two partial converses of this proposition corresponding to those given in the last note. Lemma 4 (VII. p. 824) is to the effect that, *if* AB, AC *be in one straight line*, B *and* C *being on opposite sides of* A, *the circles drawn on* AB, AC *as diameters touch externally at* A. Lemma 3 (VII. p. 822) states that, *if two circles touch externally at* A *and* BAC *is drawn through* A *cutting both circles and containing the centre of one*, BAC *will also contain the centre of the other*. The proofs, as before, use the common tangent at *A*.

Mr H. M. Taylor gets over the difficulties involved by III. 11, 12 in a manner which is most ingenious but not Euclidean. He first proves that, *if two circles meet at a point not in the same straight line with their centres, the circles intersect at that point*; this is very easily established by means of III. 7, 8 and the third similar theorem. Then he gives as a corollary the statement that, *if two circles touch, the point of contact is in the same straight line with their centres*. It is not explained how this is inferred from the substantive proposition; it seems, however, to be a *logical* inference simply. By the proposition, every *A* (circles meeting at a point not in the same straight line with the centre) is *B* (circles which intersect); therefore every not-*B* is not-*A*, i.e. circles which do not intersect do not meet at a point not in the same straight line with the centres. Now non-intersecting circles may either meet (i.e. touch) or not meet. In the former case they must meet *on* the line of centres : for, if they met at a point not in that line, they would intersect. But such a purely *logical* inference is foreign to Euclid's manner. As De Morgan says, "Euclid may have been ignorant of the identity of 'Every *X* is *Y*' and 'Every not-*Y* is not-*X*,' for anything that appears in his writings; he makes the one follow from the other by a new proof each time" (quoted in Keynes' *Formal Logic*, p. 81).

There is no difficulty in proving, by means of I. 20, Mr Taylor's next proposition that, *if two circles meet at a point which lies in the same straight line as their centres and is between the centres, the circles touch at that point, and each circle lies without the other*. But the similar proof, by means of I. 20, of the corresponding theorem for internal contact seems to be open to the same objection as Euclid's proof of III. 11 in that it assumes without proof that the circle which has its centre nearest to the point of meeting is the "inner" circle. Lastly, in order to prove that, *if two circles have a point of contact, they do not meet at any other point*, Mr Taylor uses the questionable corollary. Therefore in any case his alternative procedure does not seem preferable to Euclid's.

The alternative to Eucl. III. 11—13 which finds most favour in modern continental text-books (e.g. Legendre, Baltzer, Henrici and Treutlein, Veronese, Ingrami, Enriques and Amaldi) connects the number, position and nature of the coincidences between points on two circles with the relation in which the distance between their centres stands to the length of their radii. Enriques and Amaldi, whose treatment of the different cases is typical, give the following propositions (Veronese gives them in the converse form).

1. *If the distance between the centres of two circles is greater than the sum of the radii, the two circles have no point common and are external to one another.*

Let O, O' be the centres of the circles (which we will call " the circles O, O' "), r, r' their radii respectively.

Since then $OO' > r + r'$, *a fortiori* $OO' > r$, and O' is therefore exterior to the circle O.

Next, the circumference of the circle O intersects OO' in a point A, and since $OO' > r + r'$, $AO' > r'$, and A is external to the circle O'.

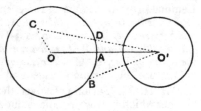

But $O'A$ is less than any straight line, as $O'B$, drawn to the circumference of the circle O [III. 8]; hence all points, as B, on the circumference of the circle O are external to the circle O'.

Lastly, if C be any point internal to the circle O, the sum of OC, $O'C$ is greater than $O'O$, and *a fortiori* greater than $r + r'$.

But OC is less than r: therefore $O'C$ is greater than r', or C is external to O.

Similarly we prove that any point on or within the circumference of the circle O' is external to the circle O.

2. *If the distance between the centres of two unequal circles is less than the difference of the radii, the two circumferences have no common point and the lesser circle is entirely within the greater.*

Let O, O' be the centres of the two circles, r, r' their radii respectively $(r < r')$.

Since $OO' < r' - r$, *a fortiori* $OO' < r'$, so that O is internal to the circle O'.

If A, A' be the points in which the straight line $O'O$ intersects respectively the circumferences of the circles O, O',

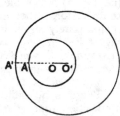

$$O'O \text{ is less than } O'A' - OA,$$

so that $O'O + OA$, or $O'A$, is less than $O'A'$,

and therefore A is internal to the circle O'.

But, of all the straight lines from O' to the circumference of the circle O, $O'A$ passing through the centre O is the greatest [III. 7];

whence all the points of the circumference of O are internal to the circle O'.

A similar argument to the preceding will show that all points within the circle O are internal to the circle O'.

3. *If the distance between the centres of two circles is equal to the sum of the radii, the two circumferences have one point common and one only, and that point is on the line of centres. Each circle is external to the other.*

Let O, O' be the centres, r, r' the radii of the circles, so that OO' is equal to $r + r'$.

Thus OO' is greater than r, so that O' is external to the circle O, and the circumference of the circle O cuts OO' in a point A.

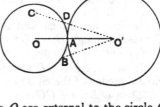

And, since OO' is equal to $r + r'$, and OA to r, it follows that $O'A$ is equal to r', so that A belongs also to the circumference of the circle O'.

The proof that all other points on, and all points within, the circumference of the circle O are external to the circle O' follows the similar proof of prop. 1 above. And similarly all points (except A) on, and all points within, the circumference of the circle O' are external to the circle O.

The two circles, having one common point only, *touch* at that point, which lies, as shown, on the line of centres. And, since the circles are external to one another, they touch *externally*.

4. *If the distance between the centres of two unequal circles is equal to the difference between the radii, the two circumferences have one point and one only in common, and that point lies on the line of centres. The lesser circle is within the other.*

The proof is that of prop. 2 above, *mutatis mutandis*.

The circles here touch *internally* at the point on the line of centres.

5. *If the distance between the centres of two circles is less than the sum, and greater than the difference, of the radii, the two circumferences have two common points symmetrically situated with respect to the line of centres but not lying on that line.*

Let O, O' be the centres of the two circles, r, r' their radii, r' being the greater, so that

$$r' - r < OO' < r + r'.$$

It follows that in any case $OO' + r > r'$, so that, if OM be taken on $O'O$ produced equal to r (so that M is on the circumference of the circle O), M is external to the circle O'.

We have to use the same Postulate as in Eucl. I. 1 that

An arc of a circle which has one extremity within and the other without a given circle has one point common with the latter and only one; from which it follows, if we consider two such arcs making a complete circumference, that, *if a circumference of a circle passes through one point internal to, and one point external to a given circle, it cuts the latter circle in two points.*

We have then to prove that the circle O, besides having one point M of its circumference external to the circle O', has one other point of its circumference (L) internal to the latter circle.

Three cases have to be distinguished according as OO' is greater than, equal to, or less than, the radius r of the lesser circle.

(1) $OO' > r$.   (See the preceding figure.)

Measure OL along OO' equal to r, so that L lies on the circumference of the circle O.

Then, since $OO' < r + r'$, $O'L$ will be less than r', so that L is within the circle O'.

(2) $OO' = r$.

In this case the circumference of the circle O passes through O', or L coincides with O'.

(3) $OO' < r$.

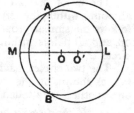

If we measure OL along OO' equal to r, the point L will lie on the circumference of the circle O.

Then $O'L = r - OO'$,

so that $O'L < r$, and *a fortiori* $O'L < r'$, so that L lies within the circle O'.

Thus, in all three cases, since the circumference of O passes through one point (M) external to, and one point (L) internal to, the circle O', the two circumferences intersect in two points A, B [Post.]

And A, B cannot lie on the line of centres OO', since this straight line intersects the circle O in L, M only, and of these points one is inside, the other outside, the circle O'.

Since AB is a common chord of both circles, the straight line bisecting it at right angles passes through both centres, i.e. is identical with OO'.

And again by means of III. 7, 8 we prove that all points except A, B on the arc ALB lie within the circle O', and all points except A, B on the arc AMB outside that circle; and so on.

PROPOSITION 13.

A circle does not touch a circle at more points than one, whether it touch it internally or externally.

For, if possible, let the circle $ABDC$ touch the circle $EBFD$, first internally, at more points than one, namely D, B.

Let the centre G of the circle $ABDC$, and the centre H of $EBFD$, be taken.

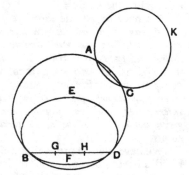

Therefore the straight line joined from G to H will fall on B, D.   [III. 11]

Let it so fall, as $BGHD$.

Then, since the point G is the centre of the circle $ABCD$,

BG is equal to GD;

therefore BG is greater than HD;

therefore BH is much greater than HD.

Again, since the point H is the centre of the circle $EBFD$,

20
$$BH \text{ is equal to } HD;$$

but it was also proved much greater than it: which is impossible.

Therefore a circle does not touch a circle internally at more points than one.

25   I say further that neither does it so touch it externally.

For, if possible, let the circle ACK touch the circle $ABDC$ at more points than one, namely A, C,

and let AC be joined.

Then, since on the circumference of each of the circles
30 $ABDC$, ACK two points A, C have been taken at random, the straight line joining the points will fall within each circle;   [III. 2]

but it fell within the circle $ABCD$ and outside ACK [III. Def. 3]: which is absurd.

35   Therefore a circle does not touch a circle externally at more points than one.

And it was proved that neither does it so touch it internally.

Therefore etc.   Q. E. D.

3, 7, 14, 27, 30, 33. **ABDC.** Euclid writes $ABCD$ (here and in the next proposition), notwithstanding the order in which the points are placed in the figure.

25, 37. **does it so touch it.** It is necessary to supply these words which the Greek (ὅτι οὐδὲ ἐκτός and ὅτι οὐδὲ ἐντός) leaves to be understood.

The difficulties which have been felt in regard to the proofs of this proposition need not trouble us now, because they have already been disposed of in the discussion of the more crucial propositions III. 11, 12.

Euclid's proof of the first part of the proposition differs from Simson's; and we will deal with Euclid's first. On this Camerer remarks that it is assumed that the supposed second point of contact lies on the line of centres *produced beyond the centre of the "outer" circle*, whereas all that is proved in III. 11 is that the line of centres *produced beyond the centre of the "inner" circle* passes through a point of contact. But, by the same argument as that given on III. 11, we show that the circles cannot have a point of contact, or even any common point, outside the line of centres, because, if there were such a point, there would be a corresponding common point on the other side of the line, and the circles would have *three* common points. Hence the only hypothesis left is that the second point of contact may be *on* the line of centres but in the direction of the centre of the "*outer*" circle; and Euclid's proof disposes of this hypothesis.

Heron (in an-Nairīzi, ed. Curtze, pp. 122—4), curiously enough, does not question Euclid's assumption that the line of centres passes through both points of contact (if double contact is possible) ; but he devotes some space to proving that the centre of the "outer" circle must lie within the "inner" circle, a fact which he represents Euclid as asserting ("sicut dixit Euclides"), though there is no such assertion in our text. The proof of the fact is of course easy. If the line of centres passes through *both* points of contact, and the centre of the "outer" circle lies either on or outside the "inner" circle, the line of centres must cut the "inner" circle in *three* points in all: which is impossible, as Heron shows by the lemma, which he places here (and proves by I. 16), that *a straight line cannot cut the circumference of a circle in more points than two.*

Simson's proof is as follows (there is no real need for giving two figures as he does).

"If it be possible, let the circle *EBF* touch the circle *ABC* in more points than one, and first on the inside, in the points *B*, *D*; join *BD*, and draw *GH* bisecting *BD* at right angles.

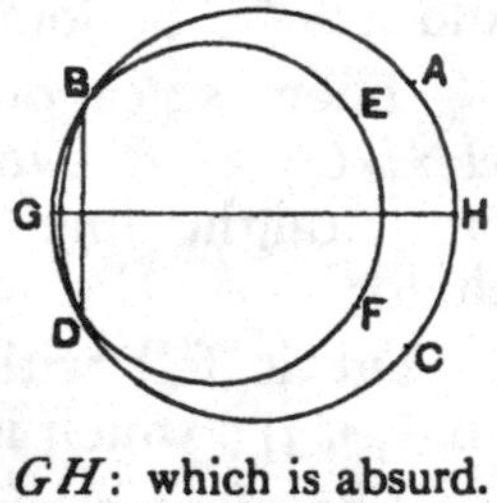

Therefore, because the points *B*, *D* are in the circumference of each of the circles, the straight line *BD* falls within each of them : And their centres are in the straight line *GH* which bisects *BD* at right angles :

Therefore *GH* passes through the point of contact [III. 11]; but it does not pass through it, because the points *B*, *D* are without the straight line *GH*: which is absurd.

Therefore one circle cannot touch another on the inside in more points than one."

On this Camerer remarks that, unless III. 11 be more completely elucidated than it is by Euclid's demonstration, which Simson has, it is not sufficiently clear that, besides the point of contact in which *GH* meets the circles, they cannot have another point of contact either (1) on *GH* or (2) outside it. Here again the latter supposition (2) is rendered impossible because in that case there would be a third common point on the opposite side of *GH*; and the former supposition (1) is that which Euclid's proof destroys.

Simson retains Euclid's proof of the second part of the proposition, though his own proof of the first part would apply to the second part also if a reference to III. 12 were substituted for the reference to III. 11. Euclid might also have proved the second part by the same method as that which he employs for the first part.

PROPOSITION 14.

In a circle equal straight lines are equally distant from the centre, and those which are equally distant from the centre are equal to one another.

Let *ABDC* be a circle, and let *AB*, *CD* be equal straight lines in it;

I say that *AB*, *CD* are equally distant from the centre.

For let the centre of the circle *ABDC* be taken [III. 1],

and let it be E; from E let EF, EG be drawn perpendicular to AB, CD, and let AE, EC be joined.

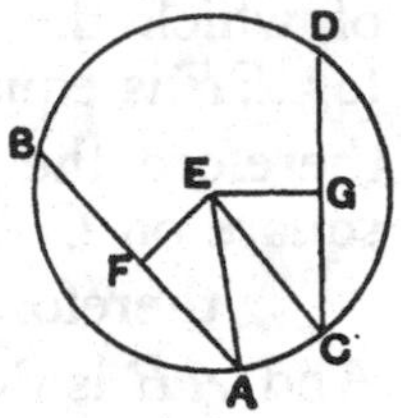

Then, since a straight line EF through the centre cuts a straight line AB not through the centre at right angles, it also bisects it.

[III. 3]

Therefore AF is equal to FB;
therefore AB is double of AF.

For the same reason

CD is also double of CG;
and AB is equal to CD;

therefore AF is also equal to CG.

And, since AE is equal to EC,
the square on AE is also equal to the square on EC.

But the squares on AF, EF are equal to the square on AE, for the angle at F is right;

and the squares on EG, GC are equal to the square on EC, for the angle at G is right;   [I. 47]

therefore the squares on AF, FE are equal to the squares on CG, GE,

of which the square on AF is equal to the square on CG, for AF is equal to CG;

therefore the square on FE which remains is equal to the square on EG,

therefore EF is equal to EG.

But in a circle straight lines are said to be equally distant from the centre when the perpendiculars drawn to them from the centre are equal;   [III. Def. 4]

therefore AB, CD are equally distant from the centre.

Next, let the straight lines AB, CD be equally distant from the centre; that is, let EF be equal to EG.

I say that AB is also equal to CD.

For, with the same construction, we can prove, similarly, that AB is double of AF, and CD of CG.

And, since AE is equal to CE,
the square on AE is equal to the square on CE.

But the squares on EF, FA are equal to the square on AE, and the squares on EG, GC equal to the square on CE. [I. 47]

Therefore the squares on EF, FA are equal to the squares on EG, GC,

of which the square on EF is equal to the square on EG, for EF is equal to EG;

therefore the square on AF which remains is equal to the square on CG;

therefore AF is equal to CG.

And AB is double of AF, and CD double of CG;

therefore AB is equal to CD.

Therefore etc.

Q. E. D.

Heron (an-Nairīzī, pp. 125—7) has an elaborate addition to this proposition in which he proves, first by *reductio ad absurdum*, and then directly, that the centre of the circle falls between the two chords.

PROPOSITION 15.

Of straight lines in a circle the diameter is greatest, and of the rest the nearer to the centre is always greater than the more remote.

Let $ABCD$ be a circle, let AD be its diameter and E the centre; and let BC be nearer to the diameter AD, and FG more remote;

I say that AD is greatest and BC greater than FG.

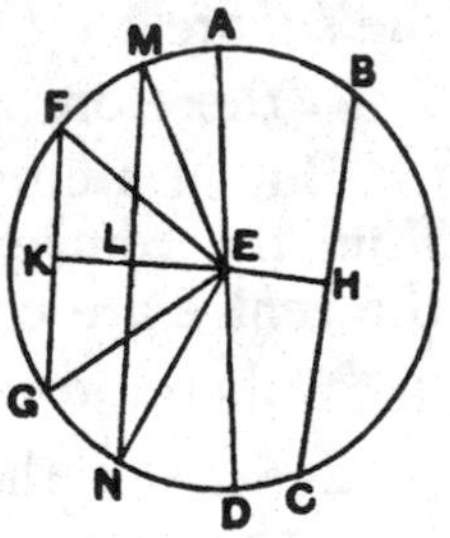

For from the centre E let EH, EK be drawn perpendicular to BC, FG.

Then, since BC is nearer to the centre and FG more remote, EK is greater than EH.　　　　　[III. Def. 5]

Let EL be made equal to EH, through L let LM be drawn at right angles to EK and carried through to N, and let ME, EN, FE, EG be joined.

Then, since EH is equal to EL,

　　　　BC is also equal to MN.　　　　　[III. 14]

Again, since AE is equal to EM, and ED to EN,

　　　　AD is equal to ME, EN.

But *ME*, *EN* are greater than *MN*,   [1. 20]
and *MN* is equal to *BC*;

therefore *AD* is greater than *BC*.

And, since the two sides *ME*, *EN* are equal to the two sides *FE*, *EG*,

and the angle *MEN* greater than the angle *FEG*,

therefore the base *MN* is greater than the base *FG*.   [1. 24]

But *MN* was proved equal to *BC*.

Therefore the diameter *AD* is greatest and *BC* greater than *FG*.

Therefore etc.   Q. E. D.

1. **Of straight lines.** The Greek leaves these words to be understood.
5. **Nearer to the diameter AD.** As *BC*, *FG* are not in general parallel to *AD*, Euclid should have said "nearer to the centre."

It will be observed that Euclid's proof differs from that given in our text-books (which is Simson's) in that Euclid introduces another line *MN*, which is drawn so as to be equal to *BC* but at right angles to *EK* and therefore parallel to *FG*. Simson dispenses with *MN* and bases his proof on a similar proof by Theodosius (*Sphaerica* I. 6). He proves that the sum of the squares on *EH*, *HB* is equal to the sum of the squares on *EK*, *KF*; whence he infers that, since the square on *EH* is less than the square on *EK*, the square on *BH* is greater than the square on *FK*. It may be that Euclid would have regarded this as too complicated an inference to make without explanation or without an increase in the number of his axioms. But, on the other hand, Euclid himself assumes that the angle subtended at the centre by *MN* is greater than the angle subtended by *FG*, or, in other words, that *M*, *N* both fall outside the triangle *FEG*. This is a similar assumption to that made in III. 7, 8, as already noticed; and its truth is obvious because *EM*, *EN*, being radii of the circle, are greater than the distances from *E* to the points in which *MN* cuts *EF*, *EG*, and therefore the latter points are nearer than *M*, *N* are to *L*, the foot of the perpendicular from *E* to *MN*.

Simson adds the converse of the proposition, proving it in the same way as he proves the proposition itself.

PROPOSITION 16.

The straight line drawn at right angles to the diameter of a circle from its extremity will fall outside the circle, and into the space between the straight line and the circumference another straight line cannot be interposed; further the angle of the semicircle is greater, and the remaining angle less, than any acute rectilineal angle.

Let *ABC* be a circle about *D* as centre and *AB* as diameter;

I say that the straight line drawn from A at right angles to AB from its extremity will fall outside the circle.

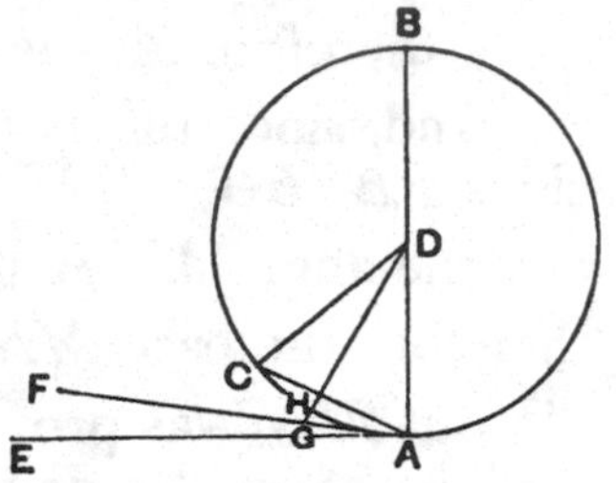

For suppose it does not, but, if possible, let it fall within as CA, and let DC be joined.

Since DA is equal to DC,

the angle DAC is also equal to the angle ACD.   [I. 5]

But the angle DAC is right;

    therefore the angle ACD is also right:

thus, in the triangle ACD, the two angles DAC, ACD are equal to two right angles: which is impossible.   [I. 17]

Therefore the straight line drawn from the point A at right angles to BA will not fall within the circle.

Similarly we can prove that neither will it fall on the circumference;

    therefore it will fall outside.

Let it fall as AE;

I say next that into the space between the straight line AE and the circumference CHA another straight line cannot be interposed.

For, if possible, let another straight line be so interposed, as FA, and let DG be drawn from the point D perpendicular to FA.

Then, since the angle AGD is right,

    and the angle DAG is less than a right angle,

                AD is greater than DG.   [I. 19]

But DA is equal to DH;

    therefore DH is greater than DG, the less than the greater: which is impossible.

Therefore another straight line cannot be interposed into the space between the straight line and the circumference.

I say further that the angle of the semicircle contained by the straight line BA and the circumference CHA is greater than any acute rectilineal angle,

and the remaining angle contained by the circumference CHA and the straight line AE is less than any acute rectilineal angle.

For, if there is any rectilineal angle greater than the angle contained by the straight line BA and the circumference

CHA, and any rectilineal angle less than the angle contained by the circumference *CHA* and the straight line *AE*, then into the space between the circumference and the straight line *AE* a straight line will be interposed such as will make an angle contained by straight lines which is greater than the angle contained by the straight line *BA* and the circumference *CHA*, and another angle contained by straight lines which is less than the angle contained by the circumference *CHA* and the straight line *AE*.

But such a straight line cannot be interposed;

therefore there will not be any acute angle contained by straight lines which is greater than the angle contained by the straight line *BA* and the circumference *CHA*, nor yet any acute angle contained by straight lines which is less than the angle contained by the circumference *CHA* and the straight line *AE*.—

PORISM. From this it is manifest that the straight line drawn at right angles to the diameter of a circle from its extremity touches the circle.

Q. E. D.

4. cannot be interposed, literally "will not fall in between" (οὐ παρεμπεσεῖται).

This proposition is historically interesting because of the controversies to which the last part of it gave rise from the 13th to the 17th centuries. History was here repeating itself, for it is certain that, in ancient Greece, both before and after Euclid's time, there had been a great deal of the same sort of contention about the nature of the "angle of a semicircle" and the "remaining angle" between the circumference of the semicircle and the tangent at its extremity. As we have seen (note on I. Def. 8), the latter angle had a recognised name, κερατοειδὴς γωνία, *horn-like* or *cornicular* angle; though this term does not appear in Euclid, it is often used by Proclus, evidently as a term well understood. While it is from Proclus that we get the best idea of the ancient controversies on this subject, we may, I think, infer their prevalence in Euclid's time from this solitary appearance of the two "angles" in the *Elements*. Along with the definition of the angle *of* a segment, it seems to show that, although these angles are only mentioned to be dropped again immediately, and are of no use in elementary geometry, or even at all, Euclid thought that an allusion to them would be expected of him; it is as if he merely meant to guard himself against appearing to ignore a subject which the geometers of his time regarded with interest. If this conjecture is right, the mention of these angles would correspond to the insertion of definitions of which he makes no use, e.g. those of a rhombus and a rhomboid.

Proclus has no hesitation in speaking of the "angle of a semicircle" and the "horn-like angle" as true *angles*. Thus he says that "angles are contained by a straight line and a circumference in two ways; for they are either contained by a straight line and a convex circumference, like that of the semi-

circle, or by a straight line and a concave circumference, like the κερατοειδής "
(p. 127, 11—14). "There are *mixed* lines, as spirals, and angles, as the angle
of a semicircle and the κερατοειδής" (p. 104, 16—18). The difficulty which
the ancients felt arose from the very fact which Euclid embodies in this
proposition. Since an angle can be divided by a line, it would seem to be a
magnitude; "but if it is a magnitude, and all homogeneous magnitudes which
are finite have a ratio to one another, then all homogeneous angles, or rather
all those on surfaces, will have a ratio to one another, so that the *cornicular*
will also have a ratio to the rectilineal. But things which have a ratio to one
another can, if multiplied, exceed one another. Therefore the *cornicular*
angle will also sometime exceed the rectilineal; which is impossible, for it is
proved that the former is less than any rectilineal angle" (Proclus, p. 121,
24—122, 6). The nature of contact between straight lines and circles was
also involved in the question, and that this was the subject of controversy
before Euclid's time is clear from the title of a work attributed to Democritus
(fl. 420—400 B.C.) περὶ διαφορῆς γνώμονος ἢ περὶ ψαύσιος κύκλου καὶ σφαίρης,
On a difference in a gnomon or on contact of a circle and a sphere. There is,
however, another reading of the first words of this title as given by Diogenes
Laertius (IX. 47), namely περὶ διαφορῆς γνώμης, *On a difference of opinion*, etc.
May it not be that neither reading is correct, but that the words should be
περὶ διαφορῆς γωνίης ἢ περὶ ψαύσιος κύκλου καὶ σφαίρης, *On a difference in an
angle or on contact with a circle and a sphere?* There would, of course,
hardly be any "angle" in connexion with the sphere; but I do not think that
this constitutes any difficulty, because the sphere might easily be tacked on as
a kindred subject to the circle. A curiously similar collocation of words
appears in a passage of Proclus, though this may be an accident. He says
(p. 50, 4) πῶς δὲ γωνιῶν διαφορὰς λέγομεν καὶ αὐξήσεις αὐτῶν ... and then, in
the next line but one, πῶς δὲ τὰς ἀφὰς τῶν κύκλων ἢ τῶν εὐθειῶν, "In what
sense do we speak of *differences of angles and of increases of them* ... and in
what sense of the *contacts* (or meetings) *of circles* or of straight lines?"
I cannot help thinking that this subject of *cornicular* angles would have had
a fascination for Democritus as being akin to the question of infinitesimals,
and very much of the same character as the other question which Plutarch
(*On Common Notions,* XXXIX. 3) says that he raised, namely that of the
relation between the base of a cone and a section of it by a plane parallel to
the base and apparently, to judge by the context, infinitely near to it: "if
a cone were cut by a plane parallel to its base, what must we think of the
surfaces of the sections, that they are equal or unequal? For, if they are
unequal, they will make the cone irregular, as having many indentations like
steps, and unevennesses; but, if they are equal, the sections will be equal,
and the cone will appear to have the property of the cylinder, as being made
up of equal and not unequal circles, which is the height of absurdity."

The contributions by Democritus to such investigations are further attested
by a passage in the *Method* of Archimedes discovered by Heiberg in 1906
(*Archimedes*, ed. Heiberg, Vol. II. 1913, p. 430; T. L. Heath, *The* Method
of Archimedes, 1912, p. 13), which says that, though Eudoxus was the first to
discover the scientific proof of the propositions (attributed to him) that the
cone and the pyramid are one-third of the cylinder and prism respectively
which have the same base and equal height, they were first *stated*, without
proof, by Democritus.

A full history of the later controversies about the cornicular "angle"
cannot be given here; more on the subject will be found in Camerer's
Euclid (Excursus IV. on III. 16) or in Cantor's *Geschichte der Mathematik.*

Vol. II. (see *Contingenzwinkel* in the index). But the following short note about the attitude of certain well-known mathematicians to the question will perhaps not be out of place. Johánnes Campanus, who edited Euclid in the 13th century, inferred from III. 16 that there was a flaw in the principle that *the transition from the less to the greater, or vice versâ, takes place through all intermediate quantities and therefore through the equal.* If a diameter of a circle, he says, be moved about its extremity until it takes the position of the tangent to that circle, then, as long as it cuts the circle, it makes an acute angle *less* than the "angle of a semicircle"; but the moment it ceases to cut, it makes a right angle *greater* than the same "angle of a semicircle." The rectilineal angle is never, during the transition, *equal* to the "angle of a semicircle." There is therefore an apparent inconsistency with X. 1, and Campanus could only observe (as he does on that proposition), in explanation of the paradox, that "these are not angles in the same sense (univoce), for the curved and the straight are not things of the same kind without qualification (simpliciter)." The argument assumes, of course, that the right angle *is* greater than the "angle of a semicircle."

Very similar is the statement of the paradox by Cardano (1501—1576), who observed that *a quantity may continually increase without limit, and another diminish without limit; and yet the first, however increased, may be less than the second, however diminished.* The first quantity is of course the *angle of contact*, as he calls it, which may be "increased" indefinitely by drawing smaller and smaller circles touching the same straight line at the same point, but will always be less than any acute rectilineal angle however small.

We next come to the French geometer, Peletier (Peletarius), who edited the *Elements* in 1557, and whose views on this subject seem to mark a great advance. Peletier's opinions and arguments are most easily accessible in the account of them given by Clavius (Christoph Klau [?], 1537—1612) in the 1607 edition of his Euclid. The violence of the controversy between the two will be understood from the fact that the arguments and counter-arguments (which sometimes run into other matters than the particular question at issue) cover, in that book, 26 pages of small print. Peletier held that the "angle of contact" was not an angle at all, that the "contact of two circles," i.e. the "angle" between the circumferences of two circles touching one another internally or externally, is not a *quantity*, and that the "contact of a straight line with a circle" is not a *quantity* either; that angles contained by a diameter and a circumference whether inside or outside the circle are *right angles* and equal to rectilineal right angles, and that angles contained by a diameter and the circumference in *all* circles are *equal.* The proof which Peletier gave of the latter proposition in a letter to Cardano is sufficiently ingenious. If a greater and a less semicircle be placed with their diameters terminating at a common point and lying in a straight line, then (1) the *angle of* the larger obviously cannot be *less* than the *angle of* the smaller. Neither (2) can the former be *greater* than the latter; for, if it were, we could obtain another *angle of* a semicircle greater still by drawing a still larger semicircle, and so on, until we should ultimately have an *angle of* a semicircle greater than a right angle: which is impossible. Hence the *angles of* semicircles must all be *equal*, and the differences between them *nothing.* Having satisfied himself that all *angles of contact* are *not*-angles, *not*-quantities, and therefore *nothings*, Peletier holds the difficulty about X. 1 to be at an end. He adds the interesting remark that the essence of an angle is in *cutting*, not contact, and that a tangent is not *inclined* to the circle at the point of contact but is, as it were, *immersed* in it at that point, just as much as if the circle did not diverge from it on either side.

The reply of Clavius need not detain us.   He argues, evidently appealing
to the eye, that the angle of contact can be *divided* by the arc of a circle
greater than the given one, that the angles of two semicircles of different sizes
cannot be equal, since they do not coincide if they are applied to one another,
that there is nothing to prevent *angles of contact* from being *quantities*, it being
only necessary, in view of x. 1, to admit that they are not of the same kind as
rectilineal angles ; lastly that, if the angle of contact had been a *nothing*,
Euclid would not have given himself so much trouble to prove that it is less
than any acute angle.   (The word is *desudasset*, which is certainly an
exaggeration as applied to what is little more than an *obiter dictum* in III. 16.)

Vieta (1540—1603) ranged himself on the side of Peletier, maintaining
that the *angle of contact* is no angle ; only he uses a new method of proof.
The circle, he says, may be regarded as a plane figure with an infinite number
of sides and angles ; but *a straight line touching a straight line, however short
it may be, will coincide with that straight line and will not make an angle.*
Never before, says Cantor (II₁, p. 540), had it been so plainly declared what
exactly was to be understood by *contact*.

Galileo Galilei (1564—1642) seems to have held the same view as Vieta
and to have supported it by a very similar argument derived from the comparison of the circle and an inscribed polygon with an infinite number of
sides.

The last writer on the question who must be mentioned is John Wallis
(1616—1703).   He published in 1656 a paper entitled *De angulo contactus et
semicirculi tractatus* in which he also maintained that the so-called angle was
not a true angle, and was not a *quantity*.   Vincent Leotaud (1595—1672)
took up the cudgels for Clavius in his *Cyclomathia* which appeared in 1663.
This brought a reply from Wallis in a letter to Leotaud dated 17 February,
1667, but not apparently published till it appeared in *A defense of the treatise
of the angle of contact* which, with a separate title-page, and date 1684, was
included in the English edition of his Algebra dated 1685.   The essence of
Wallis' position may be put as follows.   According to Euclid's definition, a
plane angle is an *inclination* of two lines ; therefore two lines forming an angle
must *incline* to one another, and, if two lines meet without being *inclined* to
one another at the point of meeting (which is the case when a circumference
is touched by a straight line), the lines do not form an *angle*.   The "angle of
contact" is therefore no angle, because *at the point of contact* the straight line
is not inclined to the circle but lies on it ἀκλινῶς, or is coincident with it.
Again, as a point is not a line but a *beginning* of a line, and a line is not a
surface but a *beginning* of a surface, so an angle is not the distance between
two lines, but their initial tendency towards separation : *Angulus (seu gradus
divaricationis) Distantia non est sed Inceptivus distantiae.*   How far lines, which
at their point of meeting do not form an angle, separate from one another as
they pass on depends on the *degree of curvature* (gradus curvitatis), and it is
the latter which has to be compared in the case of two lines so meeting.   The
arc of a smaller circle is more curved as having as much curvature in a lesser
length, and is therefore curved in a greater degree.   Thus what Clavius called
angulus contactus becomes with Wallis *gradus curvitatis*, the use of which
expression shows that curvature and curvature can be compared according to
one and the same standard.   A straight line has the least possible curvature ;
but of the "angle" made by it with a curve which it touches we cannot say that
it is greater or less than the "angle" which a second curve touching the same
straight line at the same point makes with the first curve ; for in both cases
there is no true angle at all (cf. Cantor III₁, p. 24).

The words usually given as a part of the corollary "and that a straight line touches a circle at one point only, since in fact the straight line meeting it in two points was proved to fall within it" are omitted by Heiberg as being an undoubted addition of Theon's. It was Simson who added the further remark that "it is evident that there can be but one straight line which touches the circle at the same point."

PROPOSITION 17.

From a given point to draw a straight line touching a given circle.

Let *A* be the given point, and *BCD* the given circle; thus it is required to draw from the point *A* a straight line touching the circle *BCD*.

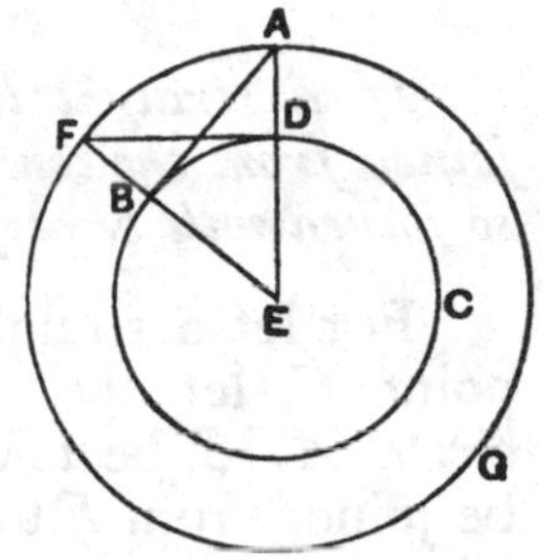

For let the centre *E* of the circle be taken;   [III. 1] let *AE* be joined, and with centre *E* and distance *EA* let the circle *AFG* be described;

from *D* let *DF* be drawn at right angles to *EA*,

and let *EF*, *AB* be joined;

I say that *AB* has been drawn from the point *A* touching the circle *BCD*.

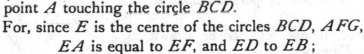

For, since *E* is the centre of the circles *BCD*, *AFG*,

      EA is equal to *EF*, and *ED* to *EB*;

therefore the two sides *AE*, *EB* are equal to the two sides *FE*, *ED*;

and they contain a common angle, the angle at *E*;

      therefore the base *DF* is equal to the base *AB*,

      and the triangle *DEF* is equal to the triangle *BEA*,

      and the remaining angles to the remaining angles;   [I. 4]

      therefore the angle *EDF* is equal to the angle *EBA*.

But the angle *EDF* is right;

      therefore the angle *EBA* is also right.

Now *EB* is a radius;

and the straight line drawn at right angles to the diameter of a circle, from its extremity, touches the circle; [III. 16, Por.]

      therefore *AB* touches the circle *BCD*.

Therefore from the given point *A* the straight line *AB* has been drawn touching the circle *BCD*.   Q. E. F.

The construction shows, of course, that two straight lines can be drawn from a given external point to touch a given circle; and it is equally obvious that these two straight lines are equal in length and equally inclined to the straight line joining the external point to the centre of the given circle. These facts are given by Heron (an-Nairīzī, p. 130).

It is true that Euclid leaves out the case where the given point lies *on* the circumference of the circle, doubtless because the construction is so directly indicated by III. 16, Por. as to be scarcely worth a separate statement.

An easier solution is of course possible as soon as we know (III. 31) that the angle in a semicircle is a right angle; for we have only to describe a circle on *AE* as diameter, and this circle cuts the given circle in the two points of contact.

PROPOSITION 18.

If a straight line touch a circle, and a straight line be joined from the centre to the point of contact, the straight line so joined will be perpendicular to the tangent.

For let a straight line *DE* touch the circle *ABC* at the point *C*, let the centre *F* of the circle *ABC* be taken, and let *FC* be joined from *F* to *C*;
I say that *FC* is perpendicular to *DE*.

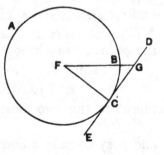

For, if not, let *FG* be drawn from *F* perpendicular to *DE*.

Then, since the angle *FGC* is right,
the angle *FCG* is acute; [I. 17]
and the greater angle is subtended by the greater side;   [I. 19]
therefore *FC* is greater than *FG*.

But *FC* is equal to *FB*;
therefore *FB* is also greater than *FG*,
the less than the greater: which is impossible.

Therefore *FG* is not perpendicular to *DE*.

Similarly we can prove that neither is any other straight line except *FC*;
therefore *FC* is perpendicular to *DE*.

Therefore etc.

Q. E. D.

3. the tangent, ἡ ἐφαπτομένη.

Just as III. 3 contains two *partial* converses of the Porism to III. 1, so the present proposition and the next give two partial converses of the corollary to III. 16. We may show their relation thus : suppose three things, (1) a tangent at a point of a circle, (2) a straight line drawn from the centre to the point of contact, (3) right angles made at the point of contact [with (1) or (2) as the case may be]. Then the corollary to III. 16 asserts that (2) and (3) together give (1), III. 18 that (1) and (2) give (3), and III. 19 that (1) and (3) give (2), i.e. that the straight line drawn from the point of contact at right angles to the tangent passes through the centre.

PROPOSITION 19.

If a straight line touch a circle, and from the point of contact a straight line be drawn at right angles to the tangent, the centre of the circle will be on the straight line so drawn.

For let a straight line DE touch the circle ABC at the point C, and from C let CA be drawn at right angles to DE ;
I say that the centre of the circle is on AC.

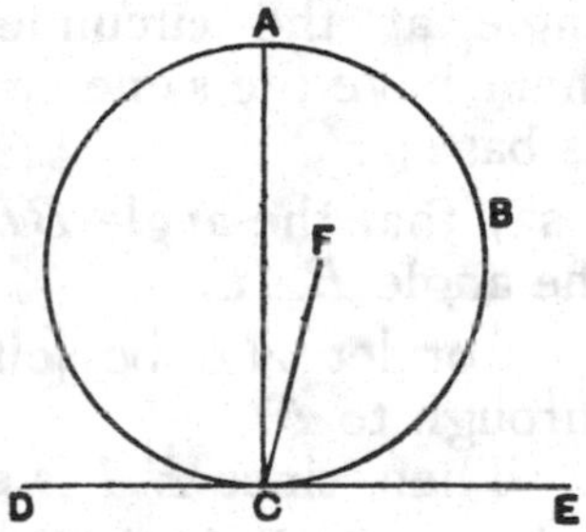

For suppose it is not, but, if possible, let F be the centre, and let CF be joined.
Since a straight line DE touches the circle ABC,
and FC has been joined from the centre to the point of contact,

FC is perpendicular to DE ;   [III. 18]
therefore the angle FCE is right.

But the angle ACE is also right ;
therefore the angle FCE is equal to the angle ACE,

the less to the greater : which is impossible.
Therefore F is not the centre of the circle ABC.
Similarly we can prove that neither is any other point except a point on AC.
Therefore etc.

Q. E. D.

We may also regard III. 19 as a partial converse of III. 18. Thus suppose (1) a straight line through the centre, (2) a straight line through the point of contact, and suppose (3) to mean perpendicular to the tangent; then III. 18 asserts that (1) and (2) combined produce (3), and III. 19 that (2) and (3) produce (1); while again we may enunciate a second partial converse of III. 18, corresponding to the statement that (1) and (3) produce (2), to the effect that a straight line drawn through the centre perpendicular to the tangent passes through the point of contact.

We may add at this point, or even after the Porism to III. 16, the theorem that *two circles which touch one another internally or externally have a common tangent at their point of contact.* For the line joining their centres, produced if necessary, passes through their point of contact, and a straight line drawn through that point at right angles to the line of centres is a tangent to both circles.

PROPOSITION 20.

In a circle the angle at the centre is double of the angle at the circumference, when the angles have the same circumference as base.

Let *ABC* be a circle, let the angle *BEC* be an angle
5 at its centre, and the angle *BAC* an angle at the circumference, and let them have the same circumference *BC* as base;

I say that the angle *BEC* is double of
10 the angle *BAC*.

For let *AE* be joined and drawn through to *F*.

Then, since *EA* is equal to *EB*, the angle *EAB* is also equal to the
15   angle *EBA*;   [I. 5]
therefore the angles *EAB*, *EBA* are double of the angle *EAB*.

But the angle *BEF* is equal to the angles *EAB*, *EBA*;
                                                        [I. 32]
therefore the angle *BEF* is also double of the angle
20 *EAB*.

For the same reason
    the angle *FEC* is also double of the angle *EAC*.

Therefore the whole angle *BEC* is double of the whole angle *BAC*.

25   Again let another straight line be inflected, and let there be another angle *BDC*; let *DE* be joined and produced to *G*.

Similarly then we can prove that the angle *GEC* is double of the angle *EDC*,

30   of which the angle *GEB* is double of the angle *EDB*;

therefore the angle *BEC* which remains is double of the angle *BDC*.

Therefore etc.   Q. E. D.

25. **let another straight line be inflected,** κεκλάσθω δὴ πάλιν (without εὐθεῖα). The verb κλάω (to *break off*) was the regular technical term for drawing from a point a (broken) straight line which first meets another straight line or curve and is then *bent back* from it to another point, or (in other words) for drawing straight lines from two points meeting at a point on a curve or another straight line. κεκλάσθαι is one of the geometrical terms the definition of which must according to Aristotle be assumed (*Anal. Post.* I. 10, 76 b 9).

The early editors, Tartaglia, Commandinus, Peletarius, Clavius and others, gave the extension of this proposition to the case where the segment is less than a semicircle, and where accordingly the "angle" corresponding to Euclid's "angle at the centre" is greater than two right angles. The convenience of the extension is obvious, and the proof of it is the same as the first part of Euclid's proof. By means of the extension III. 21 is demonstrated without making two cases; III. 22 will follow immediately from the fact that the sum of the "angles at the centre" for two segments making up a whole circle is equal to four right angles; also III. 31 follows immediately from the extended proposition.

But all the editors referred to were forestalled in this matter by Heron, as we now learn from the commentary of an-Nairīzī (ed. Curtze, p. 131 sqq.). Heron gives the extension of Euclid's proposition which, he says, it had been left for him to make, but which is necessary in order that the caviller may not be able to say that the next proposition (about the equality of the angles in any segment) is not established generally, i.e. in the case of a segment less than a semicircle as well as in the case of a segment greater than a semicircle, inasmuch as III. 20, as given by Euclid, only enables us to prove it in the latter case. Heron's enunciation is important as showing how he describes what we should now call an "angle" greater than two right angles. (The language of Gherard's translation is, in other respects, a little obscure; but the meaning is made clear by what follows.)

"The angle," Heron says, "which is at the centre of any circle is double of the angle which is at the circumference of it *when one arc is the base of both angles*; and *the remaining angles which are at the centre, and fill up the four right angles*, are double of the angle at the circumference of the arc which is subtended by the [original] angle which is at the centre."

Thus the "angle greater than two right angles" is for Heron the sum of certain "angles" in the Euclidean sense of angles less than two right angles. The particular method of splitting up which Heron adopts will be seen from his proof, which is in substance as follows.

Let *CDB* be an *angle* at the centre, *CAB* that at the circumference.

Produce *BD*, *CD* to *F*, *G*;

take any point *E* on *BC*, and join *BE*, *EC*, *ED*.

Then any angle in the segment *BAC* is half of the angle *BDC*; and *the sum of the angles* BDG, GDF, FDC *is double of any angle in the segment* BEC.

Proof. Since *CD* is equal to *ED*,

the angles *DCE, DEC* are equal.

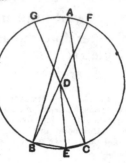

Therefore the exterior angle *GDE* is equal to twice the angle *DEC*.

Similarly the exterior angle *FDE* is equal to twice the angle *DEB*.

By addition, the angles *GDE, FDE* are double of the angle *BEC*.

But

the angle *BDC* is equal to the angle *FDG*,

therefore *the sum of the angles* BDG, GDF, FDC *is double of the angle* BEC.

And Euclid has proved the first part of the proposition, namely that the angle *BDC* is double of the angle *BAC*.

Now, says Heron, *BAC* is *any* angle in the segment *BAC*, and therefore *any* angle in the segment *BAC* is half of the angle *BDC*.

Therefore all the angles in the segment *BAC* are equal.

Again, *BEC* is *any* angle in the segment *BEC* and is equal to *half the sum of the angles* BDG, GDF, FDC.

Therefore all the angles in the segment *BEC* are equal.

Hence III. 21 is proved *generally*.

Lastly, says Heron,

since *the sum of the angles* BDG, GDF, FDC is double of the angle *BEC*,

and the angle *BDC* is double of the angle *BAC*,

therefore, by addition, the *sum of four right angles* is double of the sum of the angles *BAC, BEC*.

Hence the angles *BAC, BEC* are together equal to two right angles, and III. 22 is proved.

The above notes of Heron show conclusively, if proof were wanted, that Euclid had no idea of III. 20 applying *in terms* (either as a matter of enunciation or proof) to the case where the angle at the circumference, or the angle in the segment, is *obtuse*. He would not have recognised the "angle" greater than two right angles or the so-called "straight angle" as being an *angle* at all. This is indeed clear from his definition of an angle as the *inclination* κ.τ.ἑ., and from the language used by other later Greek mathematicians where there would be an opportunity for introducing the extension. Thus Proclus' notion of a "four-sided triangle" (cf. the note above on the definition of a triangle) shows that he did not count a re-entrant angle as an angle, and Zenodorus' application to the same figure of the word "hollow-angled" shows that in that case it was the exterior angle only which he would have called an angle. Further it would have been inconvenient to have introduced at the beginning of the *Elements* an "angle" equal to or greater than two right angles, because other definitions, e.g. that of a *right angle*, would have needed a qualification. If an "angle" might be equal to two right angles, one straight line in a straight line with another would have satisfied Euclid's definition of a right angle. This is noticed by Dodgson (p. 160), but it is practically brought out by Proclus on I. 13. "For he did not merely say that 'any straight line standing on a straight line either makes two right angles or angles equal to two right angles' but '*if it make angles*.'

If it stand on the straight line at its extremity and make one angle, is it possible for this to be equal to two right angles? It is of course impossible; *for every rectilineal angle is less than two right angles*, as every solid angle is less than four right angles (p. 292, 13—20)." [It is true that it has been generally held that the meaning of "angle" is tacitly extended in VI. 33, but there is no real ground for this view. See the note on the proposition.]

It will be observed that, following his usual habit, Euclid omits the demonstration of the case which some editors, e.g. Clavius, have thought it necessary to give separately, the case namely where one of the lines forming the angle in the segment passes through the centre. Euclid's proof gives so obviously the means of proving this that it is properly left out.

Todhunter observes, what Clavius had also remarked, that there are two assumptions in the proof of III. 20, namely that, if A is double of B and C double of D, then the sum, or difference, of A and C is equal to double the sum, or difference, of B and D respectively, the assumptions being particular cases of V. 1 and V. 5. But of course it is easy to satisfy ourselves of the correctness of the assumption without any recourse to Book V.

PROPOSITION 21.

In a circle the angles in the same segment are equal to one another.

Let $ABCD$ be a circle, and let the angles BAD, BED be angles in the same segment $BAED$;
I say that the angles BAD, BED are equal to one another.

For let the centre of the circle $ABCD$ be taken, and let it be F; let BF, FD be joined.

Now, since the angle BFD is at the centre,

and the angle BAD at the circumference,

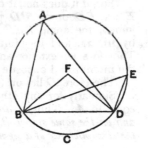

and they have the same circumference BCD as base, therefore the angle BFD is double of the angle BAD. [III. 20]

For the same reason

the angle BFD is also double of the angle BED;

therefore the angle BAD is equal to the angle BED.

Therefore etc.

Q. E. D.

Under the restriction that the "angle at the centre" used in III. 20 must be less than two right angles, Euclid's proof of this proposition only applies to the case of a segment greater than a semicircle, and the case of a segment equal to or less than a semicircle has to be considered separately. The simplest proof, of many, seems to be that of Simson.

"But, if the segment *BAED* be not greater than a semicircle, let *BAD*, *BED* be angles in it: these also are equal to one another.

Draw *AF* to the centre, and produce it to *C*, and join *CE*.

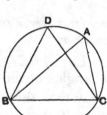

Therefore the segment *BADC* is greater than a semicircle, and the angles in it *BAC, BEC* are equal, by the first case.

For the same reason, because *CBED* is greater than a semicircle,

the angles *CAD, CED* are equal.

Therefore the whole angle *BAD* is equal to the whole angle *BED*."

We can prove, by means of *reductio ad absurdum*, the important converse of this proposition, namely that, *if there be any two triangles on the same base and on the same side of it, and with equal vertical angles, the circle passing through the extremities of the base and the vertex of one triangle will pass through the vertex of the other triangle also.* That a circle can be thus described about a triangle is clear from Euclid's construction in III. 9, which shows how to draw a circle passing through any three points, though it is in IV. 5 only that we have the problem stated. Now,

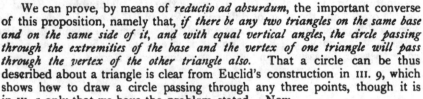

suppose a circle *BAC* drawn through the angular points of a triangle *BAC*, and let *BDC* be another triangle with the same base *BC* and on the same side of it, and having its vertical angle *D* equal to the angle *A*. Then shall the circle pass through *D*.

For, if it does not, it must pass through some point *E* on *BD* or on *BD* produced. If then *EC* be joined, the angle *BEC* is equal to the angle *BAC*, by III. 21, and therefore equal to the angle *BDC*. Therefore an exterior angle of a triangle is equal to the interior and opposite angle: which is impossible, by I. 16.

Therefore *D* lies on the circle *BAC*.

Similarly for any other triangle on the base *BC* and with vertical angle equal to *A*. Thus, *if any number of triangles be constructed on the same base and on the same side of it, with equal vertical angles, the vertices will all lie on the circumference of a segment of a circle.*

A useful theorem derivable from III. 21 is given by Serenus (*De sectione coni*, Props. 52, 53).

If *ADB* be any segment of a circle, and *C* be such a point on the circumference that *AC* is equal to *CB*, and if there be described with *C* as centre and radius *CA* or *CB* the circle *AHB*, then, *ADB* being any other angle in the segment *ACB*, and *BD* being produced to meet the outer segment in *E*, the sum of *AD, DB* is equal to *BE*.

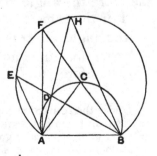

If *BC* be produced to meet the outer segment in *F*, and *FA* be joined,

CA, CB, CF are by hypothesis equal.

Therefore the angle *FAC* is equal to the angle *AFC*.

Also, by III. 21, the angles *ACB, ADB* are equal;

therefore their supplements, the angles *ACF*, *ADE*, are equal.

Further, by III. 21, the angles *AEB*, *AFB* are equal.

Hence in the triangles *ACF*, *ADE* two angles are respectively equal;
therefore the third angles *EAD*, *FAC* are equal.

But the angle *FAC* is equal to the angle *AFC*, and therefore equal to the angle *AED*.

Therefore the angles *AED*, *EAD* are equal, or the triangle *DEA* is isosceles,

and *AD* is equal to *DE*.

Adding *BD* to both, we see that

BE is equal to the sum of *AD* and *DB*.

Now, *BF* being a diameter of the circle of which the outer segment is a part,

BF is greater than *BE*;

therefore *AC*, *CB* are together greater than *AD*, *DB*.

And, generally, *of all triangles on the same base and on the same side of it which have equal vertical angles, the isosceles triangle is that which has the greatest perimeter, and of the others that has the lesser perimeter which is further from being isosceles.*

The theorem of Serenus gives us the means of solving the following problem given in Todhunter's Euclid, p. 324.

To find a point in the circumference of a given segment of a circle such that the straight lines which join the point to the extremities of the straight line on which the segment stands may be together equal to a given straight line (the length of which is of course subject to limits).

Let *ACB* in the above figure be the given segment. Find, by bisecting *AB* at right angles, a point *C* on it such that *AC* is equal to *CB*.

Then with centre *C* and radius *CA* or *CB* describe the segment of a circle *AHB* on the same side of *AB*.

Lastly, with *A* or *B* as centre and radius equal to the given straight line describe a circle. This circle will, if the given straight line be greater than *AB* and less than twice *AC*, meet the outer segment in two points, and if we join those points to the centre of the circle last drawn (whether *A* or *B*), the joining straight lines will cut the inner segment in points satisfying the given condition. If the given straight line be *equal* to twice *AC*, *C* is of course the required point. If the given straight line be greater than twice *AC*, there is no possible solution.

PROPOSITION 22.

The opposite angles of quadrilaterals in circles are equal to two right angles.

Let *ABCD* be a circle, and let *ABCD* be a quadrilateral in it;

I say that the opposite angles are equal to two right angles.

Let *AC*, *BD* be joined.

Then, since in any triangle the three angles are equal to two right angles,   [I. 32] the three angles CAB, ABC, BCA of the triangle ABC are equal to two right angles.

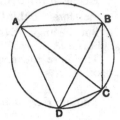

But the angle CAB is equal to the angle BDC, for they are in the same segment $BADC$;   [III. 21]

and the angle ACB is equal to the angle ADB, for they are in the same segment $ADCB$;

therefore the whole angle ADC is equal to the angles BAC, ACB.

Let the angle ABC be added to each;

therefore the angles ABC, BAC, ACB are equal to the angles ABC, ADC.

But the angles ABC, BAC, ACB are equal to two right angles;

therefore the angles ABC, ADC are also equal to two right angles.

Similarly we can prove that the angles BAD, DCB are also equal to two right angles.

Therefore etc.

Q. E. D.

As Todhunter remarks, the converse of this proposition is true and very important: *if two opposite angles of a quadrilateral be together equal to two right angles, a circle may be circumscribed about the quadrilateral.* We can, by the method of III. 9, or by IV. 5, circumscribe a circle about the triangle ABC; and we can then prove, by *reductio ad absurdum*, that the circle passes through the fourth angular point D.

PROPOSITION 23.

On the same straight line there cannot be constructed two similar and unequal segments of circles on the same side.

For, if possible, on the same straight line AB let two similar and unequal segments of circles ACB, ADB be constructed on the same side;

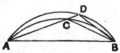

let ACD be drawn through, and let CB, DB be joined.

Then, since the segment ACB is similar to the segment ADB, and similar segments of circles are those which admit equal angles,   [III. Def. 11]

the angle ACB is equal to the angle ADB, the exterior to the interior: which is impossible.   [I. 16]

Therefore etc.

Q. E. D.

1. **cannot be constructed**, οὐ συσταθήσεται, the same phrase as in I. 7.

Clavius and the other early editors point out that, while the words "on the same side" in the enunciation are necessary for Euclid's proof, it is equally true that neither can there be two similar and unequal segments on *opposite* sides of the same straight line; this is at once made clear by causing one of the segments to revolve round the base till it is on the same side with the other.

Simson observes with reason that, while Euclid in the following proposition, III. 24, thinks it necessary to dispose of the hypothesis that, if two similar segments on equal bases are applied to one another with the bases coincident, the segments cannot cut in any other point than the extremities of the base (since otherwise two circles would cut one another in more points than two), this remark is an equally necessary preliminary to III. 23, in order that we may be justified in drawing the segments as being one inside the other. Simson accordingly begins his proof of III. 23 thus:

"Then, because the circle ACB cuts the circle ADB in the two points A, B, they cannot cut one another in any other point:

One of the segments must therefore fall within the other.

Let ACB fall within ADB and draw the straight line ACD, etc."

Simson has also substituted "not coinciding with one another" for "unequal" in Euclid's enunciation.

Then in III. 24 Simson leaves out the words referring to the hypothesis that the segment AEB when applied to the other CFD may be "otherwise placed as CGD"; in fact, after stating that AB must coincide with CD, he merely adds words quoting the result of III. 23: "Therefore, the straight line AB coinciding with CD, the segment AEB must coincide with the segment CFD, and is therefore equal to it."

PROPOSITION 24.

Similar segments of circles on equal straight lines are equal to one another.

For let AEB, CFD be similar segments of circles on equal straight lines AB, CD;

5 I say that the segment AEB is equal to the segment CFD.

For, if the segment AEB be applied to CFD, and if the point A be placed on C and the straight line AB on CD, the point B will also coincide with the point D, because AB is equal to CD;

10 and, AB coinciding with CD,

the segment AEB will also coincide with CFD.

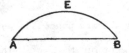 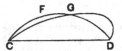

For, if the straight line AB coincide with CD but the segment AEB do not coincide with CFD,

it will either fall within it, or outside it;

15 or it will fall awry, as CGD, and a circle cuts a circle at more points than two : which is impossible.   [III. 10]

Therefore, if the straight line AB be applied to CD, the segment AEB will not fail to coincide with CFD also;

therefore it will coincide with it and will be equal to it.

20 Therefore etc.

Q. E. D.

15. **fall awry**, παραλλάξει, the same word as used in the like case in I. 8. The word implies that the applied figure will partly fall short of, and partly overlap, the figure to which it is applied.

Compare the note on the last proposition. I have put a semicolon instead of the comma which the Greek text has after "outside it," in order the better to indicate that the inference "and a circle cuts a circle in more points than two" only refers to the third hypothesis that the applied segment is "otherwise placed (παραλλάξει) as CGD." The first two hypotheses are disposed of by a *tacit* reference to the preceding proposition III. 23.

PROPOSITION 25.

Given a segment of a circle, to describe the complete circle of which it is a segment.

Let ABC be the given segment of a circle;

thus it is required to describe the complete circle belonging to the segment ABC, that is, of which it is a segment.

For let AC be bisected at D, let DB be drawn from the point D at right angles to AC, and let AB be joined;

the angle *ABD* is then greater than, equal to, or less than the angle *BAD*.

First let it be greater;

and on the straight line *BA*, and at the point *A* on it, let the angle *BAE* be constructed equal to the angle *ABD*; let *DB* be drawn through to *E*, and let *EC* be joined.

Then, since the angle *ABE* is equal to the angle *BAE*,

the straight line *EB* is also equal to *EA*.   [I. 6]

And, since *AD* is equal to *DC*, and *DE* is common,

the two sides *AD, DE* are equal to the two sides *CD, DE* respectively;

and the angle *ADE* is equal to the angle *CDE*, for each is right;

therefore the base *AE* is equal to the base *CE*.

But *AE* was proved equal to *BE*;

therefore *BE* is also equal to *CE*;

therefore the three straight lines *AE, EB, EC* are equal to one another.

Therefore the circle drawn with centre *E* and distance one of the straight lines *AE, EB, EC* will also pass through the remaining points and will have been completed.   [III. 9]

Therefore, given a segment of a circle, the complete circle has been described.

And it is manifest that the segment *ABC* is less than a semicircle, because the centre *E* happens to be outside it.

Similarly, even if the angle *ABD* be equal to the angle *BAD*,

AD being equal to each of the two *BD, DC*,

the three straight lines *DA, DB, DC* will be equal to one another,

D will be the centre of the completed circle,

and *ABC* will clearly be a semicircle.

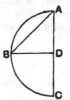

But, if the angle ABD be less than the angle BAD, and if we construct, on the straight line BA and at the point A on it, an angle equal to the angle ABD, the centre will fall on DB within the segment ABC, and the segment ABC will clearly be greater than a semicircle.

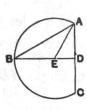

Therefore, given a segment of a circle, the complete circle has been described.

Q. E. F.

1. to describe the complete circle, προσαναγράψαι τὸν κύκλον, literally "to describe the circle *on to it.*'

It will be remembered that Simson takes first the case in which the angles ABD, BAD are equal to one another, and then takes the other two cases together, telling us to "produce BD, if necessary." This is a little shorter than Euclid's procedure, though Euclid does not repeat the proof of the first case in giving the third, but only refers to it as equally applicable.

Campanus, Peletarius and others give the solution of this problem in which we take two chords not parallel and bisect each at right angles by straight lines, which must meet in the centre, since each contains the centre and they only intersect in one point. Clavius, Billingsley, Barrow and others give the rather simpler solution in which the two chords have one extremity common (cf. Euclid's proofs of III. 9, 10). This method De Morgan favours, and (as noted on III. 1 above) would make III. 1, this proposition, and IV. 5 all *corollaries* of the theorem that "the line which bisects a chord perpendicularly must contain the centre." Mr H. M. Taylor practically adopts this order and method, though he finds the centre of a circle by means of any two non-parallel chords; but he finds *the centre of the circle of which a given arc is a part* (his proposition corresponding to III. 25) by bisecting at right angles first the base and then the chord joining one extremity of the base to the point in which the line bisecting the base at right angles meets the circumference of the segment. Under De Morgan's alternative the relation between Euclid III. 1 and the Porism to it would be reversed, and Euclid's notion of a Porism or *corollary* would have to be considerably extended.

If the problem is solved after the manner of IV. 5, it is still desirable to state, as Euclid does, after proving AE, EB, EC to be all equal, that "the circle drawn with centre E and distance one of the straight lines AE, EB, EC will also pass through the remaining points of the segment" [III. 9], in order to show that part of the circle described actually coincides with the given segment. This is not so clear if the centre is determined as the intersection of the straight lines bisecting at right angles chords which join pairs of *four* different points.

Proposition 26.

In equal circles equal angles stand on equal circumferences, whether they stand at the centres or at the circumferences.

Let *ABC*, *DEF* be equal circles, and in them let there be equal angles, namely at the centres the angles *BGC*, *EHF*, and at the circumferences the angles *BAC*, *EDF*; I say that the circumference *BKC* is· equal to the circumference *ELF*.

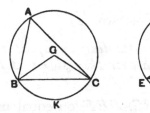

For let *BC*, *EF* be joined.

Now, since the circles *ABC*, *DEF* are equal,

the radii are equal.

Thus the two straight lines *BG*, *GC* are equal to the two straight lines *EH*, *HF*;

and the angle at *G* is equal to the angle at *H*;

therefore the base *BC* is equal to the base *EF*.   [I. 4]

And, since the angle at *A* is equal to the angle at *D*,

the segment *BAC* is similar to the segment *EDF*;

[III. Def. 11]

and they are upon equal straight lines.

But similar segments of circles on equal straight lines are equal to one another;   [III. 24]

therefore the segment *BAC* is equal to *EDF*.

But the whole circle *ABC* is also equal to the whole circle *DEF*;

therefore the circumference *BKC* which remains is equal to the circumference *ELF*.

Therefore etc.   Q. E. D.

As in III. 21, if Euclid's proof is to cover all cases, it requires us to take cognisance of "angles at the centre" which are equal to or greater than two right angles. Otherwise we must deal separately with the cases where the angle at the circumference is equal to or greater than a right angle. The case of an *obtuse* angle at the circumference can of course be reduced by means of III. 22 to the case of an acute angle at the circumference; and, in case the angle at the circumference is right, it is readily proved, by drawing the radii to the vertex of the angle and to the other extremities of the lines containing it, that the latter two radii are in a straight line, whence they make equal bases in the two circles as in Euclid's proof.

Lardner has another way of dealing with the right angle or obtuse angle at the circumference. In either case, he says, "bisect them, and the halves of them are equal, and it can be proved, as above, that the arcs upon which these halves stand are equal, whence it follows that the arcs on which the given angles stand are equal."

PROPOSITION 27.

In equal circles angles standing on equal circumferences are equal to one another, whether they stand at the centres or at the circumferences.

For in equal circles ABC, DEF, on equal circumferences BC, EF, let the angles BGC, EHF stand at the centres G, H, and the angles BAC, EDF at the circumferences;

I say that the angle BGC is equal to the angle EHF,

and the angle BAC is equal to the angle EDF.

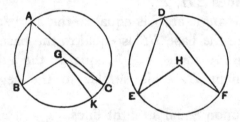

For, if the angle BGC is unequal to the angle EHF,

one of them is greater.

Let the angle BGC be greater: and on the straight line BG, and at the point G on it, let the angle BGK be constructed equal to the angle EHF.   [I. 23]

Now equal angles stand on equal circumferences, when they are at the centres;   [III. 26]

therefore the circumference BK is equal to the circumference EF.

But EF is equal to BC;

therefore BK is also equal to BC, the less to the greater: which is impossible.

Therefore the angle BGC is not unequal to the angle EHF;

therefore it is equal to it.

And the angle at A is half of the angle BGC,

and the angle at D half of the angle EHF;   [III. 20]

therefore the angle at A is also equal to the angle at D.

Therefore etc.

Q. E. D.

This proposition is the converse of the preceding one, and the remarks about the method of treating the different cases apply here also.

PROPOSITION 28.

In equal circles equal straight lines cut off equal circumferences, the greater equal to the greater and the less to the less.

Let ABC, DEF be equal circles, and in the circles let AB, DE be equal straight lines cutting off ACB, DFE as greater circumferences and AGB, DHE as lesser;

I say that the greater circumference ACB is equal to the greater circumference DFE, and the less circumference AGB to DHE.

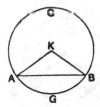 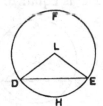

For let the centres K, L of the circles be taken, and let AK, KB, DL, LE be joined.

Now, since the circles are equal,

the radii are also equal;

therefore the two sides AK, KB are equal to the two sides DL, LE;

and the base AB is equal to the base DE;

therefore the angle AKB is equal to the angle DLE.

[I. 8]

But equal angles stand on equal circumferences, when they are at the centres;   [III. 26]

therefore the circumference AGB is equal to DHE.

And the whole circle ABC is also equal to the whole circle DEF;

therefore the circumference ACB which remains is also equal to the circumference DFE which remains.

Therefore etc.

Q. E. D.

Euclid's proof does not in terms cover the particular case in which the chord in one circle passes through its centre; but indeed this was scarcely worth giving, as the proof can easily be supplied. Since the chord in one circle passes through its centre, the chord in the second circle must also be a diameter of that circle, for equal circles are those which have equal diameters, and all other chords in any circle are less than its diameter [III. 15]; hence the segments cut off in each circle are semicircles, and these must be equal because the circles are equal.

PROPOSITION 29.

In equal circles equal circumferences are subtended by equal straight lines.

Let ABC, DEF be equal circles, and in them let equal circumferences BGC, EHF be cut off; and let the straight lines BC, EF be joined;

I say that BC is equal to EF.

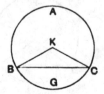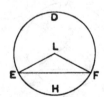

For let the centres of the circles be taken, and let them be K, L; let BK, KC, EL, LF be joined.

Now, since the circumference BGC is equal to the circumference EHF,

the angle BKC is also equal to the angle ELF. [III. 27]

And, since the circles ABC, DEF are equal,

the radii are also equal;

therefore the two sides BK, KC are equal to the two sides EL, LF; and they contain equal angles;

therefore the base BC is equal to the base EF.   [I. 4]

Therefore etc.

Q. E. D.

The particular case of this converse of III. 28 in which the given arcs are arcs of semicircles is even easier than the corresponding case of III. 28 itself.

The propositions III. 26—29 are of course equally true if the *same* circle is taken instead of *two equal* circles.

PROPOSITION 30.

To bisect a given circumference.

Let *ADB* be the given circumference;

thus it is required to bisect the circumference *ADB*.

Let *AB* be joined and bisected at *C*; from the point *C* let *CD* be drawn at right angles to the straight line *AB*, and let *AD*, *DB* be joined.

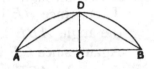

Then, since *AC* is equal to *CB*, and *CD* is common,

the two sides *AC, CD* are equal to the two sides *BC, CD*;

and the angle *ACD* is equal to the angle *BCD*, for each is right;

therefore the base *AD* is equal to the base *DB*.   [I. 4]

But equal straight lines cut off equal circumferences, the greater equal to the greater, and the less to the less;   [III. 28]

and each of the circumferences *AD, DB* is less than a semicircle;

therefore the circumference *AD* is equal to the circumference *DB*.

Therefore the given circumference has been bisected at the point *D*.

                                                Q. E. F.

PROPOSITION 31.

In a circle the angle in the semicircle is right, that in a greater segment less than a right angle, and that in a less segment greater than a right angle; and further the angle of the greater segment is greater than a right angle, and the angle of the less segment less than a right angle.

Let $ABCD$ be a circle, let BC be its diameter, and E its centre, and let BA, AC, AD, DC be joined;

I say that the angle BAC in the semicircle BAC is right,

the angle ABC in the segment ABC greater than the semicircle is less than a right angle,

and the angle ADC in the segment ADC less than the semicircle is greater than a right angle.

Let AE be joined, and let BA be carried through to F.

Then, since BE is equal to EA,

the angle ABE is also equal to the angle BAE.   [I. 5]

Again, since CE is equal to EA,

the angle ACE is also equal to the angle CAE.   [I. 5]

Therefore the whole angle BAC is equal to the two angles ABC, ACB.

But the angle FAC exterior to the triangle ABC is also equal to the two angles ABC, ACB;   [I. 32]

therefore the angle BAC is also equal to the angle FAC;

therefore each is right;   [I. Def. 10]

therefore the angle BAC in the semicircle BAC is right.

Next, since in the triangle ABC the two angles ABC, BAC are less than two right angles,   [I. 17]

and the angle BAC is a right angle,

the angle ABC is less than a right angle;

and it is the angle in the segment ABC greater than the semicircle.

Next, since $ABCD$ is a quadrilateral in a circle,

and the opposite angles of quadrilaterals in circles are equal to two right angles,   [III. 22]

while the angle ABC is less than a right angle,

therefore the angle ADC which remains is greater than a right angle;

and it is the angle in the segment ADC less than the semicircle.

I say further that the angle of the greater segment, namely that contained by the circumference ABC and the straight line AC, is greater than a right angle ;

and the angle of the less segment, namely that contained by the circumference ADC and the straight line AC, is less than a right angle.

This is at once manifest.

For, since the angle contained by the straight lines BA, AC is right,

the angle contained by the circumference ABC and the straight line AC is greater than a right angle.

Again, since the angle contained by the straight lines AC, AF is right,

the angle contained by the straight line CA and the circumference ADC is less than a right angle.

Therefore etc.　　　　　　　　　　　　　　　Q. E. D.

As already stated, this proposition is immediately deducible from III. 20 if that theorem is extended so as to include the case where the segment is equal to or less than a semicircle, and where consequently the "angle at the centre" is equal to two right angles or greater than two right angles respectively.

There are indications in Aristotle that the proof of the first part of the theorem in use before Euclid's time proceeded on different lines. Two passages of Aristotle refer to the proposition that the angle in a semicircle is a right angle. The first passage is *Anal. Post.* II. 11, 94 a 28 : "Why is the angle in a semicircle a right angle? Or what makes it a right angle? (τίνος ὄντος ὀρθή;) Suppose A to be a right angle, B half of two right angles, C the angle in a semicircle. Then B is the cause of A, the right angle, being an attribute of C, the angle in the semicircle. For B is equal to A, and C to B; for C is half of two right angles. Therefore it is in virtue of B being half of two right angles that A is an attribute of C; and the latter means the fact that the angle in a semicircle is right." Now this passage by itself would be consistent with a proof like Euclid's or the alternative interpolated proof next to be mentioned. But the second passage throws a different light on the subject. This is *Metaph.* 1051 a 26 : "Why is the angle in a semicircle a right angle invariably (καθόλου)? Because, *if there be three straight lines, two forming the base, and the third set up at right angles at its middle point*, the fact is obvious by simple inspection to any one who knows the property referred to" (ἐκεῖνο is the property that the angles of a triangle are together equal to two right angles, mentioned two lines before). That is to say, the angle *at the middle point* of the circumference of the semicircle was taken and proved, by means of the two isosceles right-angled triangles, to be the sum of two angles each equal to one-fourth of the sum of the angles of the large triangle in the figure, or of two right angles; and the proof

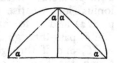

must have been completed by means of the theorem of III. 21 (that angles in the same segment are equal), which Euclid's more general proof does not need.

In the Greek texts before that of August there is an alternative proof that the angle BAC (in a semicircle) is right. August and Heiberg relegate it to an Appendix.

"Since the angle AEC is double of the angle BAE (for it is equal to the two interior and opposite angles), while the angle AEB is also double of the angle EAC,

the angles AEB, AEC are double of the angle BAC.

But the angles AEB, AEC are equal to two right angles;

therefore the angle BAC is right."

Lardner gives a slightly different proof of the second part of the theorem.

If ABC be a segment greater than a semicircle, draw the diameter AD, and join CD, CA.

Then, in the triangle ACD, the angle ACD is right (being the angle in a semicircle);

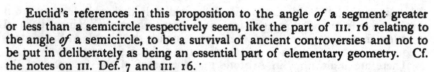

therefore the angle ADC is acute.

But the angle ADC is equal to the angle ABC in the same segment;

therefore the angle ABC is acute.

Euclid's references in this proposition to the angle *of* a segment greater or less than a semicircle respectively seem, like the part of III. 16 relating to the angle *of* a semicircle, to be a survival of ancient controversies and not to be put in deliberately as being an essential part of elementary geometry. Cf. the notes on III. Def. 7 and III. 16.

The corollary ordinarily attached to this proposition is omitted by Heiberg as an interpolation of date later than Theon. It is to this effect: "From this it is manifest that, if one angle of a triangle be equal to the other two, the first angle is right because the exterior angle to it is also equal to the same angles, and if the adjacent angles be equal, they are right." No doubt the corollary is rightly suspected, because there is no necessity for it here, and the words ὅπερ ἔδει δεῖξαι come before it, not after it, as is usual with Euclid. But, on the other hand, as the fact stated does appear in the proof of III. 31, the Porism would be a Porism after the usual type, and I do not quite follow Heiberg's argument that, "if Euclid had wished to add it, he ought to have placed it after I. 32."

It has already been mentioned above (p. 44) that this proposition supplies us with an alternative construction for the problem in III. 17 of drawing the two tangents to a circle from an external point.

Two theorems of some historical interest which follow directly from III. 31 may be mentioned.

The first is a lemma of Pappus on "the 24th problem" of the second Book of Apollonius' lost treatise on νεύσεις (Pappus VII. p. 812) and is to this effect. If a circle, as DEF, pass through D, the centre of a circle ABC, and if through F, the other point in which the line of centres meets the circle DEF, any straight line be drawn (and produced if necessary) meeting the circle DEF in E and the circle ABC in B, G,

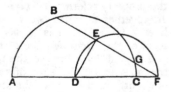

then E is the middle point of BG. For, if DE be joined, the angle DEF (in a semicircle) is a right angle [III. 31]; and DE, being at right angles to the chord BG of the circle ABC, also bisects it [III. 3].

The second is a proposition in the *Liber Assumptorum*, attributed (no doubt erroneously as regards much of it) to Archimedes, which has reached us through the Arabic (Archimedes, ed. Heiberg, II. pp. 520—521).

If two chords AB, CD *in a circle intersect at right angles in a point* O, *then the sum of the squares on* AO, BO, CO, DO *is equal to the square on the diameter.*

For draw the diameter CE, and join AC, CB, AD, BE.

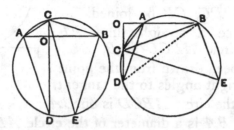

Then the angle CAO is equal to the angle CEB. (This follows, in the first figure, from III. 21 and, in the second, from I. 13 and III. 22.) Also the angle COA, being right, is equal to the angle CBE which, being the angle in a semicircle, is also right [III. 31].

Therefore the triangles AOC, EBC have two angles equal respectively; whence the third angles ACO, ECB are equal. (In the second figure the angle ACO is, by I. 13 and III. 22, equal to the angle ABD, and therefore the angles ABD, ECB are equal.)

Therefore, in both figures, the arcs AD, BE, and consequently the chords AD, BE subtended by them, are equal.   [III. 26, 29]

Now the squares on AO, DO are equal to the square on AD [I. 47], that is, to the square on BE.

And the squares on CO, BO are equal to the square on BC.

Therefore, by addition, the squares on AO, BO, CO, DO are equal to the squares on EB, BC, i.e. to the square on CE.   [I. 47]

PROPOSITION 32.

If a straight line touch a circle, and from the point of contact there be drawn across, in the circle, a straight line cutting the circle, the angles which it makes with the tangent will be equal to the angles in the alternate segments of the circle.

For let a straight line EF touch the circle $ABCD$ at the point B, and from the point B let there be drawn across, in the circle $ABCD$, a straight line BD cutting it;

I say that the angles which BD makes with the tangent EF will be equal to the angles in the alternate segments of the circle, that is, that the angle *FBD* is equal to the angle constructed in the segment *BAD*, and the angle *EBD* is equal to the angle constructed in the segment *DCB*.

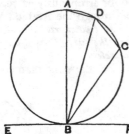

For let *BA* be drawn from *B* at right angles to *EF*,

let a point *C* be taken at random on the circumference *BD*,

and let *AD*, *DC*, *CB* be joined.

Then, since a straight line *EF* touches the circle *ABCD* at *B*, and *BA* has been drawn from the point of contact at right angles to the tangent, the centre of the circle *ABCD* is on *BA*.   [III. 19]

Therefore *BA* is a diameter of the circle *ABCD*;

therefore the angle *ADB*, being an angle in a semicircle, is right.   [III. 31]

Therefore the remaining angles *BAD*, *ABD* are equal to one right angle.   [I. 32]

But the angle *ABF* is also right;

therefore the angle *ABF* is equal to the angles *BAD*, *ABD*.

Let the angle *ABD* be subtracted from each;

therefore the angle *DBF* which remains is equal to the angle *BAD* in the alternate segment of the circle.

Next, since *ABCD* is a quadrilateral in a circle,

its opposite angles are equal to two right angles.   [III. 22]

But the angles *DBF*, *DBE* are also equal to two right angles;

therefore the angles *DBF*, *DBE* are equal to the angles *BAD*, *BCD*,

of which the angle *BAD* was proved equal to the angle *DBF*;

therefore the angle *DBE* which remains is equal to the angle *DCB* in the alternate segment *DCB* of the circle.

Therefore etc.   Q. E. D.

The converse of this theorem is true, namely that, *If a straight line drawn through one extremity of a chord of a circle make with that chord angles equal respectively to the angles in the alternate segments of the circle, the straight line so drawn touches the circle.*

This can, as Camerer and Todhunter remark, be proved indirectly; or we may prove it, with Clavius, directly. Let BD be the given chord, and let EF be drawn through B so that it makes with BD angles equal to the angles in the alternate segments of the circle respectively.

Let BA be the diameter through B, and let C be any point on the circumference of the segment DCB which does not contain A. Join AD, DC, CB.

Then, since, by hypothesis, the angle FBD is equal to the angle BAD, let the angle ABD be added to both;

therefore the angle ABF is equal to the angles ABD, BAD.

But the angle BDA, being the angle in a semicircle, is a right angle;

therefore the remaining angles ABD, BAD in the triangle ABD are equal to a right angle.

Therefore the angle ABF is right;

hence, since BA is the diameter through B,

EF touches the circle at B.   [III. 16, Por.]

Pappus assumes in one place (IV. p. 196) the consequence of this proposition that, *If two circles touch, any straight line drawn through the point of contact and terminated by both circles cuts off segments in each which are respectively similar.* Pappus also shows how to prove this (VII. p. 826) by drawing the common tangent at the point of contact and using this proposition, III. 32.

PROPOSITION 33.

On a given straight line to describe a segment of a circle admitting an angle equal to a given rectilineal angle.

Let AB be the given straight line, and the angle at C the given rectilineal angle;

thus it is required to describe on the given straight line AB a segment of a circle admitting an angle equal to the angle at C.

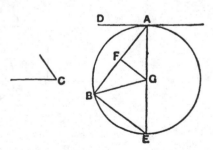

The angle at C is then acute, or right, or obtuse.

First let it be acute,

and, as in the first figure, on the straight line AB, and at the point A, let the angle BAD be constructed equal to the angle at C;

therefore the angle BAD is also acute.

Let AE be drawn at right angles to DA, let AB be

bisected at F, let FG be drawn from the point F at right angles to AB, and let GB be joined.

Then, since AF is equal to FB,
and FG is common,
the two sides AF, FG are equal to the two sides BF, FG;
and the angle AFG is equal to the angle BFG;
　　　　therefore the base AG is equal to the base BG.　　[I. 4]

Therefore the circle described with centre G and distance GA will pass through B also.

Let it be drawn, and let it be ABE;
let EB be joined.

Now, since AD is drawn from A, the extremity of the diameter AE, at right angles to AE,
therefore AD touches the circle ABE.　　　　[III. 16, Por.]

Since then a straight line AD touches the circle ABE,
and from the point of contact at A a straight line AB is drawn across in the circle ABE,
the angle DAB is equal to the angle AEB in the alternate segment of the circle.　　　　[III. 32]

But the angle DAB is equal to the angle at C;
therefore the angle at C is also equal to the angle AEB.

Therefore on the given straight line AB the segment AEB of a circle has been described admitting the angle AEB equal to the given angle, the angle at C.

Next let the angle at C be right;

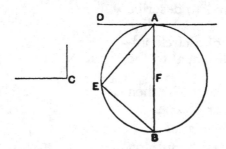

and let it be again required to describe on AB a segment of a circle admitting an angle equal to the right angle at C.

Let the angle BAD be constructed equal to the right angle at C, as is the case in the second figure;

let AB be bisected at F, and with centre F and distance either FA or FB let the circle AEB be described.

Therefore the straight line AD touches the circle ABE, because the angle at A is right.   [III. 16, Por.]

And the angle BAD is equal to the angle in the segment AEB, for the latter too is itself a right angle, being an angle in a semicircle.   [III. 31]

But the angle BAD is also equal to the angle at C.

Therefore the angle AEB is also equal to the angle at C.

Therefore again the segment AEB of a circle has been described on AB admitting an angle equal to the angle at C.

Next, let the angle at C be obtuse;

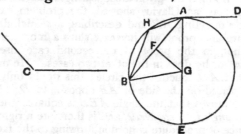

and on the straight line AB, and at the point A, let the angle BAD be constructed equal to it, as is the case in the third figure;

let AE be drawn at right angles to AD, let AB be again bisected at F, let FG be drawn at right angles to AB, and let GB be joined.

Then, since AF is again equal to FB,
and FG is common,
the two sides AF, FG are equal to the two sides BF, FG;
and the angle AFG is equal to the angle BFG;
therefore the base AG is equal to the base BG.   [I. 4]

Therefore the circle described with centre G and distance GA will pass through B also; let it so pass, as AEB.

Now, since AD is drawn at right angles to the diameter AE from its extremity,
AD touches the circle AEB.   [III. 16, Por.]

And AB has been drawn across from the point of contact at A;

therefore the angle BAD is equal to the angle constructed in the alternate segment AHB of the circle.   [III. 32]

But the angle BAD is equal to the angle at C.

Therefore the angle in the segment AHB is also equal to the angle at C:

Therefore on the given straight line AB the segment AHB of a circle has been described admitting an angle equal to the angle at C.

Q. E. F.

Simson remarks truly that the first and third cases, those namely in which the given angle is acute and obtuse respectively, have exactly the same construction and demonstration, so that there is no advantage in repeating them. Accordingly he deals with the cases as one, merely drawing two different figures. It is also true, as Simson says, that the demonstration of the second case in which the given angle is a right angle "is done in a round-about way," whereas, as Clavius showed, the problem can be more easily solved by merely bisecting AB and describing a semicircle on it. A glance at Euclid's figure and proof will however show a more curious fact, namely that he does not, in the proof of the second case, use the *angle in the alternate segment*, as he does in the other two cases. He might have done so after proving that AD touches the circle; this would only have required his point E to be placed on the side of AB opposite to D. Instead of this, he uses III. 31, and proves that the angle AEB is equal to the angle C, because the former is an *angle in a semicircle*, and is therefore a right angle as C is.

The difference of procedure is no doubt owing to the fact that he has not, in III. 32, distinguished the case in which the cutting and touching straight lines are at right angles, i.e. in which the two alternate segments are semicircles. To prove this case would also have required III. 31, so that nothing would have been gained by stating it separately in III. 32 and then quoting the result as part of III. 32, instead of referring directly to III. 31.

It is assumed in Euclid's proof of the first and third cases that AE and FG will meet; but of course there is no difficulty in satisfying ourselves of this.

Proposition 34.

From a given circle to cut off a segment admitting an angle equal to a given rectilineal angle.

Let ABC be the given circle, and the angle at D the given rectilineal angle;

thus it is required to cut off from the circle ABC a segment admitting an angle equal to the given rectilineal angle, the angle at D.

Let EF be drawn touching ABC at the point B, and on the straight line FB, and at the point B on it, let the angle FBC be constructed equal to the angle at D.   [I. 23]

Then, since a straight line EF touches the circle ABC,

and *BC* has been drawn across from the point of contact at *B*,

the angle *FBC* is equal to the angle constructed in the alternate segment *BAC*.   [III. 32]

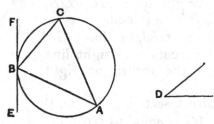

But the angle *FBC* is equal to the angle at *D*;

therefore the angle in the segment *BAC* is equal to the angle at *D*.

Therefore from the given circle *ABC* the segment *BAC* has been cut off admitting an angle equal to the given rectilineal angle, the angle at *D*.

Q. E. F.

An alternative construction here would be to make an "angle at the centre" (in the extended sense, if necessary) double of the given angle; and, if the given angle is right, it is only necessary to draw a diameter of the circle.

PROPOSITION 35.

If in a circle two straight lines cut one another, the rectangle contained by the segments of the one is equal to the rectangle contained by the segments of the other.

For in the circle *ABCD* let the two straight lines *AC*, *BD* cut one another at the point *E*;

I say that the rectangle contained by *AE*, *EC* is equal to the rectangle contained by *DE*, *EB*.

If now *AC*, *BD* are through the centre, so that *E* is the centre of the circle *ABCD*, it is manifest that, *AE*, *EC*, *DE*, *EB* being equal,

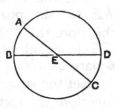

the rectangle contained by *AE*, *EC* is also equal to the rectangle contained by *DE*, *EB*.

Next let *AC*, *DB* not be through the centre ;
let the centre of *ABCD* be taken, and
let it be *F* ;
from *F* let *FG*, *FH* be drawn perpendicular to the straight lines *AC*, *DB*,
and let *FB*, *FC*, *FE* be joined.

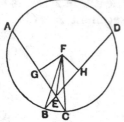

Then, since a straight line *GF*
through the centre cuts a straight line
AC not through the centre at right
angles,

            it also bisects it ;   [III. 3]

    therefore *AG* is equal to *GC*.

Since, then, the straight line *AC* has been cut into equal
parts at *G* and into unequal parts at *E*,
the rectangle contained by *AE*, *EC* together with the square
on *EG* is equal to the square on *GC* ;   [II. 5]

Let the square on *GF* be added ;
therefore the rectangle *AE*, *EC* together with the squares
on *GE*, *GF* is equal to the squares on *CG*, *GF*.

But the square on *FE* is equal to the squares on *EG*, *GF*,
and the square on *FC* is equal to the squares on *CG*, *GF* ;

                           [I. 47]

therefore the rectangle *AE*, *EC* together with the square
on *FE* is equal to the square on *FC*.

And *FC* is equal to *FB* ;
therefore the rectangle *AE*, *EC* together with the square on
EF is equal to the square on *FB*.

For the same reason, also,
the rectangle *DE*, *EB* together with the square on *FE* is
equal to the square on *FB*.

But the rectangle *AE*, *EC* together with the square on
FE was also proved equal to the square on *FB* ;
therefore the rectangle *AE*, *EC* together with the square on
FE is equal to the rectangle *DE*, *EB* together with the
square on *FE*.

Let the square on *FE* be subtracted from each ;
therefore the rectangle contained by *AE*, *EC* which remains
is equal to the rectangle contained by *DE*, *EB*.

Therefore etc.

                               Q. E. D.

In addition to the two cases in Euclid's text, Simson (following Campanus) gives two intermediate cases, namely (1) that in which one chord passes through the centre and bisects the other which does not pass through the centre at right angles, and (2) that in which one passes through the centre and cuts the other which does not pass through the centre but not at right angles. Simson then reduces Euclid's second case, the most general one, to the second of the two intermediate cases by drawing the diameter through E. His note is as follows: "As the 25th and 33rd propositions are divided into more cases, so this 35th is divided into fewer cases than are necessary. Nor can it be supposed that Euclid omitted them because they are easy; as he has given the case which by far is the easiest of them all, viz. that in which both the straight lines pass through the centre: And in the following proposition he separately demonstrates the case in which the straight line passes through the centre, and that in which it does not pass through the centre: So that it seems Theon, or some other, has thought them too long to insert: But cases that require different demonstrations should not be left out in the Elements, as was before taken notice of: These cases are in the translation from the Arabic and are now put into the text." Notwithstanding the ingenuity of the argument based on the separate mention by Euclid of the simplest case of all, I think the conclusion that Euclid himself gave *four* cases is unsafe; in fact, in giving the simplest and most difficult cases only, he seems to be following quite consistently his habit of avoiding *too great* multiplicity of cases, while not ignoring their existence.

The deduction from the next proposition (III. 36) which Simson, following Clavius and others, gives as a corollary to it, namely that, *If from any point without a circle there be drawn two straight lines cutting it, the rectangles contained by the whole lines and the parts of them without the circle are equal to one another*, can of course be combined with III. 35 in one enunciation.

As remarked by Todhunter, a large portion of the proofs of III. 35, 36 amounts to proving the proposition, *If any point be taken on the base, or the base produced, of an isosceles triangle, the rectangle contained by the segments of the base (i.e. the respective distances of the ends of the base from the point) is equal to the difference between the square on the straight line joining the point to the vertex and the square on one of the equal sides of the triangle.* This is of course an immediate consequence of I. 47 combined with II. 5 or II. 6.

The converse of III. 35 and Simson's corollary to III. 36 may be stated thus. *If two straight lines AB, CD, produced if necessary, intersect at O, and if the rectangle AO, OB be equal to the rectangle CO, OD, the circumference of a circle will pass through the four points A, B, C, D.* The proof is indirect. We describe a circle through three of the points, as *A, B, C* (by the method used in Euclid's proofs of III. 9, 10), and then we prove, by the aid of III. 35 and the corollary to III. 36, that the circle cannot but pass through *D* also.

PROPOSITION 36.

If a point be taken outside a circle and from it there fall on the circle two straight lines, and if one of them cut the circle and the other touch it, the rectangle contained by the whole of the straight line which cuts the circle and the straight

*line intercepted on it outside between the point and the convex
circumference will be equal to the square on the tangent.*

For let a point *D* be taken outside the circle *ABC*,
and from *D* let the two straight lines *DCA*,
DB fall on the circle *ABC*; let *DCA* cut
the circle *ABC* and let *BD* touch it;
I say that the rectangle contained by *AD*,
DC is equal to the square on *DB*.

Then *DCA* is either through the centre
or not through the centre.

First let it be through the centre, and
let *F* be the centre of the circle *ABC*;
let *FB* be joined;

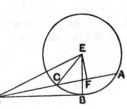

       therefore the angle *FBD* is right.   [III. 18]

And, since *AC* has been bisected at *F*, and *CD* is added
to it,
the rectangle *AD*, *DC* together with the square on *FC* is
equal to the square on *FD*.   [II. 6]

But *FC* is equal to *FB*;
therefore the rectangle *AD*, *DC* together with the square on
FB is equal to the square on *FD*.

And the squares on *FB*, *BD* are equal to the square on
FD;   [I. 47]
therefore the rectangle *AD*, *DC* together with the square on
FB is equal to the squares on *FB*, *BD*.

Let the square on *FB* be subtracted from each;
therefore the rectangle *AD*, *DC* which remains is equal to
the square on the tangent *DB*.

Again, let *DCA* not be through the centre of the circle
ABC;
let the centre *E* be taken, and from *E*
let *EF* be drawn perpendicular to *AC*;
let *EB*, *EC*, *ED* be joined.

Then the angle *EBD* is right.
           [III. 18]

And, since a straight line *EF*
through the centre cuts a straight line
AC not through the centre at right angles,

           it also bisects it;   [III. 3]

    therefore *AF* is equal to *FC*.

Now, since the straight line AC has been bisected at the point F, and CD is added to it,
the rectangle contained by AD, DC together with the square on FC is equal to the square on FD.　　　　[II. 6]

Let the square on FE be added to each;
therefore the rectangle AD, DC together with the squares on CF, FE is equal to the squares on FD, FE.

But the square on EC is equal to the squares on CF, FE, for the angle EFC is right;　　　　[I. 47]
and the square on ED is equal to the squares on DF, FE;
therefore the rectangle AD, DC together with the square on EC is equal to the square on ED.

And EC is equal to EB;
therefore the rectangle AD, DC together with the square on EB is equal to the square on ED.

But the squares on EB, BD are equal to the square on ED, for the angle EBD is right;　　　　[I. 47]
therefore the rectangle AD, DC together with the square on EB is equal to the squares on EB, BD.

Let the square on EB be subtracted from each;
therefore the rectangle AD, DC which remains is equal to the square on DB.

Therefore etc.　　　　Q. E. D.

Cf. note on the preceding proposition. Observe that, whereas it would be natural with us to prove first that, if A is an external point, and two straight lines AEB, AFC cut the circle in E, B and F, C respectively, the rectangle BA, AE is equal to the rectangle CA, AF, and thence that, the tangent from A being *a straight line like* AEB *in its limiting position when* E *and* B *coincide*, either rectangle is equal to the square on the tangent (cf. Mr H. M. Taylor, p. 253), Euclid and the Greek geometers generally did not allow themselves to infer the truth of a proposition in a *limiting case* directly from the general case including it, but preferred a separate proof of the limiting case (cf. *Apollonius of Perga*, p. 40, 139—140). This accounts for the form of III. 36.

Proposition 37.

If a point be taken outside a circle and from the point there fall on the circle two straight lines, if one of them cut the circle, and the other fall on it, and if further the rectangle contained by the whole of the straight line which cuts the circle and the straight line intercepted on it outside between the point and the convex circumference be equal to the square on the straight line which falls on the circle, the straight line which falls on it will touch the circle.

For let a point D be taken outside the circle ABC; from D let the two straight lines DCA, DB fall on the circle ACB; let DCA cut the circle and DB fall on it; and let the rectangle AD, DC be equal to the square on DB.

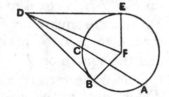

I say that DB touches the circle ABC.

For let DE be drawn touching ABC; let the centre of the circle ABC be taken, and let it be F; let FE, FB, FD be joined.

Thus the angle FED is right. 　　　　　　　　[III. 18]

Now, since DE touches the circle ABC, and DCA cuts it, the rectangle AD, DC is equal to the square on DE. [III. 36]

But the rectangle AD, DC was also equal to the square on DB;

therefore the square on DE is equal to the square on DB;

therefore DE is equal to DB.

And FE is equal to FB;

therefore the two sides DE, EF are equal to the two sides DB, BF;

and FD is the common base of the triangles;

therefore the angle DEF is equal to the angle DBF.

　　　　　　　　　　　　　　　　　　　　[I. 8]

But the angle DEF is right;

therefore the angle DBF is also right.

And FB produced is a diameter;

and the straight line drawn at right angles to the diameter of a circle, from its extremity, touches the circle; [III. 16, Por.]

therefore DB touches the circle.

Similarly this can be proved to be the case even if the centre be on AC.

Therefore etc. 　　　　　　　　　　　Q. E. D.

De Morgan observes that there is here the same defect as in I. 48, i.e. an apparent avoidance of indirect demonstration by drawing the tangent DE on the opposite side of *DF* from *DB*. The case is similar to the *apparently* direct proof which Campanus gave. He drew the straight line from *D* passing through the centre, and then (without drawing a second tangent) proved by the aid of II. 6 that the square on *DF* is equal to the sum of the squares on *DB*, *BF*; whence (by I. 48) the angle *DBF* is a right angle. But this proof uses I. 48, the very proposition to which De Morgan's original remark relates.

The undisguised indirect proof is easy. If *DB* does not touch the circle, it must cut it if produced, and it follows that the square on *DB* must be equal to the rectangle contained by *DB* and a longer line: which is absurd.

BOOK IV.

DEFINITIONS.

1. A rectilineal figure is said to be **inscribed in a rectilineal figure** when the respective angles of the inscribed figure lie on the respective sides of that in which it is inscribed.

2. Similarly a figure is said to be **circumscribed about a figure** when the respective sides of the circumscribed figure pass through the respective angles of that about which it is circumscribed.

3. A rectilineal figure is said to be **inscribed in a circle** when each angle of the inscribed figure lies on the circumference of the circle.

4. A rectilineal figure is said to be **circumscribed about a circle**, when each side of the circumscribed figure touches the circumference of the circle.

5. Similarly a circle is said to be **inscribed in a figure** when the circumference of the circle touches each side of the figure in which it is inscribed.

6. A circle is said to be **circumscribed about a figure** when the circumference of the circle passes through each angle of the figure about which it is circumscribed.

7. A straight line is said to be **fitted into a circle** when its extremities are on the circumference of the circle.

DEFINITIONS 1—7.

I append, as usual, the Greek text of the definitions.

1. Σχῆμα εὐθύγραμμον εἰς σχῆμα εὐθύγραμμον ἐγγράφεσθαι λέγεται, ὅταν ἑκάστη τῶν τοῦ ἐγγραφομένου σχήματος γωνιῶν ἑκάστης πλευρᾶς τοῦ, εἰς ὃ ἐγγράφεται, ἅπτηται.

2. Σχῆμα δὲ ὁμοίως περὶ σχῆμα περιγράφεσθαι λέγεται, ὅταν ἑκάστη πλευρὰ τοῦ περιγραφομένου ἑκάστης γωνίας τοῦ, περὶ ὃ περιγράφεται, ἅπτηται.

3. Σχῆμα εὐθύγραμμον εἰς κύκλον ἐγγράφεσθαι λέγεται, ὅταν ἑκάστη γωνία τοῦ ἐγγραφομένου ἅπτηται τῆς τοῦ κύκλου περιφερείας.

4. Σχῆμα δὲ εὐθύγραμμον περὶ κύκλον περιγράφεσθαι λέγεται, ὅταν ἑκάστη πλευρὰ τοῦ περιγραφομένου ἐφάπτηται τῆς τοῦ κύκλου περιφερείας.

5. Κύκλος δὲ εἰς σχῆμα ὁμοίως ἐγγράφεσθαι λέγεται, ὅταν ἡ τοῦ κύκλου περιφέρεια ἑκάστης πλευρᾶς τοῦ, εἰς ὃ ἐγγράφεται, ἅπτηται.

6. Κύκλος δὲ περὶ σχῆμα περιγράφεσθαι λέγεται, ὅταν ἡ τοῦ κύκλου περιφέρεια ἑκάστης γωνίας τοῦ, περὶ ὃ περιγράφεται, ἅπτηται.

7. Εὐθεῖα εἰς κύκλον ἐναρμόζεσθαι λέγεται, ὅταν τὰ πέρατα αὐτῆς ἐπὶ τῆς περιφερείας ᾖ τοῦ κύκλου.

In the first two definitions an English translation, if it is to be clear, must depart slightly from the exact words used in the Greek, where "each side" of one figure is said to pass through "each angle" of another, or "each angle" (i.e. angular point) of one lies on "each side" of another (ἑκάστη πλευρά, ἑκάστη γωνία).

It is also necessary, in the five definitions 1, 2, 3, 5 and 6, to translate the same Greek word ἅπτηται in three different ways. It was observed on III. Def. 2 that the usual meaning of ἅπτεσθαι in Euclid is to *meet*, in contra-distinction to ἐφάπτεσθαι, which means to *touch*. Exceptionally, as in Def. 5, ἅπτεσθαι has the meaning of *touch*. But two new meanings of the word appear, the first being to *lie on*, as in Deff. 1 and 3, the second to *pass through*, as in Deff. 2 and 6; "each angle" lies on (ἅπτεται) a side or on a circle, and "each side," or a circle, passes through (ἅπτεται) an angle or "each angle." The first meaning of *lying on* is exemplified in the phrase of Pappus ἅψεται τὸ σημεῖον θέσει δεδομένης εὐθείας, "will lie on a straight line given in position"; the meaning of *passing through* seems to be much rarer (I have not seen it in Archimedes or Pappus), but, as pointed out on III. Def. 2, Aristotle uses the compound ἐφάπτεσθαι in this sense.

Simson proposed to read ἐφάπτηται in the case (Def. 5) where ἅπτηται means *touches*. He made the like suggestion as regards the Greek text of III. 11, 12, 13, 18, 19; in the first four of these cases there seems to be MS. authority for the compound verb, and in the fifth Heiberg adopts Simson's correction.

BOOK IV. PROPOSITIONS

PROPOSITION 1.

Into a given circle to fit a straight line equal to a given straight line which is not greater than the diameter of the circle.

Let ABC be the given circle, and D the given straight line not greater than the diameter of the circle;

thus it is required to fit into the circle ABC a straight line equal to the straight line D.

Let a diameter BC of the circle ABC be drawn.

Then, if BC is equal to D, that which was enjoined will have been done; for BC has been fitted into the circle ABC equal to the straight line D.

But, if BC is greater than D,

let CE be made equal to D, and with centre C and distance CE let the circle EAF be described;

let CA be joined.

Then, since the point C is the centre of the circle EAF,

CA is equal to CE.

But CE is equal to D;

therefore D is also equal to CA.

Therefore into the given circle ABC there has been fitted CA equal to the given straight line D.

Q. E. F.

Of this problem as it stands there are of course an infinite number of solutions; and, if a particular point be chosen as one extremity of the chord to be "fitted in," there are two solutions. More difficult cases of "fitting into" a circle a chord of given length are arrived at by adding some further condition, e.g. (1) that the chord is to be parallel to a given straight line, or (2) that the chord, produced if necessary, shall pass through a given point. The former problem is solved by Pappus (III. p. 132); instead of drawing the chord as a tangent to a circle concentric with the given circle and having as radius a straight line the square on which is equal to the difference between the squares on the radius of the given circle and on half the given length, he merely draws the diameter of the circle which is parallel to the given direction, measures from the centre along it in each direction a length equal to half the given length, and then draws, on one side of the diameter, perpendiculars to it through the two points so determined.

The second problem of drawing a chord of given length, being less than the diameter of the circle, and passing through a given point, is more important as having been one of the problems discussed by Apollonius in his work entitled νεύσεις, now lost. Pappus states the problem thus (VII. p. 670): "A circle being given in position, to fit into it a straight line given in magnitude and verging (νεύουσαν) towards a given (point)." To do this we

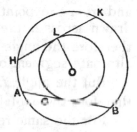

have only to place any chord *HK* in the given circle (with centre *O*) equal to the given length, take *L* the middle point of it, with *O* as centre and *OL* as radius describe a circle, and lastly through the given point *C* draw a tangent to this circle meeting the given circle in *A, B. AB* is then one of *two* chords which can be drawn satisfying the given conditions, if *C* is outside the inner circle; if *C* is on the inner circle, there is one solution only; and, if *C* is within the inner circle, there is no solution. Thus, if *C* is within the outer (given) circle, besides the condition that the given length must not be greater than the diameter of the circle, there is another necessary condition of the possibility of a solution, viz. that the given length must not be *less* than double of the straight line the square on which is equal to the difference between the squares (1) on the radius of the given circle and (2) on the distance between its centre and the given point.

PROPOSITION 2.

In a given circle to inscribe a triangle equiangular with a given triangle.

Let *ABC* be the given circle, and *DEF* the given triangle;

thus it is required to inscribe in the circle *ABC* a triangle equiangular with the triangle *DEF*.

Let *GH* be drawn touching the circle *ABC* at *A* [III. 16, Por.];

on the straight line *AH*, and at the point *A* on it, let the
angle *HAC* be constructed equal to the angle *DEF*,

and on the straight line *AG*, and at the point *A* on it, let
the angle *GAB* be constructed equal to the angle *DFE* ;

[I. 23]

let *BC* be joined.

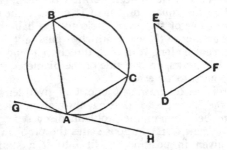

Then, since a straight line *AH* touches the circle *ABC*,
and from the point of contact at *A* the straight line *AC* is
drawn across in the circle,

therefore the angle *HAC* is equal to the angle *ABC* in the
alternate segment of the circle.   [III. 32]

But the angle *HAC* is equal to the angle *DEF* ;
therefore the angle *ABC* is also equal to the angle *DEF*.

For the same reason

the angle *ACB* is also equal to the angle *DFE* ;

therefore the remaining angle *BAC* is also equal to the
remaining angle *EDF*.   [I. 32]

Therefore in the given circle there has been inscribed a
triangle equiangular with the given triangle.   Q. E. F.

Here again, since any point on the circle may be taken as an angular
point of the triangle, there are an infinite number of solutions.   Even when a
particular point has been chosen to form one angular point, the required
triangle may be constructed in six ways.   For any one of the three angles
may be placed at the point ; and, whichever is placed there, the positions of
the two others relatively to it may be interchanged.   The sides of the triangle
will, in all the different solutions, be of the same length respectively ; only
their relative positions will be different.

This problem can of course be reduced (as it was by Borelli) to III. 34,
namely the problem of cutting off from a given circle a segment containing an
angle equal to a given angle.   It can also be solved by the alternative method
applicable to III. 34 of drawing "angles at the centre" equal to double the
angles of the given triangle respectively ; and by this method we can easily
solve this problem, or III. 34, with the further condition that one side of the

required triangle, or the base of the required segment, respectively, shall be parallel to a given straight line.

As a particular case, we can, by the method of this proposition, describe an *equilateral* triangle in any circle after we have first constructed any equilateral triangle by the aid of I. I. The possibility of this is assumed in IV. 16. It is of course equivalent to dividing the circumference of a circle into *three equal parts*. As De Morgan says, the idea of dividing a revolution into equal parts should be kept prominent in considering Book IV.; this aspect of the construction of regular polygons is obvious enough, and the reason why the division of the circle into *three* equal parts is not given by Euclid is that it happens to be as easy to divide the circle into three parts which are in the ratio of the angles of any triangle as to divide it into three equal parts.

PROPOSITION 3.

About a given circle to circumscribe a triangle equiangular with a given triangle.

Let *ABC* be the given circle, and *DEF* the given triangle;

5 thus it is required to circumscribe about the circle *ABC* a triangle equiangular with the triangle *DEF*.

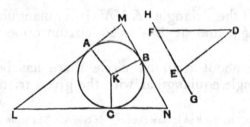

Let *EF* be produced in both directions to the points *G, H,*

let the centre *K* of the circle *ABC* be taken [III. 1], and let 10 the straight line *KB* be drawn across at random;

on the straight line *KB*, and at the point *K* on it, let the angle *BKA* be constructed equal to the angle *DEG,*

and the angle *BKC* equal to the angle *DFH*;   [I. 23]

and through the points *A, B, C* let *LAM, MBN, NCL* be 15 drawn touching the circle *ABC*.   [III. 16, Por.]

Now, since *LM, MN, NL* touch the circle *ABC* at the points *A, B, C,*

and *KA, KB, KC* have been joined from the centre *K* to the points *A, B, C,*

20 therefore the angles at the points *A*, *B*, *C* are right.   [III. 18]

And, since the four angles of the quadrilateral *AMBK* are equal to four right angles, inasmuch as *AMBK* is in fact divisible into two triangles,

and the angles *KAM*, *KBM* are right,

25 therefore the remaining angles *AKB*, *AMB* are equal to two right angles.

But the angles *DEG*, *DEF* are also equal to two right angles ;   [I. 13]

therefore the angles *AKB*, *AMB* are equal to the angles 30 *DEG*, *DEF*,

of which the angle *AKB* is equal to the angle *DEG* ;

therefore the angle *AMB* which remains is equal to the angle *DEF* which remains.

Similarly it can be proved that the angle *LNB* is also 35 equal to the angle *DFE* ;

therefore the remaining angle *MLN* is equal to the angle *EDF*.   [I. 32]

Therefore the triangle *LMN* is equiangular with the triangle *DEF*; and it has been circumscribed about the 40 circle *ABC*.

Therefore about a given circle there has been circumscribed a triangle equiangular with the given triangle.

Q. E. F.

10. **at random**, literally " as it may chance," ὡς ἔτυχεν. The same expression is used in III. 1 and commonly.

22. **is in fact divisible**, καὶ διαιρεῖται, literally " is actually divided."

The remarks as to the number of ways in which Prop. 2 can be solved apply here also.

Euclid leaves us to satisfy ourselves that the three tangents *will* meet and form a triangle. This follows easily from the fact that each of the angles *AKB*, *BKC*, *CKA* is less than two right angles. The first two are so by construction, being the supplements of two angles of the given triangle respectively, and, since all three angles round *K* are together equal to four right angles, it follows that the third, the angle *AKC*, is equal to the sum of the two angles *E*, *F* of the triangle, i.e. to the supplement of the angle *D*, and is therefore less than two right angles.

Peletarius and Borelli gave an alternative solution, first inscribing a triangle equiangular to the given triangle, by IV. 2, and then drawing tangents to the circle parallel to the sides of the inscribed triangle respectively. This method will of course give two solutions, since two tangents can be drawn parallel to each of the sides of the inscribed triangle.

If the three pairs of parallel tangents be drawn and produced far enough, they will form *eight* triangles, two of which are the triangles *circumscribed* to the circle in the manner required in the proposition. The other six triangles are so related to the circle that the circle touches two of the sides in each *produced*, i.e. the circle is an *escribed* circle to each of the six triangles.

PROPOSITION 4.

In a given triangle to inscribe a circle.

Let ABC be the given triangle;
thus it is required to inscribe a circle in the triangle ABC.

Let the angles ABC, ACB
5 be bisected by the straight lines
BD, CD [I. 9], and let these meet
one another at the point D;
from D let DE, DF, DG be
drawn perpendicular to the straight
10 lines AB, BC, CA.

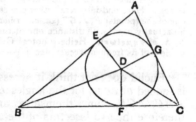

Now, since the angle ABD
is equal to the angle CBD,
and the right angle BED is also equal to the right angle BFD,
15 EBD, FBD are two triangles having two angles equal to two angles and one side equal to one side, namely that subtending one of the equal angles, which is BD common to the triangles;
therefore they will also have the remaining sides equal to
20 the remaining sides: [I. 26]
therefore DE is equal to DF.
For the same reason
DG is also equal to DF.
Therefore the three straight lines DE, DF, DG are equal
25 to one another;
therefore the circle described with centre D and distance one of the straight lines DE, DF, DG will pass also through the remaining points, and will touch the straight lines AB, BC, CA, because the angles at the points E, F, G
30 are right.

For, if it cuts them, the straight line drawn at right angles to the diameter of the circle from its extremity will be found to fall within the circle: which was proved absurd; [III. 16]

therefore the circle described with centre D and distance
35 one of the straight lines DE, DF, DG will not cut the
straight lines AB, BC, CA;

therefore it will touch them, and will be the circle inscribed
in the triangle ABC.   [IV. Def. 5]

Let it be inscribed, as FGE.

40   Therefore in the given triangle ABC the circle EFG has
been inscribed.

                                                    Q. E. F.

26, 34. **and distance one of the (straight lines D)E, (D)F, (D)G.** The words
and letters here shown in brackets are put in to fill out the rather careless language of the
Greek. Here and in several other places in Book IV. Euclid says literally "and with distance
one of the (points) E, F, G" (καὶ διαστήματι ἑνὶ τῶν E, Z, H) and the like. In one case (IV. 13)
he actually has "with distance one of the *points* G, H, K, L, M" (διαστήματι ἑνὶ τῶν H, Θ,
K, Λ, M σημείων). Heiberg notes "Graecam locutionem satis miram et negligentem," but,
in view of its frequent occurrence in good MSS., does not venture to correct it.

Euclid does not think it necessary to prove that BD, CD *will* meet; this
is indeed obvious, for the angles DBC, DCB are together half of the angles
ABC, ACB, which themselves are together less than two right angles, and
therefore the two bisectors of the angles B, C must meet, by Post. 5.

It follows from the proof of this proposition that, if the bisectors of two
angles B, C of a triangle meet in D, the line joining D to A also bisects the
third angle A, or the bisectors of the three angles of a triangle meet in
a point.

It will be observed that Euclid uses the *indirect* form of proof when
showing that the circle touches the three sides of the triangle. Simson proves
it directly, and points out that Euclid does the same in III. 17, 33 and 37,
whereas in IV. 8 and 13 as well as here he uses the *indirect* form. The
difference is unimportant, being one of form and not of substance; the
indirect proof refers back to III. 16, whereas the direct refers back to the
Porism to that proposition.

We may state this problem in the more general form: *To describe a circle
touching three given straight lines which do not all meet in one point, and of
which not more than two are parallel.*

In the case (1) where two of the straight lines are parallel and the third
cuts them, two pairs of interior angles are formed, one on each side of the
third straight line. If we bisect each of the interior angles on one side, the
bisectors will meet in a point, and this point will be the centre of a circle
which can be drawn touching each of the three straight lines, its radius being
the perpendicular from the point on any one of the three. Since the *alternate*
angles are equal, two equal circles can be drawn in this manner satisfying the
given condition.

In the case (2) where the three straight lines form a triangle, suppose each
straight line produced indefinitely. Then each straight line will make two
pairs of interior angles with the other two, one pair forming two angles of the
triangle, and the other pair being their supplements. By bisecting each angle
of either pair we obtain, in the manner of the proposition, two circles
satisfying the conditions, one of them being the inscribed circle of the triangle
and the other being a circle *escribed* to it, i.e. touching one side and the other

two sides *produced*. Next, taking the pairs of interior angles formed by a second side with the other two produced indefinitely, we get two circles satisfying the conditions, one of which is the same inscribed circle that we had before, while the other is a second escribed circle. Similarly with the third side. Hence we have the inscribed circle, and three escribed circles (one opposite each angle of the triangle), i.e. four circles in all, satisfying the conditions of the problem.

It may perhaps not be inappropriate to give at this point Heron's elegant proof of the formula for the area of a triangle in terms of the sides, which we usually write thus :

$$\triangle = \sqrt{s(s-a)(s-b)(s-c)},$$

although it requires the theory of proportions and uses some ungeometrical expressions, e.g. the product of two areas and the "side" of such a product, where of course the areas are so many square units of length. The proof is given in the *Metrica*, I. 8., and in the *Dioptra*, 30 (Heron, Vol. III., Teubner, 1903, pp. 20—24 and pp. 280—4, or Heron, ed. Hultsch, pp. 235—7).

Suppose the sides of the triangle ABC to be given in length.

Inscribe the circle DEF, and let G be its centre.

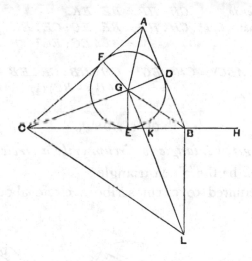

Join AG, BG, CG, DG, EG, FG.

Then
$$BC \cdot EG = 2 \cdot \triangle BGC,$$
$$CA \cdot FG = 2 \cdot \triangle ACG,$$
$$AB \cdot DG = 2 \cdot \triangle ABG.$$

Therefore, by addition,
$$p \cdot EG = 2 \cdot \triangle ABC,$$

where p is the perimeter.

Produce CB to H, so that $BH = AD$.

Then, since $AD = AF$, $DB = BE$, $FC = CE$,
$$CH = \tfrac{1}{2}p.$$

Hence
$$CH \cdot EG = \triangle ABC.$$

But $CH.EG$ is the "side" of the product $CH^2.EG^2$, that is $\sqrt{CH^2.EG^2}$;

therefore $\qquad\qquad (\triangle ABC)^2 = CH^2.EG^2.$

Draw GL at right angles to CG, and BL at right angles to CB, meeting at L. Join CL.

Then, since each of the angles CGL, CBL is right, $CGBL$ is a quadrilateral in a circle.

Therefore the angles CGB, CLB are equal to two right angles.

Now the angles CGB, AGD are equal to two right angles, since AG, BG, CG bisect the angles at G, and the angles CGB, AGD are equal to the angles AGC, DGB, while the sum of all four is equal to four right angles.

Therefore the angles AGD, CLB are equal.

So are the right angles ADG, CBL.

Therefore the triangles AGD, CLB are similar.

Hence $\qquad\qquad\qquad BC:BL = AD:DG$
$\qquad\qquad\qquad\qquad\qquad = BH:EG,$

and, alternately, $\qquad CB:BH = BL:EG$
$\qquad\qquad\qquad\qquad\qquad = BK:KE,$

whence, *componendo*, $\quad CH:HB = BE:EK.$

It follows that $\quad CH^2:CH.HB = BE.EC:CE.EK$
$\qquad\qquad\qquad\qquad\qquad = BE.EC:EG^2$

Therefore

$\qquad (\triangle ABC)^2 = CH^2.EG^2 = CH.HB.CE.EB$
$\qquad\qquad\qquad\qquad = \tfrac{1}{2}p\,(\tfrac{1}{2}p - BC)\,(\tfrac{1}{2}p - AB)\,(\tfrac{1}{2}p - AC).$

PROPOSITION 5.

About a given triangle to circumscribe a circle.

Let ABC be the given triangle;

thus it is required to circumscribe a circle about the given triangle ABC.

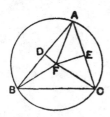 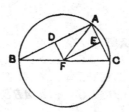 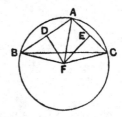

Let the straight lines AB, AC be bisected at the points D, E [I. 10], and from the points D, E let DF, EF be drawn at right angles to AB, AC;

they will then meet within the triangle ABC, or on the straight line BC, or outside BC.

First let them meet within at F, and let FB, FC, FA be joined.

Then, since AD is equal to DB,
and DF is common and at right angles,
therefore the base AF is equal to the base FB.   [I. 4]

Similarly we can prove that
$\qquad CF$ is also equal to AF;
so that FB is also equal to FC;

therefore the three straight lines FA, FB, FC are equal to one another.

Therefore the circle described with centre F and distance one of the straight lines FA, FB, FC will pass also through the remaining points, and the circle will have been circumscribed about the triangle ABC.

Let it be circumscribed, as ABC.

Next, let DF, EF meet on the straight line BC at F, as is the case in the second figure; and let AF be joined.

Then, similarly, we shall prove that the point F is the centre of the circle circumscribed about the triangle ABC.

Again, let DF, EF meet outside the triangle ABC at F, as is the case in the third figure, and let AF, BF, CF be joined.

Then again, since AD is equal to DB,
and DF is common and at right angles,
therefore the base AF is equal to the base BF.   [I. 4]

Similarly we can prove that
$\qquad CF$ is also equal to AF;
so that BF is also equal to FC;

therefore the circle described with centre F and distance one of the straight lines FA, FB, FC will pass also through the remaining points, and will have been circumscribed about the triangle ABC.

Therefore about the given triangle a circle has been circumscribed.

Q. E. F.

And it is manifest that, when the centre of the circle falls within the triangle, the angle BAC, being in a segment greater than the semicircle, is less than a right angle;

when the centre falls on the straight line BC, the angle BAC, being in a semicircle, is right;

and when the centre of the circle falls outside the triangle, the angle BAC, being in a segment less than the semicircle, is greater than a right angle.   [III. 31]

·Simson points out that Euclid does not prove that DF, EF will meet, and he inserts in the text the following argument to supply the omission.

."DF, EF produced meet one another. For, if they do not meet, they are parallel, wherefore AB, AC, which are at right angles to them, are parallel [or, he should have added, in a straight line]: which is absurd."

This assumes, of course, that straight lines which are at right angles to two parallels are themselves parallel; but this is an obvious deduction from I. 28.

On the assumption that DF, EF will meet Todhunter has this note: "It has been proposed to show this in the following way: join DE; then the angles EDF and DEF are together less than the angles ADF and AEF, that is, they are together less than two right angles; and therefore DF and EF will meet, by Axiom 12 [Post. 5]. This assumes that ADE and AED are *acute* angles; it may, however, be easily shown that DE is parallel to BC, so that the triangle ADE is equiangular to the triangle ABC; and we must therefore select the two sides AB and AC such that ABC and ACB may be acute angles."

This is, however, unsatisfactory. Euclid makes no such selection in III. 9 and III. 10, where the same assumption is tacitly made; and it is unnecessary, because it is easy to prove that the straight lines DF, EF meet in *all* cases, by considering the different possibilities separately and drawing a separate figure for each case.

Simson thinks that Euclid's demonstration had been spoiled by some unskilful hand both because of the omission to prove that the perpendicular bisectors meet, and because "without any reason he divides the proposition into three cases, whereas one and the same construction and demonstration serves for them all, as Campanus has observed." However, up to the usual words ὅπερ ἔδει ποιῆσαι there seems to be no doubt about the text. Heiberg suggests that Euclid gave separately the case where F falls on BC because, in that case, only AF needs to be drawn and not BF, CF as well.

The addition, though given in Simson and the text-books as a "corollary," has no heading πόρισμα in the best MSS.; it is an explanation like that which is contained in the penultimate paragraph of III. 25.

The Greek text has a further addition, which is rejected by Heiberg as not genuine, "So that, further, when the given angle happens to be less than a right angle, DF, EF will fall within the triangle, when it is right, on BC, and, when it is greater than a right angle, outside BC. (being) what it was required to do." Simson had already observed that the text here is vitiated "where mention is made of a given angle, though there neither is, nor can be, anything in the proposition relating to a given angle."

PROPOSITION 6.

In a given circle to inscribe a square.

Let *ABCD* be the given circle;
thus it is required to inscribe a square in the circle *ABCD*.

Let two diameters *AC, BD* of the
circle *ABCD* be drawn at right angles
to one another, and let *AB, BC, CD,
DA* be joined.

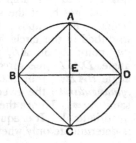

Then, since *BE* is equal to *ED*, for
E is the centre,
and *EA* is common and at right angles,
therefore the base *AB* is equal to the
base *AD*.   [I. 4]

For the same reason
each of the straight lines *BC, CD* is also equal to each of
the straight lines *AB, AD*;
therefore the quadrilateral *ABCD* is equilateral.

I say next that it is also right-angled.

For, since the straight line *BD* is a diameter of the circle
ABCD,
therefore *BAD* is a semicircle;
therefore the angle *BAD* is right.   [III. 31]

For the same reason
each of the angles *ABC, BCD, CDA* is also right;
therefore the quadrilateral *ABCD* is right-angled.

But it was also proved equilateral;
therefore it is a square;   [I. Def. 22]
and it has been inscribed in the circle *ABCD*.

Therefore in the given circle the square *ABCD* has been
inscribed.

Q. E. F.

Euclid here proceeds to consider problems corresponding to those in
Props. 2—5 with reference to figures of four or more sides, but with the
difference that, whereas he dealt with triangles of any form, he confines
himself henceforth to regular figures. It happened to be as easy to divide a
circle into *three* parts which are in the ratio of the angles, or of the supplements
of the angles, of a triangle as into three *equal* parts. But, when it is required to
inscribe in a circle a figure equiangular to a given *quadrilateral,* this can only be

done provided that the quadrilateral has either pair of opposite angles equal
to two right angles.   Moreover, in this case, the problem *may* be solved in the
same way as that of IV. 2, i.e. by simply inscribing a triangle equiangular to one
of the triangles into which the quadrilateral is divided by either diagonal, and
then drawing on the side corresponding to the diagonal as base another
triangle equiangular to the other triangle contained in the quadrilateral.   But
this is not the *only* solution; there are an infinite
number of other solutions in which the inscribed
quadrilateral will, unlike that found by this particular
method, not be of the same *form* as the given quadrilateral.   For suppose *ABCD* to be the quadrilateral
inscribed in the circle by the method of IV. 2.   Take
any point *B'* on *AB*, join *AB'*, and then make the
angle *DAD'* (measured towards *AC*) equal to the
angle *BAB'*.   Join *B'C, CD'*.   Then *AB'CD'* is also
equiangular to the given quadrilateral, but not of the
same *form*.   Hence the problem is indeterminate in the case of the general
quadrilateral.   It is equally so if the given quadrilateral is a rectangle; and it
is determinate only when the given quadrilateral is a *square*.

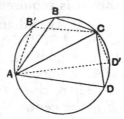

PROPOSITION 7.

About a given circle to circumscribe a square.

Let *ABCD* be the given circle;

thus it is required to circumscribe a square about the circle
ABCD.

Let two diameters *AC, BD* of the
circle *ABCD* be drawn at right angles
to one another, and through the points
A, B, C, D let *FG, GH, HK, KF* be
drawn touching the circle *ABCD*.
                                [III. 16, Por.]
Then, since *FG* touches the circle
ABCD,

and *EA* has been joined from the centre
E to the point of contact at *A*,

therefore the angles at *A* are right.   [III. 18]

For the same reason

the angles at the points *B, C, D* are also right.

Now, since the angle *AEB* is right,

and the angle *EBG* is also right,

therefore *GH* is parallel to *AC*.   [I. 28]

For the same reason

AC is also parallel to FK,

so that GH is also parallel to FK.   [I. 30]

Similarly we can prove that

each of the straight lines GF, HK is parallel to BED.

Therefore GK, GC, AK, FB, BK are parallelograms;

therefore GF is equal to HK, and GH to FK.   [I. 34]

And, since AC is equal to BD,

and AC is also equal to each of the straight lines GH, FK,

while BD is equal to each of the straight lines GF, HK,

[I. 34]

therefore the quadrilateral $FGHK$ is equilateral.

I say next that it is also right-angled.

For, since $GBEA$ is a parallelogram,

and the angle AEB is right,

therefore the angle AGB is also right.   [I. 34]

Similarly we can prove that

the angles at H, K, F are also right.

Therefore $FGHK$ is right-angled.

But it was also proved equilateral;

therefore it is a square;

and it has been circumscribed about the circle $ABCD$.

Therefore about the given circle a square has been circumscribed.

Q. E. F.

It is just as easy to describe about a given circle a polygon equiangular to any given polygon as it is to describe a square about a given circle. We have only to use the method of IV. 3, i.e. to take any radius of the circle, to measure round the centre successive angles in one and the same direction equal to the supplements of the successive angles of the given polygon and, lastly, to draw tangents to the circle at the extremities of the several radii so determined; but again the polygon would in general not be of the same *form* as the given one; it would only be so if the given polygon happened to be such that a circle could be inscribed in it. To take the case of a *quadrilateral* only: it is easy to prove that, if a quadrilateral be described about a circle, the sum of one pair of opposite sides must be equal to the sum of the other pair. It may be proved, *conversely*, that, if a quadrilateral has the sums of the pairs of opposite sides equal, a circle can be inscribed in it. If then a given quadrilateral has the sums of the pairs of opposite sides equal, a quadrilateral can be described about any given circle not only equiangular with it but having the same *form* or, in the words of Book VI., *similar* to it.

PROPOSITION 8.

In a given square to inscribe a circle.

Let *ABCD* be the given square;
thus it is required to inscribe a circle in the given square
ABCD.

Let the straight lines *AD*, *AB* be
bisected at the points *E*, *F* respectively
[I. 10],
through *E* let *EH* be drawn parallel
to either *AB* or *CD*, and through
F let *FK* be drawn parallel to either
AD or *BC*;   [I. 31]
therefore each of the figures *AK*, *KB*,
AH, *HD*, *AG*, *GC*, *BG*, *GD* is a parallelogram,
and their opposite sides are evidently equal.   [I. 34]

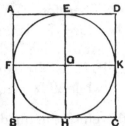

Now, since *AD* is equal to *AB*,
and *AE* is half of *AD*, and *AF* half of *AB*,
therefore *AE* is equal to *AF*,
so that the opposite sides are also equal;
therefore *FG* is equal to *GE*.

Similarly we can prove that each of the straight lines *GH*,
GK is equal to each of the straight lines *FG*, *GE*;
therefore the four straight lines *GE*, *GF*, *GH*, *GK* are
equal to one another.

Therefore the circle described with centre *G* and distance
one of the straight lines *GE*, *GF*, *GH*, *GK* will pass also
through the remaining points.

And it will touch the straight lines *AB*, *BC*, *CD*, *DA*,
because the angles at *E*, *F*, *H*, *K* are right.

For, if the circle cuts *AB*, *BC*, *CD*, *DA*, the straight
line drawn at right angles to the diameter of the circle from
its extremity will fall within the circle: which was proved
absurd;   [III. 16]
therefore the circle described with centre *G* and distance
one of the straight lines *GE*, *GF*, *GH*, *GK* will not cut
the straight lines *AB*, *BC*, *CD*, *DA*.

Therefore it will touch them, and will have been inscribed
in the square *ABCD*.

Therefore in the given square a circle has been inscribed.
                                         Q. E. F.

As was remarked in the last note, a circle can be inscribed in any *quadrilateral* which has the sum of one pair of opposite sides equal to the sum of the other pair.   In particular, it follows that a circle can be inscribed in a *square* or a *rhombus*, but not in a rectangle or a rhomboid.

PROPOSITION 9.

About a given square to circumscribe a circle.

Let $ABCD$ be the given square;

thus it is required to circumscribe a circle about the square $ABCD$.

For let AC, BD be joined, and let them cut one another at E.

Then, since DA is equal to AB, and AC is common,

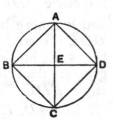

therefore the two sides DA, AC are equal to the two sides BA, AC;

and the base DC is equal to the base BC;

therefore the angle DAC is equal to the angle BAC.   [I. 8]

Therefore the angle DAB is bisected by AC.

Similarly we can prove that each of the angles ABC, BCD, CDA is bisected by the straight lines AC, DB.

Now, since the angle DAB is equal to the angle ABC, and the angle EAB is half the angle DAB, and the angle EBA half the angle ABC, therefore the angle EAB is also equal to the angle EBA; so that the side EA is also equal to EB.   [I. 6]

Similarly we can prove that each of the straight lines EA, EB is equal to each of the straight lines EC, ED.

Therefore the four straight lines EA, EB, EC, ED are equal to one another.

Therefore the circle described with centre E and distance one of the straight lines EA, EB, EC, ED will pass also through the remaining points;

and it will have been circumscribed about the square $ABCD$.

Let it be circumscribed, as $ABCD$.

Therefore about the given square a circle has been circumscribed.

Q. E. F.

PROPOSITION 10.

To construct an isosceles triangle having each of the angles at the base double of the remaining one.

Let any straight line AB be set out, and let it be cut at the point C so that the rectangle contained by AB, BC is equal to the square on CA;   [II. 11]

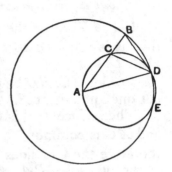

with centre A and distance AB let the circle BDE be described,

and let there be fitted in the circle BDE the straight line BD equal to the straight line AC which is not greater than the diameter of the circle BDE.   [IV. 1]

Let AD, DC be joined, and let the circle ACD be circumscribed about the triangle ACD.

[IV. 5]

Then, since the rectangle AB, BC is equal to the square on AC,

and AC is equal to BD,

therefore the rectangle AB, BC is equal to the square on BD.

And, since a point B has been taken outside the circle ACD,

and from B the two straight lines BA, BD have fallen on the circle ACD, and one of them cuts it, while the other falls on it,

and the rectangle AB, BC is equal to the square on BD,

therefore BD touches the circle ACD.   [III. 37]

Since, then, BD touches it, and DC is drawn across from the point of contact at D,

therefore the angle BDC is equal to the angle DAC in the alternate segment of the circle.   [III. 32]

Since, then, the angle BDC is equal to the angle DAC, let the angle CDA be added to each;

therefore the whole angle BDA is equal to the two angles CDA, DAC.

But the exterior angle BCD is equal to the angles CDA, DAC;   [I. 32]

therefore the angle BDA is also equal to the angle BCD.

But the angle BDA is equal to the angle CBD, since the side AD is also equal to AB;   [I. 5]

so that the angle DBA is also equal to the angle BCD.

Therefore the three angles BDA, DBA, BCD are equal to one another.

And, since the angle DBC is equal to the angle BCD,

the side BD is also equal to the side DC.   [I. 6]

But BD is by hypothesis equal to CA;

therefore CA is also equal to CD,

so that the angle CDA is also equal to the angle DAC;   [I. 5]

therefore the angles CDA, DAC are double of the angle DAC.

But the angle BCD is equal to the angles CDA, DAC;

therefore the angle BCD is also double of the angle CAD.

But the angle BCD is equal to each of the angles BDA, DBA;

therefore each of the angles BDA, DBA is also double of the angle DAB.

Therefore the isosceles triangle ABD has been constructed having each of the angles at the base DB double of the remaining one.

Q. E. F.

There is every reason to conclude that the connexion of the triangle constructed in this proposition with the regular pentagon, and the construction of the triangle itself, were the discovery of the Pythagoreans. In the first place the Scholium IV. No. 2 (Heiberg, Vol. v. p. 273) says "this Book is the discovery of the Pythagoreans." Secondly, the summary in Proclus (p. 65, 20) says that Pythagoras discovered "the construction of the cosmic figures," by which must be understood the five regular solids. Thirdly, Iamblichus (*Vit. Pyth.* c. 18, s. 88) quotes a story of Hippasus, "that he was one of the Pythagoreans but, owing to his being the first to publish and write down (the construction of) the sphere arising from the twelve pentagons (τὴν ἐκ τῶν δώδεκα πενταγώνων), perished by shipwreck for his impiety, having got credit for the discovery all the same, whereas everything belonged to HIM (ἐκείνου τοῦ ἀνδρός), for it is thus that they refer to Pythagoras, and they do not call him by his name." Cantor has (I_3, pp. 176 sqq.) collected notices which help us to form an idea how the discovery of the Euclidean construction for a regular pentagon may have been arrived at by the Pythagoreans.

Plato puts into the mouth of Timaeus a description of the formation from right angled triangles of the figures which are the faces of the first four regular solids. The face of the cube is the square which is formed from isosceles right-angled triangles by placing four of these triangles contiguously so that the four right angles are in contact at the centre. The equilateral triangle, however, which is the form of the faces of the tetrahedron, the octahedron and the icosahedron, cannot be constructed from isosceles right-angled triangles, but is constructed from a particular scalene right-angled triangle which Timaeus (54 A, B) regards as the most beautiful of all scalene right-angled triangles, namely that in which the square on one of the sides about the right angle is three times the square on the other. This is, of course, the triangle forming half of an equilateral triangle bisected by the perpendicular from one angular point on the opposite side. The Platonic Timaeus does not construct his equilateral triangle from two such triangles but from six, by placing the latter contiguously round a point so that the hypotenuses and the smaller of the sides about the right angles respectively adjoin, and all of them meet at the common centre, as shown in the figure (*Timaeus*, 54 D, E.). The probability that this exposition was Pythagorean is confirmed by the independent testimony of Proclus (pp. 304—5), who attributes to the Pythagoreans the theorem that six equilateral triangles, or three hexagons, or four squares, placed contiguously with one angular point of each at a common point, will just fill up the four right angles round that point, and that no other regular polygons in any numbers have this property.

How then would it be proposed to split up into triangles, or to make up out of triangles, the face of the remaining solid, the dodecahedron? It would easily be seen that the pentagon could not be constructed by means of the two right-angled triangles which were used for constructing the square and the equilateral triangle respectively. But attempts would naturally be made to split up the pentagon into elementary triangles, and traces of such attempts are actually forthcoming. Plutarch has in two passages spoken of the division of the faces of the dodecahedron into triangles, remarking in one place (*Quaest. Platon.* v. 1) that each of the twelve faces is made up of 30 elemen-

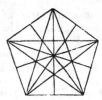

tary scalene triangles, so that, taken together, they give 360 such triangles, and in another (*De defectu oraculorum*, c. 33) that the elementary triangle of the dodecahedron must be different from that of the tetrahedron, octahedron and icosahedron. Another writer of the 2nd cent., Alcinous, has, in his introduction to the study of Plato (*De doctrina Platonis*, c. 11), spoken similarly of the 360 elements which are produced when every one of the pentagons is divided into 5 isosceles triangles, and each of the latter into 6 scalene triangles. Now, if we proceed to draw lines in a pentagon separating it into this number of small triangles as shown in the above figure, the figure which stands out most prominently in the mass of lines is the "star-pentagon," as drawn separately, which then (if the consecutive corners be joined) suggests the drawing, as part of a pentagon, of a triangle of a definite character.   Now we are expressly told by Lucian and the scholiast to the *Clouds* of Aristophanes (see Bretschneider, pp. 85—86) that the triple interwoven triangle, the pentagram (τὸ τριπλοῦν τρίγωνον, τὸ δι' ἀλλήλων, τὸ πεντάγραμμον), was used by the Pythagoreans as a symbol of recognition between the members of the same school (συμβόλῳ πρὸς τοὺς ὁμοδόξους ἐχρῶντο), and was called by them Health. There seems to be therefore no room for doubt that the construction of a pentagon by means of an isosceles triangle having each of its base angles double of the vertical angle was due to the Pythagoreans.

The construction of this triangle depends upon II. 11, or the problem of dividing a straight line so that the rectangle contained by the whole and one of the parts is equal to the square on the other part.   This problem of course appears again in Eucl. VI. 30 as the problem of cutting a given straight line *in extreme and mean ratio*, i.e. the problem of the *golden section*, which is no doubt "the section" referred to in the passage of the summary given by Proclus (p. 67, 6) which says that Eudoxus "greatly added to the number of the theorems which Plato originated regarding the section."   This idea that Plato began the study of the "golden section" as a subject in itself is not in the least inconsistent with the supposition that the problem of Eucl. II. 11 was solved by the Pythagoreans.   The very fact that Euclid places it among other propositions which are clearly Pythagorean in origin is significant, as is also the fact that its solution is effected by "applying to a straight line a rectangle equal to a given square and exceeding by a square," while Proclus says plainly (p. 419, 15) that, according to Eudemus, "the application of areas, their *exceeding* and their falling short, are ancient and discoveries of the Muse of the Pythagoreans."

We may suppose the construction of IV. 10 to have been arrived at by analysis somewhat as follows (Todhunter's Euclid, p. 325).

Suppose the problem solved, i.e. let *ABD* be an isosceles triangle having each of its base angles double of the vertical angle.

Bisect the angle *ADB* by the straight line *DC* meeting *AB* in *C*.   [I. 9]

Therefore the angle *BDC* is equal to the angle *BAD*; and the angle *CDA* is also equal to the angle *BAD*,

so that *DC* is equal to *CA*.

Again, since, in the triangles *BCD*, *BDA*,

the angle *BDC* is equal to the angle *BAD*,

and the angle *B* is common,

the third angle *BCD* is equal to the third angle *BDA*, and therefore to the angle *DBC*.

Therefore *DC* is equal to *DB*.

Now, if a circle be described about the triangle *ACD* [IV. 5], since the angle *BDC* is equal to the angle in the segment *CAD*,

BD must touch the circle [by the converse of III. 32 easily proved from it by *reductio ad absurdum*].

Hence [III. 36] the square on *BD* and therefore the square on *CD*, or *AC*, is equal to the rectangle *AB*, *BC*.

Thus the problem is reduced to that of cutting *AB* at *C* so that the rectangle *AB*, *BC* is equal to the square on *AC*.   [II. 11]

When this is done, we have only to draw a circle with centre A and radius AB and place in it a chord BD equal in length to AC.   [IV. 1]

Since each of the angles ABD, ADB is double of the angle BAD, the latter is equal to one-fifth of the sum of all three, i.e. is one-fifth of two right angles, or two-fifths of a right angle, and each of the base angles is four-fifths of a right angle.

If we bisect the angle BAD, we obtain an angle equal to one-fifth of a right angle, so that the proposition enables us *to divide a right angle into five equal parts.*

It will be observed that BD is the side of a regular *decagon* inscribed in the larger circle.

Proclus, as remarked above (Vol. I. p. 130), gives IV. 10 as an instance in which two of the six formal divisions of a proposition, the *setting-out* and the "*definition,*" are left out, and explains that they are unnecessary because there is no *datum* in the enunciation. This is however no more than formally true, because Euclid does begin his proposition by *setting out* "any straight line AB," and he constructs an isosceles triangle having AB for one of its equal sides, i.e. he does practically imply a datum in the enunciation, and a corresponding *setting-out* and "*definition*" in the proposition itself.

PROPOSITION 11.

In a given circle to inscribe an equilateral and equiangular pentagon.

Let $ABCDE$ be the given circle;
thus it is required to inscribe in the circle $ABCDE$ an equilateral and equiangular pentagon.

Let the isosceles triangle FGH be set out having each of the angles at G, H double of the angle at F;

[IV. 10]

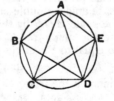

let there be inscribed in the circle $ABCDE$ the triangle ACD equiangular with the triangle FGH, so that the angle CAD is equal to the angle at F and the angles at G, H respectively equal to the angles ACD, CDA; [IV. 2] therefore each of the angles ACD, CDA is also double of the angle CAD.

Now let the angles ACD, CDA be bisected respectively by the straight lines CE, DB [I. 9], and let AB, BC, DE, EA be joined.

Then, since each of the angles ACD, CDA is double of the angle CAD,
and they have been bisected by the straight lines CE, DB,

therefore the five angles DAC, ACE, ECD, CDB, BDA are equal to one another.

But equal angles stand on equal circumferences;   [III. 26]

therefore the five circumferences AB, BC, CD, DE, EA are equal to one another.

But equal circumferences are subtended by equal straight lines;   [III. 29]

therefore the five straight lines AB, BC, CD, DE, EA are equal to one another;

therefore the pentagon $ABCDE$ is equilateral.

I say next that it is also equiangular.

For, since the circumference AB is equal to the circumference DE, let BCD be added to each;

therefore the whole circumference $ABCD$ is equal to the whole circumference $EDCB$.

And the angle AED stands on the circumference $ABCD$, and the angle BAE on the circumference $EDCB$;

therefore the angle BAE is also equal to the angle AED.
     [III. 27]

For the same reason

each of the angles ABC, BCD, CDE is also equal to each of the angles BAE, AED;

therefore the pentagon $ABCDE$ is equiangular.

But it was also proved equilateral;

therefore in the given circle an equilateral and equiangular pentagon has been inscribed.

Q. E. F.

De Morgan remarks that "the method of IV. 11 is not so natural as making a direct use of the angle obtained in the last." On the other hand, if we look at the figure and notice that it shows the whole of the *pentagram-star* except one line (that connecting B and E), I think we shall conclude that the method is nearer to that used by the Pythagoreans, and therefore of much more historical interest.

Another method would of course be to use IV. 10 to describe a *decagon* in the circle, and then to join any vertex to the next alternate one, the latter to the next alternate one, and so on.

Mr H. M. Taylor gives "a complete geometrical construction for inscribing a regular decagon or pentagon in a given circle," as follows.

" Find O the centre.

Draw two diameters AOC, BOD at right angles to one another.

Bisect OD in E.

Draw AE and cut off EF equal to OE.

Place round the circle ten chords equal to AF.

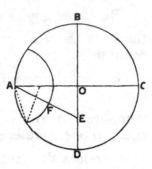

These chords will be the sides of a regular decagon. Draw the chords joining five alternate vertices of the decagon; they will be the sides of a regular pentagon."

The construction is of course only a combination of those in II. 11 and IV. 1; and the proof would have to follow that in IV. 10.

PROPOSITION 12.

About a given circle to circumscribe an equilateral and equiangular pentagon.

Let $ABCDE$ be the given circle;

thus it is required to circumscribe an equilateral and equiangular pentagon about the circle $ABCDE$.

Let A, B, C, D, E be conceived to be the angular points of the inscribed pentagon, so that the circumferences AB, BC, CD, DE, EA are equal;

[IV. 11]

through A, B, C, D, E let GH, HK, KL, LM, MG be drawn touching the circle;   [III. 16, Por.]

let the centre F of the circle $ABCDE$ be taken [III. 1], and let FB, FK, FC, FL, FD be joined.

Then, since the straight line KL touches the circle $ABCDE$ at C,

and FC has been joined from the centre F to the point of contact at C,

therefore FC is perpendicular to KL;   [III. 18]

therefore each of the angles at C is right.

For the same reason

the angles at the points B, D are also right.

And, since the angle *FCK* is right,
therefore the square on *FK* is equal to the squares on *FC, CK*.

For the same reason　　　　　　　　　　　　　　　　[I. 47]
the square on *FK* is also equal to the squares on *FB, BK* ;
so that the squares on *FC, CK* are equal to the squares on *FB, BK*,
of which the square on *FC* is equal to the square on *FB* ;
therefore the square on *CK* which remains is equal to the square on *BK*.

Therefore *BK* is equal to *CK*.

And, since *FB* is equal to *FC*,
and *FK* common,
the two sides *BF, FK* are equal to the two sides *CF, FK* ;
and the base *BK* equal to the base *CK* ;
therefore the angle *BFK* is equal to the angle *KFC*,　[I. 8]
and the angle *BKF* to the angle *FKC*.

Therefore the angle *BFC* is double of the angle *KFC*,
and the angle *BKC* of the angle *FKC*.

For the same reason
the angle *CFD* is also double of the angle *CFL*,
and the angle *DLC* of the angle *FLC*.

Now, since the circumference *BC* is equal to *CD*,
the angle *BFC* is also equal to the angle *CFD*.　　　[III. 27]

And the angle *BFC* is double of the angle *KFC*, and the angle *DFC* of the angle *LFC* ;
therefore the angle *KFC* is also equal to the angle *LFC*.

But the angle *FCK* is also equal to the angle *FCL* ;
therefore *FKC, FLC* are two triangles having two angles equal to two angles and one side equal to one side, namely *FC* which is common to them ;
therefore they will also have the remaining sides equal to the remaining sides, and the remaining angle to the remaining angle ;　　　　　　　　　　　　　　　　　　　[I. 26]
therefore the straight line *KC* is equal to *CL*,
and the angle *FKC* to the angle *FLC*.

And, since *KC* is equal to *CL*,
therefore *KL* is double of *KC*.

For the same reason it can be proved that
HK is also double of BK.

And BK is equal to KC;

therefore HK is also equal to KL.

Similarly each of the straight lines HG, GM, ML can also be proved equal to each of the straight lines HK, KL;

therefore the pentagon $GHKLM$ is equilateral.

I say next that it is also equiangular.

For, since the angle FKC is equal to the angle FLC, and the angle HKL was proved double of the angle FKC,

and the angle KLM double of the angle FLC,

therefore the angle HKL is also equal to the angle KLM.

Similarly each of the angles KHG, HGM, GML can also be proved equal to each of the angles HKL, KLM;

therefore the five angles GHK, HKL, KLM, LMG, MGH are equal to one another.

Therefore the pentagon $GHKLM$ is equiangular.

And it was also proved equilateral; and it has been circumscribed about the circle $ABCDE$.

Q. E. F.

De Morgan remarks that IV. 12, 13, 14 supply the place of the following: *Having given a regular polygon of any number of sides inscribed in a circle, to describe the same about the circle; and, having given the polygon, to inscribe and circumscribe a circle.* For the method can be applied generally, as indeed Euclid practically says in the Porism to IV. 15 about the regular hexagon and in the remark appended to IV. 16 about the regular fifteen-angled figure.

The *conclusion* of this proposition, "therefore about the given circle an equilateral and equiangular pentagon has been circumscribed," is omitted in the MSS.

PROPOSITION 13.

In a given pentagon, which is equilateral and equiangular, to inscribe a circle.

Let $ABCDE$ be the given equilateral and equiangular pentagon;

thus it is required to inscribe a circle in the pentagon $ABCDE$.

For let the angles BCD, CDE be bisected by the straight lines CF, DF respectively; and from the point F, at which the straight lines CF, DF meet one another, let the straight lines FB, FA, FE be joined.

Then, since BC is equal to CD, and CF common, the two sides BC, CF are equal to the two sides DC, CF; and the angle BCF is equal to the angle DCF;

therefore the base BF is equal to the base DF, and the triangle BCF is equal to the triangle DCF, and the remaining angles will be equal to the remaining angles, namely those which the equal sides subtend.　　　[I. 4]

Therefore the angle CBF is equal to the angle CDF.

And, since the angle CDE is double of the angle CDF, and the angle CDE is equal to the angle ABC, while the angle CDF is equal to the angle CBF; therefore the angle CBA is also double of the angle CBF;

therefore the angle ABF is equal to the angle FBC; therefore the angle ABC has been bisected by the straight line BF.

Similarly it can be proved that the angles BAE, AED have also been bisected by the straight lines FA, FE respectively.

Now let FG, FH, FK, FL, FM be drawn from the point F perpendicular to the straight lines AB, BC, CD, DE, EA.

Then, since the angle HCF is equal to the angle KCF, and the right angle FHC is also equal to the angle FKC, FHC, FKC are two triangles having two angles equal to two angles and one side equal to one side, namely FC which is common to them and subtends one of the equal angles; therefore they will also have the remaining sides equal to the remaining sides;　　　[I. 26]

therefore the perpendicular FH is equal to the perpendicular FK.

Similarly it can be proved that each of the straight lines FL, FM, FG is also equal to each of the straight lines FH, FK;

therefore the five straight lines *FG, FH, FK, FL, FM* are equal to one another.

Therefore the circle described with centre *F* and distance one of the straight lines *FG, FH, FK, FL, FM* will pass also through the remaining points;

and it will touch the straight lines *AB, BC, CD, DE, EA*, because the angles at the points *G, H, K, L, M* are right.

For, if it does not touch them, but cuts them,

it will result that the straight line drawn at right angles to the diameter of the circle from its extremity falls within the circle: which was proved absurd.   [III. 16]

Therefore the circle described with centre *F* and distance one of the straight lines *FG, FH, FK, FL, FM* will not cut the straight lines *AB, BC, CD, DE, EA*;

therefore it will touch them.

Let it be described, as *GHKLM*.

Therefore in the given pentagon, which is equilateral and equiangular, a circle has been inscribed.

Q. E. F.

PROPOSITION 14.

About a given pentagon, which is equilateral and equiangular, to circumscribe a circle.

Let *ABCDE* be the given pentagon, which is equilateral and equiangular;

thus it is required to circumscribe a circle about the pentagon *ABCDE*.

Let the angles *BCD, CDE* be bisected by the straight lines *CF, DF* respectively, and from the point *F*, at which the straight lines meet, let the straight lines *FB, FA, FE* be joined to the points *B, A, E*.

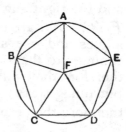

Then in manner similar to the preceding it can be proved that the angles *CBA, BAE, AED* have also been bisected by the straight lines *FB, FA, FE* respectively.

Now, since the angle *BCD* is equal to the angle *CDE*,
and the angle *FCD* is half of the angle *BCD*,
and the angle *CDF* half of the angle *CDE*,
therefore the angle *FCD* is also equal to the angle *CDF*,

so that the side *FC* is also equal to the side *FD*.   [I. 6]

Similarly it can be proved that
each of the straight lines *FB*, *FA*, *FE* is also equal to each
of the straight lines *FC*, *FD* ;
therefore the five straight lines *FA*, *FB*, *FC*, *FD*, *FE* are
equal to one another.

Therefore the circle described with centre *F* and distance
one of the straight lines *FA*, *FB*, *FC*, *FD*, *FE* will pass
also through the remaining points, and will have been
circumscribed.

Let it be circumscribed, and let it be *ABCDE*.

Therefore about the given pentagon, which is equilateral
and equiangular, a circle has been circumscribed.

Q. E. F.

PROPOSITION 15.

In a given circle to inscribe an equilateral and equiangular hexagon.

Let *ABCDEF* be the given circle ;
thus it is required to inscribe an equilateral and equiangular
hexagon in the circle *ABCDEF*.

Let the diameter *AD* of the circle
ABCDEF be drawn ;
let the centre *G* of the circle be taken, and
with centre *D* and distance *DG* let the
circle *EGCH* be described ;
let *EG*, *CG* be joined and carried through
to the points *B*, *F*,
and let *AB*, *BC*, *CD*, *DE*, *EF*, *FA* be
joined.

I say that the hexagon *ABCDEF* is
equilateral and equiangular.

For, since the point *G* is the centre of the circle *ABCDEF*,

GE is equal to *GD*.

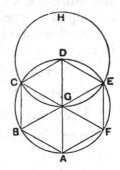

Again, since the point D is the centre of the circle GCH,

DE is equal to DG.

But GE was proved equal to GD;

therefore GE is also equal to ED;

therefore the triangle EGD is equilateral;

and therefore its three angles EGD, GDE, DEG are equal to one another, inasmuch as, in isosceles triangles, the angles at the base are equal to one another. [I. 5]

And the three angles of the triangle are equal to two right angles; [I. 32]

therefore the angle EGD is one-third of two right angles.

Similarly, the angle DGC can also be proved to be one-third of two right angles.

And, since the straight line CG standing on EB makes the adjacent angles EGC, CGB equal to two right angles,

therefore the remaining angle CGB is also one-third of two right angles.

Therefore the angles EGD, DGC, CGB are equal to one another;

so that the angles vertical to them, the angles BGA, AGF, FGE are equal. [I. 15]

Therefore the six angles EGD, DGC, CGB, BGA, AGF, FGE are equal to one another.

But equal angles stand on equal circumferences; [III. 26] therefore the six circumferences AB, BC, CD, DE, EF, FA are equal to one another.

And equal circumferences are subtended by equal straight lines; [III. 29]

therefore the six straight lines are equal to one another;

therefore the hexagon $ABCDEF$ is equilateral.

I say next that it is also equiangular.

For, since the circumference FA is equal to the circumference ED,

let the circumference $ABCD$ be added to each;

therefore the whole $FABCD$ is equal to the whole $EDCBA$;

and the angle *FED* stands on the circumference *FABCD*,
and the angle *AFE* on the circumference *EDCBA*;

therefore the angle *AFE* is equal to the angle *DEF*.

[III. 27]

Similarly it can be proved that the remaining angles of the hexagon *ABCDEF* are also severally equal to each of the angles *AFE*, *FED*;

therefore the hexagon *ABCDEF* is equiangular.

But it was also proved equilateral;
and it has been inscribed in the circle *ABCDEF*.

Therefore in the given circle an equilateral and equiangular hexagon has been inscribed.

Q. E. F.

PORISM. From this it is manifest that the side of the hexagon is equal to the radius of the circle.

And, in like manner as in the case of the pentagon, if through the points of division on the circle we draw tangents to the circle, there will be circumscribed about the circle an equilateral and equiangular hexagon in conformity with what was explained in the case of the pentagon.

And further by means similar to those explained in the case of the pentagon we can both inscribe a circle in a given hexagon and circumscribe one about it.

Q. E. F.

Heiberg, I think with good reason, considers the Porism to this proposition to be referred to in the instance which Proclus (p. 304, 2) gives of a porism following a *problem*. As the text of Proclus stands, "the (porism) found in the second Book (τὸ δὲ ἐν τῷ δευτέρῳ βιβλίῳ κείμενον) is a porism to a problem"; but this is not true of the only porism that we find in the second Book, namely the porism to II. 4. Hence Heiberg thinks that for τῷ δευτέρῳ βιβλίῳ should be read τῷ δ' βιβλίῳ, i.e. the fourth Book. Moreover Proclus speaks of *the* porism in the particular Book, from which we gather that there was only *one* porism in Book IV. as he knew it, and therefore that he did not regard as a *porism* the addition to IV. 5. Cf. note on that proposition.

It appears that Theon substituted for the first words of the Porism to IV. 15 "And in like manner as in the case of the pentagon" (ὁμοίως δὲ τοῖς ἐπὶ τοῦ πενταγώνου) the simple word "and" or "also" (καί), apparently thinking that the words had the same meaning as the similar words lower down. This is however not the case, the meaning being that "if, as in the case of the pentagon, we draw tangents, we can prove, also as was done in the case of the pentagon, that the figure so formed is a circumscribed regular hexagon."

PROPOSITION 16.

In a given circle to inscribe a fifteen-angled figure which shall be both equilateral and equiangular.

Let $ABCD$ be the given circle;

thus it is required to inscribe in the circle $ABCD$ a fifteen-angled figure which shall be both equilateral and equiangular.

In the circle $ABCD$ let there be inscribed a side AC of the equilateral triangle inscribed in it, and a side AB of an equilateral pentagon;

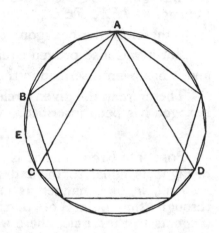

therefore, of the equal segments of which there are fifteen in the circle $ABCD$, there will be five in the circumference ABC which is one-third of the circle, and there will be three in the circumference AB which is one-fifth of the circle;

therefore in the remainder BC there will be two of the equal segments.

Let BC be bisected at E;   [III. 30]

therefore each of the circumferences BE, EC is a fifteenth of the circle $ABCD$.

If therefore we join BE, EC and fit into the circle $ABCD$ straight lines equal to them and in contiguity, a fifteen-angled figure which is both equilateral and equiangular will have been inscribed in it.

                                                    Q. E. F.

And, in like manner as in the case of the pentagon, if through the points of division on the circle we draw tangents to the circle, there will be circumscribed about the circle a fifteen-angled figure which is equilateral and equiangular.

And further, by proofs similar to those in the case of the pentagon, we can both inscribe a circle in the given fifteen-angled figure and circumscribe one about it.

                                                    Q. E. F.

Here, as in III. 10, we have the term "circle" used by Euclid in its exceptional sense of the *circumference* of a circle, instead of the "*plane figure contained by one line*" of I. Def. 15. Cf. the note on that definition (Vol. I. pp. 184—5).

Proclus (p. 269) refers to this proposition in illustration of his statement that Euclid gave proofs of a number of propositions with an eye to their use in astronomy. "With regard to the last proposition in the fourth Book in which he inscribes the side of the fifteen-angled figure in a circle, for what object does anyone assert that he propounds it except for the reference of this problem to astronomy? For, when we have inscribed the fifteen-angled figure in the circle through the poles, we have the distance from the poles both of the equator and the zodiac, since they are distant from one another by the side of the fifteen-angled figure." This agrees with what we know from other sources, namely that up to the time of Eratosthenes (*circa* 284—204 B.C.) 24 was generally accepted as the correct measurement of the obliquity of the ecliptic. This measurement, and the construction of the fifteen-angled figure, were probably due to the Pythagoreans, though it would appear that the former was not known to Oenopides of Chios (fl. *circa* 460 B.C.), as we learn from Theon of Smyrna (pp. 198—9, ed. Hiller), who gives Dercyllides as his authority, that Eudemus (fl. *circa* 320 B.C.) stated in his ἀστρολογίαι that, while Oenopides discovered certain things, and Thales, Anaximander and Anaximenes others, it was the rest (οἱ λοιποί) who added other discoveries to these and, among them, that "the axes of the fixed stars and of the planets respectively are distant from one another by the side of a fifteen-angled figure." Eratosthenes evaluated the angle to $\frac{11}{83}$rds of 180°, i.e. about 23° 51′ 20″, which measurement was apparently not improved upon in antiquity (cf. Ptolemy, *Syntaxis*, ed. Heiberg, p. 68).

Euclid has now shown how to describe regular polygons with 3, 4, 5, 6 and 15 sides. Now, when any regular polygon is given, we can construct a regular polygon with twice the number of sides by first describing a circle about the given polygon and then bisecting all the smaller arcs subtended by the sides. Applying this process any number of times, we see that we can by Euclid's methods construct regular polygons with $3 \cdot 2^n$, $4 \cdot 2^n$, $5 \cdot 2^n$, $15 \cdot 2^n$ sides, where n is zero or any positive integer.

BOOK V.

INTRODUCTORY NOTE.

The anonymous author of a scholium to Book v. (Euclid, ed. Heiberg, Vol. v. p. 280), who is perhaps Proclus, tells us that "some say" this Book, containing the general theory of proportion which is equally applicable to geometry, arithmetic, music, and all mathematical science, "is the discovery of Eudoxus, the teacher of Plato." Not that there had been no theory of proportion developed before his time; on the contrary, it is certain that the Pythagoreans had worked out such a theory with regard to *numbers*, by which must be understood commensurable and even whole numbers (a number being a "multitude made up of units," as defined in Eucl. vii). Thus we are told that the Pythagoreans distinguished three sorts of *means*, the arithmetic, the geometric and the harmonic mean, the geometric mean being called proportion (ἀναλογία) *par excellence*; and further Iamblichus speaks of the "most perfect proportion consisting of four terms and specially called *harmonic*," in other words, the proportion

$$a : \frac{a+b}{2} = \frac{2ab}{a+b} : b,$$

which was said to be a discovery of the Babylonians and to have been first introduced into Greece by Pythagoras (Iamblichus, *Comm. on Nicomachus*, p. 118). Now the principle of similitude is one which is presupposed by all the arts of design from their very beginnings; it was certainly known to the Egyptians, and it must certainly have been thoroughly familiar to Pythagoras and his school. This consideration, together with the evidence of the employment by him of the *geometric proportion*, makes it indubitable that the Pythagoreans used the theory of proportion, in the form in which it was known to them, i.e. as applicable to commensurables only, in their geometry. But the discovery, also by the Pythagoreans, of the incommensurable would of course be seen to render the proofs which depended on the theory of proportion as then understood inconclusive; as Tannery observes (*La Géométrie grecque*, p. 98), "the discovery of incommensurability must have caused a veritable logical scandal in geometry and, in order to avoid it, they were obliged to restrict as far as possible the use of the principle of similitude, pending the discovery of a means of establishing it on the basis of a theory of proportion independent of commensurability." The glory of the latter discovery belongs then most probably to Eudoxus. Certain it is that the complete theory was already familiar to Aristotle, as we shall see later.

It seems probable, as indicated by Tannery (*loc. cit.*), that the theory of proportions and the principle of similitude took, in the earliest Greek geometry, an earlier place than they do in Euclid, but that, in consequence of the discovery of the incommensurable, the treatment of the subject was fundamentally remodelled in the period between Pythagoras and Eudoxus. An indication of this is afforded by the clever device used in Euclid I. 44 for applying to a given straight line a parallelogram equal to a given triangle; the equality of the "complements" in a parallelogram is there used for doing what is practically finding a fourth proportional to three given straight lines. Thus Euclid was no doubt following for the subject-matter of Books I.—IV. what had become the traditional method, and this is probably one of the reasons why proportions and similitude are postponed till as late as Books V., VI.

It is a remarkable fact that the theory of proportions is twice treated in Euclid, in Book V. with reference to magnitudes in general, and in Book VII. with reference to the particular case of numbers. The latter exposition referring only to commensurables may be taken to represent fairly the theory of proportions at the stage which it had reached before the great extension of it made by Eudoxus. The differences between the definitions etc. in Books V. and VII. will appear as we go on; but the question naturally arises, why did Euclid not save himself so much repetition and treat numbers merely as a particular case of magnitude, referring back to the corresponding more general propositions of Book V. instead of proving the same propositions over again for numbers? It could not have escaped him that numbers fall under the conception of magnitude. Aristotle had plainly indicated that magnitudes may be numbers when he observed (*Anal. post.* I. 7, 75 b 4) that you cannot adapt the arithmetical method of proof to the properties of magnitudes if the magnitudes are not numbers. Further Aristotle had remarked (*Anal. post.* I. 5, 74 a 17) that the proposition that the terms of a proportion can be taken alternately was at one time proved separately for numbers, lines, solids and times, though it was possible to prove it for all by one demonstration; but, because there was no common name comprehending them all, namely numbers, lengths, times and solids, and their character was different, they were taken separately. Now however, he adds, the proposition is proved generally. Yet Euclid says nothing to connect the two theories of proportion even when he comes in x. 5 to a proportion two terms of which are magnitudes and two are numbers ("Commensurable magnitudes have to one another the ratio which a number has to a number"). The probable explanation of the phenomenon is that Euclid simply followed tradition and gave the two theories as he found them. This would square with the remark in Pappus (VII. p. 678) as to Euclid's fairness to others and his readiness to give them credit for their work.

DEFINITIONS.

1. A magnitude is a **part** of a magnitude, the less of the greater, when it measures the greater.

2. The greater is a **multiple** of the less when it is measured by the less.

3.   A **ratio** is a sort of relation in respect of size between two magnitudes of the same kind.

4.   Magnitudes are said to **have a ratio** to one another which are capable, when multiplied, of exceeding one another.

5.   Magnitudes are said to **be in the same ratio**, the first to the second and the third to the fourth, when, if any equimultiples whatever be taken of the first and third, and any equimultiples whatever of the second and fourth, the former equimultiples alike exceed, are alike equal to, or alike fall short of, the latter equimultiples respectively taken in corresponding order.

6.   Let magnitudes which have the same ratio be called **proportional**.

7.   When, of the equimultiples, the multiple of the first magnitude exceeds the multiple of the second, but the multiple of the third does not exceed the multiple of the fourth, then the first is said to **have a greater ratio** to the second than the third has to the fourth.

8.   A proportion in three terms is the least possible.

9.   When three magnitudes are proportional, the first is said to have to the third the **duplicate ratio** of that which it has to the second.

10.   When four magnitudes are < continuously > proportional, the first is said to have to the fourth the **triplicate ratio** of that which it has to the second, and so on continually, whatever be the proportion.

11.   The term **corresponding magnitudes** is used of antecedents in relation to antecedents, and of consequents in relation to consequents.

12.   **Alternate ratio** means taking the antecedent in relation to the antecedent and the consequent in relation to the consequent.

13.   **Inverse ratio** means taking the consequent as antecedent in relation to the antecedent as consequent.

14. **Composition of a ratio** means taking the antecedent together with the consequent as one in relation to the consequent by itself.

15. **Separation of a ratio** means taking the excess by which the antecedent exceeds the consequent in relation to the consequent by itself.

16. **Conversion of a ratio** means taking the antecedent in relation to the excess by which the antecedent exceeds the consequent.

17. A ratio **ex aequali** arises when, there being several magnitudes and another set equal to them in multitude which taken two and two are in the same proportion, as the first is to the last among the first magnitudes, so is the first to the last among the second magnitudes;

Or, in other words, it means taking the extreme terms by virtue of the removal of the intermediate terms.

18. A **perturbed proportion** arises when, there being three magnitudes and another set equal to them in multitude, as antecedent is to consequent among the first magnitudes, so is antecedent to consequent among the second magnitudes, while, as the consequent is to a third among the first magnitudes, so is a third to the antecedent among the second magnitudes.

DEFINITION I.

Μέρος ἐστὶ μέγεθος μεγέθους τὸ ἔλασσον τοῦ μείζονος, ὅταν καταμετρῇ τὸ μεῖζον.

The word *part* (μέρος) is here used in the restricted sense of a *submultiple* or an *aliquot part* as distinct from the more general sense in which it is used in the Common Notion (5) which says that "the whole is greater than the part." It is used in the same restricted sense in VII. Def. 3, which is the same definition as this with "number" (ἀριθμός) substituted for "magnitude." VII. Def. 4, keeping up the restriction, says that, when a number does not measure another number, it is *parts* (in the plural), not *a part* of it. Thus, 1, 2, or 3, is *a part* of 6, but 4 is not *a part* of 6 but *parts*. The same distinction between the restricted and the more general sense of the word *part* appears in Aristotle, *Metaph.* 1023 b 12: "In one sense a part is that into which quantity (τὸ ποσόν) can anyhow be divided; for that which is taken away from quantity, *quâ* quantity, is always called a 'part' of it, as e.g. two is said to be in a sense a part of three. But in another sense a 'part' is only *what measures* (τὰ καταμετροῦντα) such quantities. Thus two is in one sense said to be a part of three, in the other not."

DEFINITION 2.

Πολλαπλάσιον δὲ τὸ μεῖζον τοῦ ἐλάττονος, ὅταν καταμετρῆται ὑπὸ τοῦ ἐλάττονος.

DEFINITION 3.

Λόγος ἐστὶ δύο μεγεθῶν ὁμογενῶν ἡ κατὰ πηλικότητα ποιὰ σχέσις.

The best explanation of the definitions of *ratio* and *proportion* that I have seen is that of De Morgan, which will be found in the articles under those titles in the *Penny Cyclopaedia*, Vol. xix. (1841); and in the following notes I shall draw largely from these articles. Very valuable also are the notes on the definitions of Book v. given by Hankel (fragment on Euclid published as an appendix to his work *Zur Geschichte der Mathematik in Alterthum und Mittelalter*, 1874).

There has been controversy as to what is the proper translation of the word πηλικότης in the definition. σχέσις κατὰ πηλικότητα has generally been translated "relation in respect of *quantity*." Upon this De Morgan remarks that it makes nonsense of the definition; "for magnitude has hardly a different meaning from quantity, and a relation of magnitudes with respect to quantity may give a clear idea to those who want a word to convey a notion of architecture with respect to building or of battles with respect to fighting, and to no others." The true interpretation De Morgan, following Wallis and Gregory, takes to be *quantuplicity*, referring to the number of times one magnitude is contained in the other. For, he says, we cannot describe magnitude in language without quantuplicitative reference to other magnitude; hence he supposes that the definition simply conveys the fact that the mode of expressing quantity in terms of quantity is entirely based upon the notion of quantuplicity or that relation of which we take cognizance when we find how many times one is contained in the other. While all the rest of De Morgan's observations on the definition are admirable, it seems to me that on this question of the proper translation of πηλικότης he is in error. He supports his view mainly by reference (1) to the definition of a compounded ratio usually given as the 5th definition of Book vi., which speaks of the πηλικότητες of two ratios being multiplied together, and (2) to the comments of Eutocius and a scholiast on this definition. Eutocius says namely (Archimedes, ed. Heiberg, iii. p. 120) that "the term πηλικότης is evidently used of the number from which the given ratio is called, as (among others) Nicomachus says in his first book on music and Heron in his commentary on the Introduction to Arithmetic." But it now appears certain that this definition is an interpolation; it is never used, it is not found in Campanus, and Peyrard's ms. only has it in the margin. At the same time it is clear that, if the definition is admitted at all, any commentator would be obliged to explain it in the way that Eutocius does, whether the explanation was consistent with the proper meaning of πηλικότης or not. Hence we must look elsewhere for the meaning of πηλίκος and πηλικότης. If we do this, I think we shall find no case in which the words have the sense attributed to them by De Morgan. The real meaning of πηλίκος is *how great*. It is so used by Aristotle, e.g. in *Eth. Nic.* v. 10, 1134 b 11, where he speaks of a man's child being as it were a part of him so long as he is of a certain age (ἕως ἂν ᾖ πηλίκον). Again Nicomachus, to whom Eutocius appeals, himself (1. 2, 5, p. 5, ed. Hoche) distinguishes πηλίκος as referring to *magnitude*, while ποσός refers to *multitude*. So does Iamblichus in his commentary on Nicomachus (p. 8, 3—5); besides which Iamblichus distinguishes πηλίκον as the subject of geometry, being *con-*

tinuous, and ποσόν as the subject of arithmetic, being *discrete*, and speaks of a point being the origin of πηλίκον as a unit is of ποσόν, and so on. Similarly, Ptolemy (*Syntaxis*, ed. Heiberg, p. 31) speaks of the *size* (πηλικότης) of the chords in a circle (περὶ τῆς πηλικότητος τῶν ἐν τῷ κύκλῳ εὐθειῶν). Consequently I think we can only translate πηλικότης in the definition as *size*. This corresponds to Hankel's translation of it as "Grösse," though he uses this same word for a concrete "magnitude" as well; *size* seems to me to give the proper distinction between πηλικότης and μέγεθος, as *size* is the attribute, and a *magnitude* (in its ordinary mathematical sense) is the thing which possesses the attribute of size.

The view that "relation in respect of *size*" is meant by the words in the text is also confirmed, I think, by a later remark of De Morgan himself, namely that a synonym for the word *ratio* may be found in the more intelligible term *relative magnitude*. In fact σχέσις in the definition corresponds to *relative* and πηλικότης to *magnitude*. (By magnitude De Morgan here means the attribute and not the thing possessing it.)

Of the definition as a whole Simson and Hankel express the opinion that it is an interpolation. Hankel points to the fact that it is unnecessary and moreover so vague as to be of no practical use, while the very use of the expression κατὰ πηλικότητα seems to him suspicious, since the only other place in which the word πηλικότης occurs in Euclid is the 5th definition of Book VI., which is admittedly not genuine. Yet the definition of ratio appears in all the MSS., the only variation being that some add the words πρὸς ἄλληλα, "to one another," which are rejected by Heiberg as an interpolation of Theon; and on the whole there seems to be no sufficient ground for regarding it as other than genuine. The true explanation of its presence would appear to be substantially that given by Barrow (*Lectiones Cantabrig.*, London, 1684, Lect. III. of 1666), namely that Euclid inserted it for completeness' sake, more for ornament than for use, intending to give the learner a general notion of ratio by means of a metaphysical, rather than a mathematical definition; "for metaphysical it is and not, properly speaking, mathematical, since nothing depends on it or is deduced from it by mathematicians, nor, as I think, can anything be deduced." This is confirmed by the fact that there is no definition of λόγος in Book VII., and it could equally have been dispensed with here. Similarly De Morgan observes that Euclid never attempts this vague sort of definition except when, dealing with a well-known term of common life, he wishes to bring it into geometry with something like an expressed meaning which may aid the conception of the thing, though it does not furnish a perfect criterion. Thus we may compare the definition with that of a straight line, where Euclid merely calls the reader's attention to the well-known term εὐθεῖα γραμμή, tries how far he can present the conception which accompanies it in other words, and trusts for the correct use of the term to the axioms (or postulates) which the universal conception of a straight line makes self-evident.

We have now to trace as clearly as possible the development of the conception of λόγος, *ratio*, or relative magnitude. In its primitive sense λόγος was only used of a ratio between commensurables, i.e. a ratio which could be *expressed*, and the manner of expressing it is indicated in the proposition, Eucl. X. 5, which proves that *commensurable magnitudes have to one another the ratio which a number has to a number.* That this was the primitive meaning of λόγος is proved by the use of the term ἄλογος for the incommensurable, which means *irrational* in the sense of *not having a ratio* to something taken as rational (ῥητός).

Euclid himself shows us how we are to set about finding the ratio, or relative magnitude, of two commensurable magnitudes. He gives, in x. 3, practically our ordinary method of finding the greatest common measure. If A, B be two magnitudes of which B is the less, we cut off from A a part equal to B, from the remainder a part equal to B, and so on, until we leave a remainder less than B, say R_1. We measure off R_1 from B in the same way until a remainder R_2 is left which is less than R_1. We repeat the process with R_1, R_2, and so on, until we find a remainder which is contained in the preceding remainder a certain number of times exactly. If account is taken of the number of times each magnitude is contained (with something over, except at the last) in that upon which it is measured, we can calculate how many times the last remainder is contained in A and how many times the last remainder is contained in B; and we can thus express the ratio of A to B as the ratio of one number to another.

But it may happen that the two magnitudes have no common measure, i.e. are incommensurable, in which case the process described would never come to an end and the means of expression would fail; the magnitudes would then *have no ratio* in the primitive sense. But the word λόγος, ratio, acquires in Euclid, Book v., a wider sense covering the *relative magnitude* of incommensurables as well as commensurables; as stated in Euclid's 4th definition, "magnitudes are said to have a *ratio* to one another which can, when multiplied, exceed one another," and finite incommensurables have this property as much as commensurables. De Morgan explains the manner of transition from the narrower to the wider signification of *ratio* as follows. "Since the relative magnitude of two quantities is always shown by the quantuplicitative mode of expression, when that is possible, and since proportional quantities (pairs which have the same relative magnitude) are pairs which have the same mode (if possible) of expression by means of each other; in all such cases sameness of relative magnitude leads to sameness of mode of expression; or proportion is sameness of ratios (in the primitive sense). But sameness of relative magnitude may exist where quantuplicitative expression is impossible; thus the diagonal of a larger square is the same compared with *its* side as the diagonal of a smaller square compared with *its* side. It is an easy transition to speak of sameness of ratio even in this case; that is, to use the term ratio in the sense of relative magnitude, that word having originally only a reference to the mode of expressing relative magnitude, in cases which allow of a particular mode of expression. The word *irrational* (ἄλογος) does not make any corresponding change but continues to have its primitive meaning, namely, incapable of quantuplicitative expression."

It remains to consider how we are to describe the *relative magnitude* of two incommensurables of the same kind. That they have a definite relation is certain. Suppose, for precision, that S is the side of a square, D its diagonal; then, if S is given, any alteration in D or any error in D would make the figure cease to be a square. At the same time, a person altogether ignorant of the relative magnitude of D and S might say that drawing two straight lines of length S so as to form a right angle and joining the ends by a straight line, the length of which would accordingly be D, does not help him to realise the relative magnitude, but that he would like to know how many diagonals make an exact number of sides. We should have to reply that no number of diagonals whatever makes an exact number of sides; but that he may mention any fraction of the side, a hundredth, a thousandth or a millionth, and that we will then express the diagonal with an error not so great as that fraction. We then tell him that 1,000,000 diagonals exceed

1,414,213 sides but fall short of 1,414,214 sides; consequently the diagonal lies between 1·414213 and 1·414214 times the side, and these differ only by one-millionth of the side, so that the error in the diagonal is less still. To enable him to continue the process further, we show him how to perform the arithmetical operation of approximating to the value of $\sqrt{2}$. This gives the means of carrying the approximation to any degree of accuracy that may be desired. In the power, then, of carrying approximations of this kind as far as we please lies that of expressing the ratio, so far as expression is possible, and of comparing the ratio with others as accurately as if expression had been possible.

Euclid was of course aware of this, as were probably others before him; though the actual approximations to the values of ratios of incommensurables of which we find record in the works of the great Greek geometers are very few. The history of such approximations up to Archimedes is, so far as material was available, sketched in *The Works of Archimedes* (pp. lxxvii and following); and it is sufficient here to note the facts (1) that Plato, and even the Pythagoreans, were familiar with $\frac{7}{5}$ as an approximation to $\sqrt{2}$, (2) that the method of finding any number of successive approximations by the system of *side-* and *diagonal-*numbers described by Theon of Smyrna was also Pythagorean (cf. the note above on Euclid, II. 9, 10), (3) that Archimedes, without a word of preliminary explanation, gives out that

$$\frac{1351}{780} > \sqrt{3} > \frac{265}{153},$$

gives approximate values for the square roots of several large numbers, and proves that the ratio of the circumference of a circle to its diameter is less than $3\frac{1}{7}$ but greater than $3\frac{10}{71}$, (4) that the first approach to the rapidity with which the decimal system enables us to approximate to the value of surds was furnished by the method of sexagesimal fractions, which was almost as convenient to work with as the method of decimals, and which appears fully developed in Ptolemy's σύνταξις. A number consisting of a whole number and any fraction was under this system represented as so many units, so many of the fractions which we should denote by $\frac{1}{60}$, so many of those which we should write $(\frac{1}{60})^2$, $(\frac{1}{60})^3$, and so on. Theon of Alexandria shows us how to extract the square root of 4500 in this sexagesimal system, and, to show how effective it was, it is only necessary to mention that Ptolemy gives

$\frac{103}{60} + \frac{55}{60^2} + \frac{23}{60^3}$ as an approximation to $\sqrt{3}$, which approximation is equivalent

to 1·7320509 in the ordinary decimal notation and is therefore correct to 6 places.

Between Def. 3 and Def. 4 two manuscripts and Campanus insert "Proportion is the sameness of ratios" (ἀναλογία δὲ ἡ τῶν λόγων ταὐτότης), and even the best MS. has it in the margin. It would be altogether out of place, since it is not till Def. 5 that it is explained what sameness of ratios is. The words are an interpolation later than Theon (Heiberg, Vol. v. pp. xxxv, lxxxix), and are no doubt taken from arithmetical works (cf. Nicomachus and Theon of Smyrna). It is true that Aristotle says similarly, "Proportion is equality of ratios" (*Eth. Nic.* v. 6, 1131 a 31), and he appears to be quoting from the Pythagoreans; but the reference is to *numbers*.

Similarly two MSS. (inferior) insert after Def. 7 "Proportion is the similarity (ὁμοιότης) of ratios." Here too we have a mere interpolation.

DEFINITION 4.

Λόγον ἔχειν πρὸς ἄλληλα μεγέθη λέγεται, ἃ δύναται πολλαπλασιαζόμενα ἀλλήλων ὑπερέχειν.

This definition supplements the last one. De Morgan says that it amounts to saying that the magnitudes are of the same species. But this can hardly be all; the definition seems rather to be meant, on the one hand, to exclude the relation of a finite magnitude to a magnitude of the same kind which is either infinitely great or infinitely small, and, even more, to emphasise the fact that the term *ratio*, as defined in the preceding definition, and about to be used throughout the book, includes the relation between any two *incommensurable* as well as between any two commensurable finite magnitudes of the same kind. Hence, while De Morgan seems to regard the extension of the meaning of *ratio* to include the relative magnitude of incommensurables as, so to speak, taking place between Def. 3 and Def. 5, the 4th definition appears to show that it is ratio in its extended sense that is being defined in Def. 3.

DEFINITION 5.

Ἐν τῷ αὐτῷ λόγῳ μεγέθη λέγεται εἶναι πρῶτον πρὸς δεύτερον καὶ τρίτον πρὸς τέταρτον, ὅταν τὰ τοῦ πρώτου καὶ τρίτου ἰσάκις πολλαπλάσια τῶν τοῦ δευτέρου καὶ τετάρτου ἰσάκις πολλαπλασίων καθ᾽ ὁποιονοῦν πολλαπλασιασμὸν ἑκάτερον ἑκατέρου ἢ ἅμα ὑπερέχῃ ἢ ἅμα ἴσα ἢ ἅμα ἐλλείπῃ ληφθέντα κατάλληλα.

In my translation of this definition I have compromised between an attempted literal translation and the more expanded version of Simson. The difficulty in the way of an exactly literal translation is due to the fact that the words (καθ᾽ ὁποιονοῦν πολλαπλασιασμὸν) signifying that the equimultiples *in each case* are any equimultiples *whatever* occur only once in the Greek, though they apply *both* to τὰ...ἰσάκις πολλαπλάσια in the nominative and τῶν...ἰσάκις πολλαπλασίων in the genitive. I have preferred "alike" to "simultaneously" as a translation of ἅμα because "simultaneously" might suggest that time was of the essence of the matter, whereas what is meant is that any particular comparison made between the equimultiples must be made between *the same* equimultiples of the two pairs respectively, not that they need to be compared *at the same time*.

Aristotle has an allusion to a definition of "the same ratio" in *Topics* VIII. 3, 158 b 29: "In mathematics too some things appear to be not easy to prove (γράφεσθαι) for want of a definition, e.g. that the parallel to the side which cuts a plane [a parallelogram] divides the straight line [the other side] and the area similarly. But, when the definition is expressed, the said property is immediately manifest; for the areas and the straight lines *have the same* ἀνταναίρεσις, *and this is the definition of 'the same ratio.'*" Upon this passage Alexander says similarly, "This is the definition of proportionals which the ancients used: *magnitudes are proportional to one another which have* (or show) *the same* ἀνθυφαίρεσις, and Aristotle has called the latter ἀνταναίρεσις." Heiberg (*Mathematisches zu Aristoteles*, p. 22) thinks that Aristotle is alluding to the fact that the proposition referred to could not be rigorously proved so long as the Pythagorean definition applicable to commensurable magnitudes only was adhered to, and is quoting the definition belonging to the complete theory of Eudoxus; whence, in view of the positive statement of Aristotle that the definition quoted *is* the definition of "the same ratio," it would appear that the Euclidean definition (which Heiberg describes as a careful and exact paraphrase of ἀνταναίρεσις) is Euclid's own. I do not feel able to subscribe to this view, which seems to me to involve very grave difficulties. The Euclidean definition is regularly appealed to in Book v. as the criterion of magnitudes being in proportion, and the use of it would appear to constitute the whole essence of the new general theory of proportion; if then this theory is due to Eudoxus, it seems impossible to believe that the definition was not also due to him. Certainly the definition given by Aristotle would be no substitute for it; ἀνθυφαίρεσις and ἀνταναίρεσις are words almost as vague and "metaphysical" (as Barrow would say) as the words used to define *ratio*, and it is difficult to see how any mathematical facts could be deduced from such a definition. Consider for a moment the etymology of the words. ὑφαίρεσις or ἀναίρεσις means "removal," "taking away" or "destruction" of a thing; and the prefix ἀντὶ indicates that the "taking away" from one magnitude *answers to*, corresponds with, alternates with, the "taking away" from the other. So far therefore as the etymology goes, the word seems rather to suggest the "taking away" of corresponding *fractions*, and therefore to suit the old imperfect theory of proportion rather than the new one. Thus Waitz (*ad loc.*) paraphrases the definition as meaning that "as many parts as are taken from one magnitude, so many are at the same time taken from the other as well." A possible explanation would seem to be that, though Eudoxus had formulated the new definition, the old one was still current in the text-books of Aristotle's time, and was taken by him as being a good enough illustration of what he wished to bring out in the passage of the *Topics* referred to.

From the revival of learning in Europe onwards the Euclidean definition of proportion was the subject of much criticism. Campanus had failed to understand it, had in fact misinterpreted it altogether, and he may have misled others such as Ramus (1515—72), always a violently hostile critic of Euclid. Among the objectors to it was no less a person than Galileo. For particulars of the controversies on the subject down to Thomas Simpson (*Elem. of Geometry*, Lond. 1800) the reader is referred to the Excursus at the end of the second volume of Camerer's Euclid (1825). For us it is interesting to note that the unsoundness of the usual criticisms of the definition was never better exposed than by Barrow. Some of the objections, he pointed out (*Lect. Cantabr.* VII. of 1666), are due to misconception on the part of their authors as to the nature of a definition. Thus Euclid is required by these objectors (e.g. Tacquet) to do the impossible and to show that what is predicated in the definition is true of the thing defined, as if any one should be required to show that the name "circle" was applicable to those figures alone which have their radii all equal! As we are entitled to assign to such figures and such figures only the name of "circle," so Euclid is entitled ("quamvis non temere nec imprudenter at certis de causis iustis illis et idoneis") to describe a certain property which four magnitudes may have, and to call magnitudes possessing that property magnitudes "in the same ratio." Others had argued from the occurrence of the other definition of proportion in VII. Def. 20 that Euclid was dissatisfied with the present one; Barrow pointed out that, on the contrary, it was the fact that VII. Def. 20 was not adequate to cover the case of incommensurables which made Euclid adopt the present definition here. Lastly, he maintains, against those who descant on the "obscurity" of v. Def. 5, that the supposed obscurity is due, partly no doubt to the inherent difficulty of the subject of incommensurables, but also to faulty translators, and most of all to lack of effort in the learner to grasp thoroughly the meaning of words which, in themselves, are as clearly expressed as they could be.

To come now to the merits of the case, the best defence and explanation of the definition that I have seen is that given by De Morgan. He first translates it, observes that it applies equally to commensurable or incommensurable quantities because no attempt is made to measure one by an aliquot part of another, and then proceeds thus.

"The two questions which must be asked, and satisfactorily answered, previously to its [the definition's] reception, are as follows:

1. What right had Euclid, or any one else, to expect that the preceding most prolix and unwieldy statement should be received by the beginner as the definition of a relation the perception of which is one of the most common acts of his mind, since it is performed on every occasion where similarity or dissimilarity of figure is looked for or presents itself?

2. If the preceding question should be clearly answered, how can the definition of proportion ever be used; or how is it possible to compare every one of the infinite number of multiples of A with every one of the multiples of B?

To the first question we reply that not only is the test proposed by Euclid tolerably simple, when more closely examined, but that it is, or might be made to appear, an easy and natural consequence of those fundamental perceptions with which it may at first seem difficult to compare it."

To elucidate this De Morgan gives the following illustration.

Suppose there is a straight colonnade composed of equidistant columns (which we will understand to mean the vertical lines forming the axes of the columns respectively), the first of which is at a distance from a bounding wall equal to the distance between consecutive columns. In front of the colonnade let there be a straight row of equidistant railings (regarded as meaning their axes), the first being at a distance from the bounding wall equal to the distance between consecutive railings. Let the columns be numbered from the wall, and also the railings. We suppose of course that the column distance (say, C) and the railing distance (say, R) are different and that they may bear to each other any ratio, commensurable or incommensurable; i.e. that there need not go any exact number of railings to any exact number of columns.

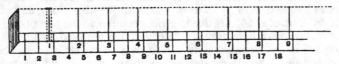

If the construction be supposed carried on to any extent, a spectator can, by mere inspection, and without measurement, compare C with R to any degree of accuracy. For example, since the 10th railing falls between the 4th and 5th columns, $10R$ is greater than $4C$ and less than $5C$, and therefore R lies between $\frac{4}{10}$ths of C and $\frac{5}{10}$ths of C. To get a more accurate notion, the ten-thousandth railing may be taken; suppose it falls between the 4674th and 4675th columns. Therefore $10,000R$ lies between $4674C$ and $4675C$, or R lies between $\frac{4674}{10000}$ and $\frac{4675}{10000}$ of C. There is no limit to the degree of accuracy thus obtainable; and the ratio of R to C is determined when the order of distribution of the railings among the columns is assigned *ad infinitum*; or, in other words, when the position of *any* given railing can be found, as to the numbers of the columns between which it lies. Any alteration, however small, in the place of the first railing must at last affect the order of distribution. Suppose e.g. that the first railing is moved from the wall by one part in a thousand of the distance between the columns; then the second railing is pushed forward by $\frac{2}{1000}C$, the third by $\frac{3}{1000}C$, and so on, so that the railings after the thousandth are pushed forward by more than C; i.e. the order with respect to the columns is disarranged.

Now let it be proposed to make a model of the preceding construction in which c shall be the column distance and r the railing distance. It needs no definition of proportion, nor anything more than the conception which we have of that term prior to definition (and with which we must show the agreement of any definition that we may adopt), to assure us that C must be to R in the same proportion as c to r if the model be truly formed. Nor is it drawing too largely on that conception of proportion to assert that the distribution of the railings among the columns in the model must be everywhere the same as in the original; for example, that the model would be *out of proportion* if its 37th railing fell between the 18th and 19th columns, while the 37th railing of the original fell between the 17th and 18th columns. Thus the dependence of Euclid's definition upon common notions is settled; for the obvious relation between the construction and its model which has just been described contains the collection of conditions, the fulfilment of which, according to Euclid, constitutes proportion. According to Euclid, whenever mC exceeds, equals, or falls short of nR, then mc must exceed, equal, or fall short of nr; and, by the most obvious property of the constructions, according as the mth column comes after, opposite to, or before the nth railing in the original, the mth column must come after, opposite to, or before the nth railing in the correct model.

Thus the test proposed by Euclid is necessary. It is also sufficient. For admitting that, to a given original with a given column-distance in the model, there is one correct model railing distance (which must therefore be that which distributes the railings among the columns as in the original), we have seen that any other railing distance, however slightly different, would at last give a different distribution; that is, the correct distance, and the correct distance only, satisfies all the conditions required by Euclid's definition.

The use of the word *distribution* having been well learnt, says De Morgan, the following way of stating the definition will be found easier than that of Euclid. "Four magnitudes, A and B of one kind, and C and D of the same or another kind, are proportional when all the multiples of A can be distributed among the multiples of B in the same intervals as the corresponding multiples of C among those of D." Or, whatever numbers m, n may be, if mA lies between nB and $(n + 1)B$, mC lies between nD and $(n + 1)D$.

It is important to note that, if the test be always satisfied from and after any given multiples of A and C, it must be satisfied before those multiples. For instance, let the test be always satisfied from and after $100A$ and $100C$; and let $5A$ and $5C$ be instances for examination. Take any multiple of 5 which will exceed 100, say 50 times five; and let it be found on examination that $250A$ lies between $678B$ and $679B$; then $250C$ lies between $678D$ and $679D$. Divide by 50, and it follows that $5A$ lies between $13\frac{28}{50}B$ and $13\frac{29}{50}B$, and *a fortiori* between $13B$ and $14B$. Similarly, $5C$ lies between $13\frac{28}{50}D$ and $13\frac{29}{50}D$, and therefore between $13D$ and $14D$. Or $5A$ lies in the same interval among the multiples of B in which $5C$ lies among the multiples of D. And so for any multiple of A, C less than $100A$, $100C$.

There remains the second question relating to the infinite character of the definition; four magnitudes A, B, C, D are not to be called proportional until it is shown that *every* multiple of A falls in the same intervals among the multiples of B in which the same multiple of C is found among the multiples of D. Suppose that the distribution of the railings among the columns should be found to agree in the model and the original as far as the millionth railing. This proves only that the railing distance of the model does not err by the millionth part of the corresponding column distance. We can thus fix limits to the disproportion, if any, and we may make those limits as small as we please, by carrying on the method of *observation*; but we cannot *observe* an infinite number of cases and *so* enable ourselves to affirm proportion absolutely. Mathematical methods however enable us to avoid the difficulty. We can take *any multiples whatever* and work with them as if they were particular multiples. De Morgan gives, as an instance to show that the definition of proportion can in practice be used, notwithstanding its infinite character, the following proof of a proposition to the same effect as Eucl. VI. 2.

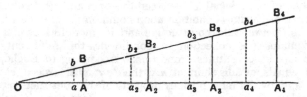

"Let OAB be a triangle to one side AB of which ab is drawn parallel, and on OA produced set off AA_2, A_2A_3 etc. equal to OA, and aa_2, a_2a_3 etc. equal to Oa.

Through every one of the points so obtained draw parallels to AB, meeting OB produced in b_2, B_2 etc.

Then it is easily proved that bb_2, b_2b_3 etc. are severally equal to Ob, and BB_2, B_2B_3 etc. to OB.

Consequently a distribution of the multiples of OA among the multiples of Oa is made on one line, and of OB among those of Ob on the other.

The examination of this distribution in all its extent (which is impossible, and hence the apparent difficulty of using the definition) is rendered unnecessary by the known property of parallel lines. For, since A_3 lies between a_3 and a_4, B_3 must lie between b_3 and b_4; for, if not, the line A_3B_3 would cut either a_3b_3 or a_4b_4.

Hence, without inquiring where A_m *does* fall, we know that, if it fall between a_n and a_{n+1}, B_m must fall between b_n and b_{n+1}; or, if $m \cdot OA$ fall in magnitude between $n \cdot Oa$ and $(n+1)Oa$, then $m \cdot OB$ must fall between $n \cdot Ob$ and $(n+1)Ob$."

Max Simon remarks (*Euclid und die sechs planimetrischen Bücher*, p. 110), after Zeuthen, that Euclid's definition of equal ratios is word for word the same as Weierstrass' definition of equal numbers. So far from agreeing in the usual view that the Greeks saw in the irrational no *number*, Simon thinks it is clear from Eucl. V. that they possessed a notion of number in all its generality as clearly defined as, nay almost identical with, Weierstrass' conception of it.

Certain it is that there is an exact correspondence, almost coincidence, between Euclid's definition of equal ratios and the modern theory of irrationals due to Dedekind. Premising the ordinal arrangement of natural numbers in ascending order, then enlarging the sphere of numbers by including (1) negative numbers as well as positive, (2) fractions, as a/b, where a, b may be any natural numbers, provided that b is not zero, and arranging the fractions ordinally among the other numbers according to the definition :

$$\text{let } \frac{a}{b} \text{ be } <=> \frac{c}{d} \text{ according as } ad \text{ is } <=> bc,$$

Dedekind arrives at the following definition of an irrational number.

An irrational number a is defined whenever a law is stated which will assign every given rational number to one and only one of two classes A and B such that (1) every number in A precedes every number in B, and (2) there is no last number in A and no first number in B; the definition of a being that it is the one number which lies between all numbers in A and all numbers in B.

Now let x/y and x'/y' be equal ratios in Euclid's sense.

Then $\dfrac{x}{y}$ will divide all rational numbers into two groups A and B ;

$\qquad \dfrac{x'}{y'} \qquad$ „ $\qquad$ „ $\qquad$ „ $\qquad A'$ and B'.

Let $\dfrac{a}{b}$ be any rational number in A, so that

$$\frac{a}{b} < \frac{x}{y}.$$

This means that $ay < bx$.

But Euclid's definition asserts that in that case $ay' < bx'$ also.

Hence also $\qquad\qquad\qquad \dfrac{a}{b} < \dfrac{x'}{y'}$;

therefore every member of group A is also a member of group A'.

Similarly every member of group B is a member of group B'.

For, if $\dfrac{a}{b}$ belong to B,

$$\frac{a}{b} > \frac{x}{y},$$

which means that $ay > bx$.

But in that case, by Euclid's definition, $ay' > bx'$;

therefore also $\qquad\qquad\qquad \dfrac{a}{b} > \dfrac{x'}{y'}.$

Thus, in other words, A and B are coextensive with A' and B' respectively ;

therefore $\dfrac{x}{y} = \dfrac{x'}{y'}$, according to Dedekind, as well as according to Euclid.

If x/y, x'/y' happen to be *rational*,

then one of the groups, say A, includes x/y,

and one of the groups, say A', includes x'/y'.

In this case $\dfrac{a}{b}$ might *coincide* with $\dfrac{x}{y}$;

that is $\qquad\qquad\qquad\qquad \dfrac{a}{b} = \dfrac{x}{y},$

which means that $\qquad\qquad\quad ay = bx.$

Therefore, by Euclid's definition, $ay' = bx'$;

so that
$$\frac{a}{b} = \frac{x'}{y'}.$$

Thus the groups are again coextensive.

In a word, Euclid's definition divides all rational numbers into two *coextensive* classes, and therefore defines equal ratios in a manner exactly corresponding to Dedekind's theory.

Alternatives for Eucl. V. Def. 5.

Saccheri records in his *Euclides ob omni naevo vindicatus* that a distinguished geometer of his acquaintance proposed to substitute for Euclid's the following definition:

"A first magnitude has to a second the same ratio that a third has to a fourth when the first contains the aliquot parts of the second, *according to any number* [i.e. with any denominator] *whatever*, the same number of times as the number of times the third contains the same aliquot parts of the fourth"; on which Saccheri remarks that he sees no advantage in this definition, which presupposes the notion of *division*, over that of Euclid which uses *multiplication* and the notions of *greater*, *equal*, and *less*.

This definition was, however, practically adopted by Faifofer (*Elementi di geometria*, 3 ed., 1882) in the following form:

"Four magnitudes taken in a certain order form a proportion when, by measuring the first and the third respectively by any equi-submultiples whatever of the second and of the fourth, equal quotients are obtained."

Ingrami (*Elementi di geometria*, 1904) takes multiples of the first and third instead of submultiples of the second and fourth:

"Given four magnitudes in predetermined order, the first two homogeneous with one another, and likewise also the last two, the magnitudes are said to form a proportion (or to be in proportion) when *any multiple* of the first contains the second the same number of times that the equimultiple of the third contains the fourth."

Veronese's definition (*Elementi di geometria*, Pt. II., 1905) is like that of Faifofer; Enriques and Amaldi (*Elementi di geometria*, 1905) adhere to Euclid's.

Proportionals of VII. Def. 20 a particular case.

It has already been observed that Euclid has nowhere proved (though the fact cannot have escaped him) that the proportion of numbers is included in the proportion of magnitudes as a special case. This is proved by Simson as being necessary to the 5th and 6th propositions of Book x. Simson's proof is contained in his propositions C and D inserted in the text of Book v. and in the notes thereon. Proposition C and the note on it prove that, *if four magnitudes are proportionals according to* VII. *Def.* 20, *they are also proportionals according to* V. *Def.* 5. Prop. D and the note prove the partial converse, namely that, if four magnitudes are proportionals according to the 5th definition of Book v., *and if the first be any multiple, or any part, or parts, of the second,* the third is the same multiple, part, or parts, of the fourth. The proofs use certain results obtained in Book v.

Prop. C is as follows:

If the first be the same multiple of the second, or the same part of it, that the third is of the fourth, the first is to the second as the third to the fourth.

Let the first A be the same *multiple* of B the second that C the third is of the fourth D;

$$A \text{ is to } B \text{ as } C \text{ is to } D.$$

```
A ───────────          E ──────────────
B ──────               G ───────────
C ─────────            F ─────────────
D ───                  H ──────
```

Take of A, C any equimultiples whatever E, F; and of B, D any equimultiples whatever G, H.

Then, because A is the same multiple of B that C is of D,

and E is the same multiple of A that F is of C,

$$E \text{ is the same multiple of } B \text{ that } F \text{ is of } D. \qquad \text{[v. 3]}$$

Therefore E, F are the same multiples of B, D.

But G, H are equimultiples of B, D;

therefore, if E be a greater multiple of B than G is, F is a greater multiple of D than H is of D;

that is, if E be greater than G, F is greater than H.

In like manner,

if E be equal to G, or less, F is equal to H, or less than it.

But E, F are equimultiples, any whatever, of A, C;

and G, H any equimultiples whatever of B, D.

Therefore A is to B as C is to D. [v. Def. 5]

Next, let the first A be the same *part* of the second B that the third C is of the fourth D:

$$A \text{ is to } B \text{ as } C \text{ is to } D.$$

```
                                        A ──────
For B is the same multiple of A that    B ──────────
D is of C;                              C ───
wherefore, by the preceding case,       D ──────────
```

$$B \text{ is to } A \text{ as } D \text{ is to } C;$$

and, *inversely*, A is to B as C is to D.

[For this last inference Simson refers to his Proposition B. That proposition is very simply proved by taking any equimultiples E, F of B, D and any equimultiples G, H of A, C and then arguing as follows:

Since A is to B as C is to D,

G, H are *simultaneously* greater than, equal to, or less than E, F respectively; so that

E, F are *simultaneously* less than, equal to, or greater than G, H respectively,

and therefore [Def. 5] B is to A as D is to C.]

We have now only to add to Prop. C the case where AB contains the *same parts* of CD that EF does of GH:

in this case likewise AB is to CD as EF to GH.

Let CK be a part of CD, and GL the same part of GH; let AB be the same multiple of CK that EF is of GL.

Therefore, by Prop. C,

$$AB \text{ is to } CK \text{ as } EF \text{ to } GL.$$

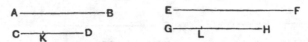

And *CD*, *GH* are equimultiples of *CK*, *GL*, the second and fourth.

Therefore *AB* is to *CD* as *EF* to *GH* [Simson's Cor. to v. 4, which however is the particular case of v. 4 in which the "equimultiples" of one pair are the pair itself, i.e. the pair multiplied by unity].

To prove the partial converse we begin with Prop. D.

If the first be to the second as the third to the fourth, and if the first be a multiple or part of the second, the third is the same multiple or the same part of the fourth.

Let *A* be to *B* as *C* is to *D*;

and, first, let *A* be a multiple of *B*;

$$C \text{ is the same multiple of } D.$$

Take *E* equal to *A*, and whatever multiple *A* or *E* is of *B*, make *F* the same multiple of *D*.

Then, because *A* is to *B* as *C* is to *D*,

and of *B* the second and *D* the fourth equimultiples have been taken *E* and *F*,

$$A \text{ is to } E \text{ as } C \text{ is to } F. \qquad\qquad \text{[v. 4, Cor.]}$$

But *A* is equal to *E*;

$$\text{therefore } C \text{ is equal to } F.$$

[In support of this inference Simson cites his Prop. A, which however we can directly deduce from v. Def. 5 by taking any, but *the same*, equimultiples of all four magnitudes.]

Now *F* is the same multiple of *D* that *A* is of *B*;

$$\text{therefore } C \text{ is the same multiple of } D \text{ that } A \text{ is of } B.$$

Next, let the first *A* be a part of the second *B*;

$$C \text{ the third is the same part of the fourth } D.$$

Because *A* is to *B* as *C* is to *D*,

inversely, *B* is to *A* as *D* is to *C*.   [Prop. B]

But *A* is a part of *B*; therefore *B* is a multiple of *A*;

and, by the preceding case, *D* is the same multiple of *C*,

$$\text{that is, } C \text{ is the same part of } D \text{ that } A \text{ is of } B.$$

We have, again, only to add to Prop. D the case where *AB* contains any parts of *CD*, and *AB* is to *CD* as *EF* to *GH*;

then shall *EF* contain the same parts of *GH* that *AB* does of *CD*.

For let *CK* be a part of *CD*, and *GL* the same part of *GH*; and let *AB* be a multiple of *CK*.

EF shall be the same multiple of *GL*.

Take *M* the same multiple of *GL* that *AB* is of *CK*;

therefore   *AB* is to *CK* as *M* is to *GL*.   [Prop. C]

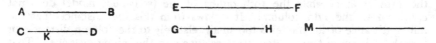

And *CD*, *GH* are equimultiples of *CK*, *GL*;

therefore   *AB* is to *CD* as *M* is to *GH*.

But, by hypothesis,   *AB* is to *CD* as *EF* is to *GH*;

therefore   *M* is equal to *EF*,   [v. 9]

and consequently *EF* is the same multiple of *GL* that *AB* is of *CK*.

DEFINITION 6.

Τὰ δὲ τὸν αὐτὸν ἔχοντα λόγον μεγέθη ἀνάλογον καλείσθω.

Ἀνάλογον, though usually written in one word, is equivalent to ἀνὰ λόγον, *in proportion*. It comes however in Greek mathematics to be used practically as an indeclinable adjective, as here; cf. αἱ τέσσαρες εὐθεῖαι ἀνάλογον ἔσονται, "the four straight lines will be proportional," τρίγωνα τὰς πλευρὰς ἀνάλογον ἔχοντα, "triangles having their sides proportional." Sometimes it is used adverbially: ἀνάλογον ἄρα ἐστὶν ὡς ἡ ΒΑ πρὸς τὴν ΑΓ, οὕτως ἡ ΗΔ πρὸς τὴν ΔΖ, "proportionally therefore, as *BA* is to *AC*, so is *GD* to *DF*"; so too, apparently, in the expression ἡ μέση ἀνάλογον (εὐθεῖα), "the mean proportional." I do not follow the objection of Max Simon (Euclid, p. 110) to "proportional" as a translation of ἀνάλογον. "We ask," he says, "in vain, what is proportional to what? We say e.g. that weight is proportional to price because double, treble etc. weight corresponds to double, treble etc. price. But here the meaning must be 'standing in a relation of proportion.'" Yet he admits that the Latin word *proportionalis* is an adequate expression. He translates by "in proportion" in the text of this definition. But I do not see that "in proportion" is better than "proportional." The fact is that both expressions are elliptical when used of four magnitudes "in proportion"; but there is surely no harm in using either when the meaning is so well understood.

The use of the word καλείσθω, "*let* magnitudes having the same ratio *be called* proportional," seems to indicate that this definition is Euclid's own.

DEFINITION 7.

Ὅταν δὲ τῶν ἰσάκις πολλαπλασίων τὸ μὲν τοῦ πρώτου πολλαπλάσιον ὑπερέχῃ τοῦ τοῦ δευτέρου πολλαπλασίου, τὸ δὲ τοῦ τρίτου πολλαπλάσιον μὴ ὑπερέχῃ τοῦ τοῦ τετάρτου πολλαπλασίου, τότε τὸ πρῶτον πρὸς τὸ δεύτερον μείζονα λόγον ἔχειν λέγεται, ἤπερ τὸ τρίτον πρὸς τὸ τέταρτον.

As De Morgan observes, the practical test of *disproportion* is simpler than that of proportion. For, whereas no examination of individual cases, however extensive, will enable an observer of the construction and its model (the illustration by means of columns and railings described above) to affirm proportion or deny disproportion, and all it enables us to do is to fix limits (as small as we please) to the disproportion (if any), a single instance may enable us to deny proportion or affirm disproportion, and also to state which way the disproportion lies. Let the 19th railing in the original fall beyond the 11th column, while the 19th railing of the (so-called) model does not come up to the 11th column. It follows from this one instance that the railing distance of the model is too small relatively to the column distance, or that the column distance is too great relatively to the railing distance. That is, the ratio of r to c is less than that of R to C, or the ratio of c to r is greater than that of C to R.

Saccheri (*op. cit.*) remarks (as Commandinus had done) that the ratio of the first magnitude to the second will also be greater than that of the third to the fourth if, while the multiple of the first is *equal* to the multiple of the second, the multiple of the third is *less* than that of the fourth: a case not mentioned in Euclid's definition. Saccheri speaks of this case being included in Clavius' interpretation of the definition. I have, however, failed to find a reference to the case in Clavius, though he adds, as a sort of corollary, in his note on the definition, that if, on the other hand, the multiple of the first is *less* than the multiple of the second, while the multiple of the third is *not less* than that of the fourth, the ratio of the first to the second is *less* than that of the third to the fourth.

Euclid presumably left out the second possible criterion for a greater ratio, and the definition of a less ratio, because he was anxious to reduce the definitions to the minimum necessary for his purpose, and to leave the rest to be inferred as soon as the development of the propositions of Book V. enabled this to be done without difficulty.

Saccheri tried to reduce the second possible criterion for a greater ratio to that given by Euclid in his definition without recourse to anything coming later in the Book, but, in order to do this, he has to use "multiples" produced by multipliers which are not integral numbers, but integral numbers *plus* proper fractions, so that Euclid's Def. 7 becomes inapplicable.

De Morgan notes that "proof should be given that the same pair of magnitudes can never offer both tests [i.e. the test in the definition for a greater ratio and the corresponding test for a less ratio, with "less" substituted for "greater" in the definition] to another pair; that is, the test of greater ratio from one set of multiples, and that of less ratio from another." In other words, if m, n, p, q are integers and A, B, C, D four magnitudes, none of the pairs of equations

$$\text{(1)} \quad mA > nB, \quad mC = \text{or} < nD,$$
$$\text{(2)} \quad mA = nB, \quad mC < nD$$

can be satisfied simultaneously with any one of the pairs of equations

$$\text{(3)} \quad pA = qB, \quad pC > qD,$$
$$\text{(4)} \quad pA < qB, \quad pC > \text{or} = qD.$$

There is no difficulty in proving this with the help of two simple assumptions which are indeed obvious.

We need only take in illustration one of the numerous cases. Suppose, if possible, that the following pairs of equations are simultaneously true:

$$\text{(1)} \quad mA > nB, \quad mC < nD$$
and $\quad \text{(2)} \quad pA < qB, \quad pC > qD.$

Multiply (1) by q and (2) by n.

(We need here to assume that, where rX, rY are any equimultiples of any magnitudes X, Y,

$$\text{according as } X \gtreqless Y, \quad rX \gtreqless rY.$$

This is contained in Simson's Axioms 1 and 3.)

We have then the pairs of equations

$$mqA > nqB, \quad mqC < nqD,$$
$$npA < nqB, \quad npC > nqD.$$

From the second equations in each pair it follows that

$$mqC < npC.$$

(We now need to assume that, if rX, sX are any multiples of X, and rY, sY the same multiples of Y, then,

$$\text{according as } rX \gtreqless sX, \quad rY \gtreqless sY.$$

Simson uses this same assumption in his proof of v. 18.)

Therefore $mqA < npA.$

But it follows from the first equations in each pair that

$$mqA > npA :$$

which is impossible.

Nor can Euclid's criterion for a greater ratio coexist with that for equal ratios.

DEFINITION 8.

Ἀναλογία δὲ ἐν τρισὶν ὅροις ἐλαχίστη ἐστίν.

This is the reading of Heiberg and Camerer (who follow Peyrard's MS.) and is that translated above. The other reading has ἐλαχίστοις, which can only be translated "consists in three terms *at least.*" Hankel regards the definition as a later interpolation, because it is superfluous, and because the word ὅρος for a *term* in a proportion is nowhere else used by Euclid, though it is common in later writers such as Nicomachus and Theon of Smyrna. The genuineness of the definition is however supported by the fact that Aristotle not only uses ὅρος in this sense (*Eth. Nic.* v. 6, 7, 1131 b 5, 9), but has a similar remark (*ibid.* 1131 a 31) that a "proportion is in *four* terms at least." The difference from Euclid is only formal; for Aristotle proceeds: "The *discrete* (διῃρημένη) (proportion) is clearly in four (terms), but so also is the *continuous* (συνεχής). For it uses one as two and mentions it twice, e.g. (in stating) that, as α is to β, so also is β to γ; thus β is mentioned twice, so that, if β be twice put down, the proportionals are four." The distinction between *discrete* and *continuous* seems to have been Pythagorean (cf. Nicomachus, II. 21, 5; 23, 2, 3; where however συνημμένη is used instead of συνεχής); Euclid does not use the words διῃρημένη and συνεχής in this connexion.

So far as they go, the first words of the next definition (9), "When three magnitudes are proportionals," which seemingly refer to Def. 8, also support the view that the latter is, at least in substance, genuine.

DEFINITIONS 9, 10.

9. Ὅταν δὲ τρία μεγέθη ἀνάλογον ᾖ, τὸ πρῶτον πρὸς τὸ τρίτον διπλασίονα λόγον ἔχειν λέγεται ἤπερ πρὸς τὸ δεύτερον.

10. Ὅταν δὲ τέσσαρα μεγέθη ἀνάλογον ᾖ, τὸ πρῶτον πρὸς τὸ τέταρτον τριπλασίονα λόγον ἔχειν λέγεται ἤπερ πρὸς τὸ δεύτερον, καὶ ἀεὶ ἑξῆς ὁμοίως, ὡς ἂν ἡ ἀναλογία ὑπάρχῃ.

Here, and in connexion with the definitions of duplicate, triplicate, etc. ratios, would be the place to expect a definition of "*compound* ratio." None such is however forthcoming, and the only "definition" of it that we find is that forming VI. Def. 5, which is an interpolation made, perhaps, even before Theon's time. According to the interpolated definition, "A ratio is said to be compounded of ratios when the sizes (πηλικότητες) of the ratios multiplied together make some (? ratio)." But the multiplication of the *sizes* (or magnitudes) of two ratios of incommensurable, and even of commensurable, magnitudes is an operation unknown to the classical Greek geometers. Eutocius (Archimedes, ed. Heiberg, III. p. 120) is driven to explain the definition by making πηλικότης mean the *number* from which the given ratio is called, or, in other words, the number which multiplied into the consequent of the ratio gives the antecedent. But he is only able to work out his idea with reference to ratios between numbers, or between *commensurable* magnitudes; and indeed the definition is quite out of place in Euclid's theory of proportion.

There is then only one statement in Euclid's text as we have it indicating what is meant by compound ratio; this is in VI. 23, where he says abruptly "But the ratio of *K* to *M* is compounded of the ratio of *K* to *L* and that of *L* to *M*." Simson accordingly gives a definition (A of Book V.) of compound ratio directly suggested by the statement in VI. 23 just quoted.

" When there are any number of magnitudes of the same kind, the first is said to have to the last of them the ratio compounded of the ratio which the first has to the second, and of the ratio which the second has to the third, and of the ratio which the third has to the fourth, and so on unto the last magnitude.

For example, if *A*, *B*, *C*, *D* be four magnitudes of the same kind, the first *A* is said to have to the last *D* the ratio compounded of the ratio of *A* to *B*, and of the ratio of *B* to *C*, and of the ratio of *C* to *D*; or the ratio of *A* to *D* is said to be compounded of the ratios of *A* to *B*, *B* to *C*, and *C* to *D*.

And if *A* has to *B* the same ratio which *E* has to *F*; and *B* to *C* the same ratio that *G* has to *H*; and *C* to *D* the same that *K* has to *L*; then, by this definition, *A* is said to have to *D* the ratio compounded of ratios which are the same with the ratios of *E* to *F*, *G* to *H*, and *K* to *L*: and the same thing is to be understood when it is more briefly expressed, by saying, *A* has to *D* the ratio compounded of the ratios of *E* to *F*, *G* to *H*, and *K* to *L*.

In like manner, the same things being supposed, if *M* has to *N* the same ratio which *A* has to *D*; then, for shortness' sake, *M* is said to have to *N* the ratio compounded of the ratios of *E* to *F*, *G* to *H*, and *K* to *L*."

De Morgan has some admirable remarks on compound ratio, which not only give a very clear view of what is meant by it but at the same time supply a plausible explanation of the *origin* of the term. "Treat ratio," says De Morgan, "as an engine of operation. Let that of *A* to *B* suggest the power of altering any magnitude in that ratio." (It is true that it is not yet proved that, *B* being any magnitude, and *P* and *Q* two magnitudes of the same kind, there does exist a magnitude *A* which is to *B* in the same ratio as *P* to *Q*. It is not till VI. 12 that this is proved, by construction, in the particular case where the three magnitudes are straight lines. The proof in the Greek text of v. 18 which assumes the truth of the more general proposition is, by reason of that assumption, open to objection; see the note on that proposition.) Now "every alteration of a magnitude is alteration in some ratio, two or more successive alterations are jointly equivalent to but one, and the ratio of the initial magnitude to the terminal one is as properly said to be the *compound* ratio of alteration as 13 to be the compound addend in lieu of 8 and 5, or 28 the compound multiple for 7 and 4. Composition is used here, as elsewhere, for the process of detecting one single alteration which produces the joint effect of two or more. The composition of the ratios of *P* to *R*, *R* to *S*, *T* to *U*, is performed by assuming *A*, altering it in the first ratio into *B*, altering *B* in the second ratio into *C*, and *C* in the third ratio into *D*. The joint effect turns *A* into *D*, and the ratio of *A* to *D* is the compounded ratio."

Another word for *compounded* ratio is συνημμένος (συνάπτω) which is common in Archimedes and later writers.

It is clear that duplicate ratio, triplicate ratio etc. defined in v. Deff. 9 and 10 are merely particular cases of compound ratio, being in fact the ratios compounded of two, three etc. *equal* ratios. The use which the Greek geometers made of compounded, duplicate, triplicate ratios etc. is well illustrated by the discovery of Hippocrates that the problem of the duplication of the cube (or, more generally, the construction of a cube which shall be to a given cube in any given ratio) reduces to that of finding "two mean proportionals in continued proportion." This amounted to seeing that, if *x, y* are two mean proportionals in continued proportion between any two lines *a, b*, in other words, if *a* is to *x* as *x* to *y*, and *x* is to *y* as *y* to *b*, then a cube with side *a* is to a cube with side *x* as *a* is to *b*; and this is equivalent to saying that *a* has to *b* the triplicate ratio of *a* to *x*.

Euclid is careful to use the forms διπλασίων, τριπλασίων, etc. to express what we translate as *duplicate, triplicate* etc. ratios; the Greek mathematicians, however, commonly used διπλάσιος λόγος, "double ratio," τριπλάσιος λόγος, "triple ratio" etc. in the sense of the ratios of 2 to 1, 3 to 1 etc. The effort, if such it was, to keep the one form for the one signification and the other for the other was only partially successful, as there are several instances of the contrary use, e.g. in Archimedes, Nicomachus and Pappus.

The expression for having the ratio which is "duplicate (triplicate) of that which it has to the second" is curious—διπλασίονα (τριπλασίονα) λόγον ἔχειν ἥπερ πρὸς τὸ δεύτερον—ἥπερ being used as if διπλασίονα or τριπλασίονα were a sort of comparative, in the same way as it is used after μείζονα or ἐλάσσονα. Another way of expressing the same thing is to say λόγος διπλασίων (τριπλασίων) τοῦ, ὃν ἔχει... the ratio "duplicate of that (ratio) which..." The explanation of both constructions would seem to be that διπλάσιος or διπλασίων is, as Hultsch translates it in his edition of Pappus (cf. p. 59, 17), *duplo maior*, where the ablative *duplo* implies not a difference but a proportion.

The four magnitudes in Def. 10 must of course be in *continued* proportion (κατὰ τὸ συνεχές). The Greek text as it stands does not state this.

Definition 11.

Ὁμόλογα μεγέθη λέγεται τὰ μὲν ἡγούμενα τοῖς ἡγουμένοις τὰ δὲ ἑπόμενα τοῖς ἑπομένοις.

It is difficult to express the meaning of the Greek in as few words. A translation more literal, but conveying less, would be, "Antecedents are called *corresponding magnitudes* to antecedents, and consequents to consequents."

I have preferred to translate ὁμόλογος by "corresponding" rather than by "homologous." I do not agree with Max Simon when he says (Euclid, p. 111) that the technical term "homologous" is not the adjective ὁμόλογος, and does not mean "corresponding," "agreeing," but "like in respect of the proportion" ("ähnlich in Bezug auf das Verhältniss"). The definition seems to me to be for the purpose of appropriating to a technical use precisely the ordinary adjective ὁμόλογος, "agreeing" or "corresponding."

Antecedents, ἡγούμενα, are literally "*leading* (terms)," and *consequents*, ἑπόμενα, "*following* (terms)."

Definition 12.

Ἐναλλὰξ λόγος ἐστὶ λῆψις τοῦ ἡγουμένου πρὸς τὸ ἡγούμενον καὶ τοῦ ἑπομένου πρὸς τὸ ἑπόμενον.

We now come to a number of expressions for the transformation of ratios or proportions. The first is ἐναλλάξ, *alternately*, which would be better described with reference to a proportion of four terms than with reference to a ratio. But probably Euclid defined all the terms in Deff. 12—16 with reference to *ratios* because to define them with reference to proportions would look like assuming what ought to be proved, namely the legitimacy of the various transformations of proportions (cf. v. 16, 7 Por., 18, 17, 19 Por.). The word ἐναλλάξ is of course a common term which has no exclusive reference to mathematics. But this same use of it with reference to proportions already occurs in Aristotle: *Anal. post.* 1. 5, 74 a 18, καὶ τὸ ἀνάλογον ὅτι ἐναλλάξ, "and that a proportion (is true) alternately, or *alternando*." Used with λόγος, as here, the adverb ἐναλλάξ has the sense of an adjective, "alternate"; we have already had it similarly used of "alternate angles" (αἱ ἐναλλὰξ γωνίαι) in the theory of parallels.

Definition 13.

Ἀνάπαλιν λόγος ἐστὶ λῆψις τοῦ ἑπομένου ὡς ἡγουμένου πρὸς τὸ ἡγούμενον ὡς ἑπόμενον.

Ἀνάπαλιν, "inversely," "the other way about," is also a general term with no exclusive reference to mathematics. For this use of it with reference to proportion cf. Aristotle, *De Caelo* 1. 6, 273 b 32 τὴν ἀναλογίαν ἣν τὰ βάρη ἔχει, οἱ χρόνοι ἀνάπαλιν ἕξουσιν, "the proportion which the weights have, the times will have *inversely*." As here used with λόγος, ἀνάπαλιν is, exceptionally, adjectival.

Definition 14.

Σύνθεσις λόγου ἐστὶ λῆψις τοῦ ἡγουμένου μετὰ τοῦ ἑπομένου ὡς ἑνὸς πρὸς αὐτὸ τὸ ἑπόμενον.

The *composition* of a ratio is to be distinguished from the *compounding* of ratios and *compounded ratio* (συγκείμενος λόγος) as explained above in the note on Deff. 9, 10.   The fact is that συντίθημι and what serves for the passive of it (σύγκειμαι) are used for *adding* as well as *compounding* in the sense of compounding ratios.   In order to distinguish the two senses, I have always used the word *componendo* where the sense is that of this definition, though this requires a slight departure from the literal rendering of some passages. Thus the enunciation of v. 17 says, literally, "if magnitudes compounded be in proportion they will also be in proportion separated" (ἐὰν συγκείμενα μεγέθη ἀνάλογον ᾖ, καὶ διαιρεθέντα ἀνάλογον ἔσται).   This practically means that, if $A + B$ is to B as $C + D$ is to D, then A is to B as C is to D. I have accordingly translated as follows: "if magnitudes be proportional *componendo*, they will also be proportional *separando*."   (It will be observed that *separando*, a term explained in the next note, is here used, not relatively to the proportion A is to B as C is to D, but relatively to the proportion *componendo*, viz. $A + B$ is to B as $C + D$ is to D.)   The corresponding term for *componendo* in the Greek mathematicians is συνθέντι, literally "to one who has compounded," i.e. "if we compound."   (For this absolute use of the dative of the participle cf. Nicomachus I. 8, 9 ἀπὸ μονάδος...κατὰ τὸν διπλάσιον λόγον προχωροῦντι μέχρις ἀπείρου, ὅσοι καὶ ἂν γένωνται, οὗτοι πάντες ἀρτιάκις ἀρτιοί εἰσιν.   A very good instance from Aristotle is *Eth. Nic.* I. 5, 1097 b 12 ἐπεκτείνοντι γὰρ ἐπὶ τοὺς γονεῖς καὶ τοὺς ἀπογόνους καὶ τῶν φίλων τοὺς φίλους εἰς ἄπειρον πρόεισιν.)   A variation for συνθέντι found in Archimedes is κατὰ σύνθεσιν.   Perhaps the more exclusive use of the form συνθέντι by geometers later than Euclid to denote the *composition of a ratio*, as compared with Euclid's more general use of σύνθεσις and other parts of the verb συντίθημι or σύγκειμαι, may point to a desire to get rid of ambiguity of terms and to make the terminology of geometry more exact.

DEFINITION 15.

Διαίρεσις λόγου ἐστὶ λῆψις τῆς ὑπεροχῆς, ᾗ ὑπερέχει τὸ ἡγούμενον τοῦ ἑπομένου, πρὸς αὐτὸ τὸ ἑπόμενον.

As *composition of a ratio* means the transformation, e.g., of the ratio of A to B into the ratio of $A + B$ to B, so the *separation of a ratio* indicates the transformation of it into the ratio of $A - B$ to B.   Thus, as the new antecedent is in one case got by *adding* the original antecedent to the original consequent, so the antecedent in the other case is obtained by *subtracting* the original consequent from the original antecedent (it being assumed that the latter is greater than the former).   Hence the literal translations of διαίρεσις λόγου, "division of a ratio," and of διελόντι (the corresponding term to συνθέντι) as *dividendo*, scarcely give a sufficiently obvious explanation of the meaning.   Heiberg accordingly translates by "subtractio rationis," which again may be thought to depart too far from the Greek.   Perhaps "separation" and *separando* may serve as a compromise.

DEFINITION 16.

Ἀναστροφὴ λόγου ἐστὶ λῆψις τοῦ ἡγουμένου πρὸς τὴν ὑπεροχήν, ᾗ ὑπερέχει τὸ ἡγούμενον τοῦ ἑπομένου.

Conversion of a ratio means taking, e.g., instead of the ratio of A to B, the ratio of A to $A - B$ (A being again supposed greater than B).   As ἀναστροφή is used for *conversion*, so ἀναστρέψαντι is used for *convertendo* (corresponding to the terms συνθέντι and διελόντι).

DEFINITION 17.

Δι' ἴσου λόγος ἐστὶ πλειόνων ὄντων μεγεθῶν καὶ ἄλλων αὐτοῖς ἴσων τὸ πλῆθος σύνδυο λαμβανομένων καὶ ἐν τῷ αὐτῷ λόγῳ, ὅταν ᾖ ὡς ἐν τοῖς πρώτοις μεγέθεσι τὸ πρῶτον πρὸς τὸ ἔσχατον, οὕτως ἐν τοῖς δευτέροις μεγέθεσι τὸ πρῶτον πρὸς τὸ ἔσχατον· ἢ ἄλλως· Λῆψις τῶν ἄκρων καθ' ὑπεξαίρεσιν τῶν μέσων.

δι' ἴσου, *ex aequali*, must apparently mean *ex aequali distantia*, at an equal distance or interval, i.e. after an equal number of intervening terms. The wording of the definition suggests that it is rather a *proportion ex aequali* than a *ratio ex aequali* which is being defined (cf. Def. 12). The meaning is clear enough. If *a, b, c, d...* be one set of magnitudes, and *A, B, C, D...* another set of magnitudes, such that

$$a \text{ is to } b \text{ as } A \text{ is to } B,$$
$$b \text{ is to } c \text{ as } B \text{ is to } C,$$

and so on, the last proportion being, e.g.,

$$k \text{ is to } l, \text{ as } K \text{ is to } L,$$

then the inference *ex aequali* is that

$$a \text{ is to } l \text{ as } A \text{ is to } L.$$

The *fact* that this is so, or the *truth* of the inference from the hypothesis, is not proved until v. 22. The definition is therefore merely verbal; it gives a convenient *name* to a certain inference which is of constant application in mathematics. But *ex aequali* could not be intelligibly defined except with reference to two sets of ratios respectively equal.

DEFINITION 18.

Τεταραγμένη δὲ ἀναλογία ἐστίν, ὅταν τριῶν ὄντων μεγεθῶν καὶ ἄλλων αὐτοῖς ἴσων τὸ πλῆθος γίνηται ὡς μὲν ἐν τοῖς πρώτοις μεγέθεσιν ἡγούμενον πρὸς ἑπόμενον, οὕτως ἐν τοῖς δευτέροις μεγέθεσιν ἡγούμενον πρὸς ἑπόμενον, ὡς δὲ ἐν τοῖς πρώτοις μεγέθεσιν ἑπόμενον πρὸς ἄλλο τι, οὕτως ἐν τοῖς δευτέροις ἄλλο τι πρὸς ἡγούμενον.

Though the words δι' ἴσου, *ex aequali*, are not in this definition, it gives a description of a case in which the inference *ex aequali* is still true, as will be hereafter proved in v. 23. A *perturbed proportion* is an expression for the case when, there being three magnitudes *a, b, c* and three others *A, B, C*,

$$a \text{ is to } b \text{ as } B \text{ is to } C,$$

and
$$b \text{ is to } c \text{ as } A \text{ is to } B.$$

Another description of this case is found in Archimedes, "*the ratios being dissimilarly ordered*" (ἀνομοίως τεταγμένων τῶν λόγων). The full description of the *inference* in this case (as proved in v. 23), namely that

$$a \text{ is to } c \text{ as } A \text{ is to } C,$$

is *ex aequali in perturbed proportion* (δι' ἴσου ἐν τεταραγμένῃ ἀναλογίᾳ). Archimedes sometimes omits the δι' ἴσου, first giving the two proportions and proceeding thus: "therefore, the proportions being dissimilarly ordered, *a* has to *c* the same ratio as *A* has to *C*."

The fact that Def. 18 describes a particular case in which the inference δι' ἴσου will be proved true seems to have suggested to some one after Theon's time the interpolation of another definition between 17 and 18 to describe the *ordinary* case where the argument *ex aequali* holds good. The interpolated definition runs thus: "an *ordered proportion* (τεταγμένη ἀναλογία) arises when, as antecedent is to consequent, so is antecedent to consequent, and, as consequent is to something else, so is consequent to something else." This case needed no description after Def. 17 itself; and the supposed definition is never used.

After the definitions of Book v. Simson supplies the following *axioms*.

1. Equimultiples of the same or of equal magnitudes are equal to one another.

2. Those magnitudes of which the same or equal magnitudes are equimultiples are equal to one another.

3. A multiple of a greater magnitude is greater than the same multiple of a less.

4. That magnitude of which a multiple is greater than the same multiple of another is greater than that other magnitude.

BOOK V. PROPOSITIONS.

PROPOSITION I.

If there be any number of magnitudes whatever which are, respectively, equimultiples of any magnitudes equal in multitude, then, whatever multiple one of the magnitudes is of one, that multiple also will all be of all.

Let any number of magnitudes whatever AB, CD be respectively equimultiples of any magnitudes E, F equal in multitude;

I say that, whatever multiple AB is of E, that multiple will AB, CD also be of E, F.

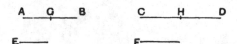

For, since AB is the same multiple of E that CD is of F, as many magnitudes as there are in AB equal to E, so many also are there in CD equal to F.

Let AB be divided into the magnitudes AG, GB equal to E,

and CD into CH, HD equal to F;

then the multitude of the magnitudes AG, GB will be equal to the multitude of the magnitudes CH, HD.

Now, since AG is equal to E, and CH to F,

therefore AG is equal to E, and AG, CH to E, F.

For the same reason

GB is equal to E, and GB, HD to E, F;

therefore, as many magnitudes as there are in AB equal to E, so many also are there in AB, CD equal to E, F;

therefore, whatever multiple AB is of E, that multiple will AB, CD also be of E, F.

Therefore etc.

Q. E. D.

De Morgan remarks of v. 1—6 that they are "simple propositions of concrete arithmetic, covered in language which makes them unintelligible to modern ears. The first, for instance, states no more than that *ten* acres and *ten* roods make *ten* times as much as one acre and one rood." One aim therefore of notes on these as well as the other propositions of Book v. should be to make their purport clearer to the learner by setting them side by side with the same truths expressed in the much shorter and more familiar modern (algebraical) notation. In doing so, we shall express magnitudes by the first letters of the alphabet, *a*, *b*, *c* etc., adopting small instead of capital letters so as to avoid confusion with Euclid's lettering; and we shall use the small letters *m*, *n*, *p* etc. to represent integral numbers. Thus *ma* will always mean *m* times *a* or the *m*th multiple of *a* (counting 1 . *a* as the first, 2 . *a* as the second multiple, and so on).

Prop. 1 then asserts that, if *ma*, *mb*, *mc* etc. be any equimultiples of *a*, *b*, *c* etc., then

$$ma + mb + mc + \ldots = m\,(a + b + c + \ldots).$$

PROPOSITION 2.

If a first magnitude be the same multiple of a second that a third is of a fourth, and a fifth also be the same multiple of the second that a sixth is of the fourth, the sum of the first and fifth will also be the same multiple of the second that the sum of the third and sixth is of the fourth.

Let a first magnitude, AB, be the same multiple of a second, C, that a third, DE, is of a fourth, F, and let a fifth, BG, also be the same multiple of the second, C, that a sixth, EH, is of the fourth F;

I say that the sum of the first and fifth, AG, will be the same multiple of the second, C, that the sum of the third and sixth, DH, is of the fourth, F.

For, since AB is the same multiple of C that DE is of F, therefore, as many magnitudes as there are in AB equal to C, so many also are there in DE equal to F.

For the same reason also,

as many as there are in BG equal to C, so many are there also in EH equal to F;

therefore, as many as there are in the whole AG equal to C, so many also are there in the whole DH equal to F.

Therefore, whatever multiple AG is of C, that multiple also is DH of F.

Therefore the sum of the first and fifth, AG, is the same multiple of the second, C, that the sum of the third and sixth, DH, is of the fourth, F.

Therefore etc.

Q. E. D.

To find the corresponding formula for the result of this proposition, we may suppose a to be the "second" magnitude and b the "fourth." If now the "first" magnitude is ma, the "third" is, by hypothesis, mb; and, if the "fifth" magnitude is na, the "sixth" is nb. The proposition then asserts that $ma + na$ is the same multiple of a that $mb + nb$ is of b.

More generally, if pa, qa... and pb, qb... be any further equimultiples of a, b respectively, $ma + na + pa + qa + \ldots$ is the same multiple of a that $mb + nb + pb + qb + \ldots$ is of b. This extension is stated in Simson's corollary to v. 2 thus:

"From this it is plain that, if any number of magnitudes AB, BG, GH be multiples of another C; and as many DE, EK, KL be the same multiples of F, each of each; the whole of the first, viz. AH, is the same multiple of C that the whole of the last, viz. DL, is of F."

The course of the proof, which separates m into its units and also n into its units, practically tells us that the multiple of a arrived at by adding the *two* multiples is the $(m + n)$th multiple; or practically we are shown that

$$ma + na = (m + n)\, a,$$

or, more generally, that

$$ma + na + pa + \ldots = (m + n + p + \ldots)\, a.$$

Proposition 3.

If a first magnitude be the same multiple of a second that a third is of a fourth, and if equimultiples be taken of the first and third, then also ex aequali *the magnitudes taken will be equimultiples respectively, the one of the second and the other of the fourth.*

Let a first magnitude A be the same multiple of a second B that a third C is of a fourth D, and let equimultiples EF, GH be taken of A, C;

I say that EF is the same multiple of B that GH is of D.

For, since EF is the same multiple of A that GH is of C, therefore, as many magnitudes as there are in EF equal to A, so many also are there in GH equal to C.

Let *EF* be divided into the magnitudes *EK*, *KF* equal to *A*, and *GH* into the magnitudes *GL*, *LH* equal to *C*; then the multitude of the magnitudes *EK*, *KF* will be equal to the multitude of the magnitudes *GL*, *LH*.

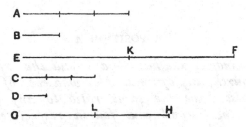

And, since *A* is the same multiple of *B* that *C* is of *D*, while *EK* is equal to *A*, and *GL* to *C*, therefore *EK* is the same multiple of *B* that *GL* is of *D*.

For the same reason

KF is the same multiple of *B* that *LH* is of *D*.

Since, then, a first magnitude *EK* is the same multiple of a second *B* that a third *GL* is of a fourth *D*, and a fifth *KF* is also the same multiple of the second *B* that a sixth *LH* is of the fourth *D*, therefore the sum of the first and fifth, *EF*, is also the same multiple of the second *B* that the sum of the third and sixth, *GH*, is of the fourth *D*.   [v. 2]

Therefore etc.

Q. E. D.

Heiberg remarks of the use of *ex aequali* in the enunciation of this proposition that, strictly speaking, it has no reference to the definition (17) of a *ratio ex aequali*. But the uses of the expression here and in the definition are, I think, sufficiently parallel, as may be seen thus. The proposition asserts that, if

$$na, nb \text{ are equimultiples of } a, b,$$

and if $m \cdot na, m \cdot nb$ are equimultiples of *na*, *nb*,

then $m \cdot na$ is the same multiple of *a* that $m \cdot nb$ is of *b*. Clearly the proposition can be extended by taking further equimultiples of the last equimultiples and so on; and we can prove that

$$p \cdot q \ldots m \cdot na \text{ is the same multiple of } a \text{ that } p \cdot q \ldots m \cdot nb \text{ is of } b,$$

where the series of numbers $p \cdot q \ldots m \cdot n$ is exactly the same in both expressions;

and *ex aequali* (δι᾽ ἴσου) expresses the fact that the equimultiples are *at the same distance* from *a*, *b* in the series *na*, $m \cdot na$... and *nb*, $m \cdot nb$... respectively.

Here again the proof breaks m into its units, and then breaks n into its units; and we are practically shown that the multiple of a arrived at, viz. $m . na$, is the multiple denoted by the product of the numbers m, n, i.e. the (mn)th multiple, or in other words that

$$m . na = mn . a.$$

Proposition 4.

If a first magnitude have to a second the same ratio as a third to a fourth, any equimultiples whatever of the first and third will also have the same ratio to any equimultiples whatever of the second and fourth respectively, taken in corresponding order.

For let a first magnitude A have to a second B the same ratio as a third C to a fourth D; and let equimultiples E, F be taken of A, C, and G, H other, chance, equimultiples of B, D;

I say that, as E is to G, so is F to H.

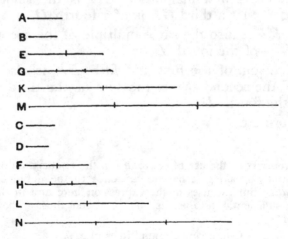

For let equimultiples K, L be taken of E, F, and other, chance, equimultiples M, N of G, H.

Since E is the same multiple of A that F is of C,

and equimultiples K, L of E, F have been taken,

therefore K is the same multiple of A that L is of C.　　[v. 3]

For the same reason

M is the same multiple of B that N is of D.

And, since, as A is to B, so is C to D,
and of A, C equimultiples K, L have been taken,
and of B, D other, chance, equimultiples M, N,
therefore, if K is in excess of M, L also is in excess of N,
if it is equal, equal, and if less, less.   [v. Def. 5]
And K, L are equimultiples of E, F,
and M, N other, chance, equimultiples of G, H;
therefore, as E is to G, so is F to H.   [v. Def. 5]
Therefore etc.

Q. E. D.

This proposition shows that, if a, b, c, d are proportionals, then

$$ma \text{ is to } nb \text{ as } mc \text{ is to } nd;$$

and the proof is as follows:

Take pma, pmc any equimultiples of ma, mc, and qnb, qnd any equimultiples of nb, nd.

Since $a : b = c : d$, it follows [v. Def. 5] that,

according as $pma > = < qnb$, $pmc > = < qnd$.

But the p- and q-equimultiples are *any* equimultiples;
therefore [v. Def. 5]

$$ma : nb = mc : nd.$$

It will be observed that Euclid's phrase for taking *any* equimultiples of A, C and *any* other equimultiples of B, D is "let there be taken equimultiples E, F of A, C, and G, H other, *chance*, equimultiples of B, D," E, F being called ἰσάκις πολλαπλάσια simply, and G, H ἄλλα, ἃ ἔτυχεν, ἰσάκις πολλαπλάσια. And similarly, when *any* equimultiples (K, L) of E, F come to be taken, and *any* other equimultiples (M, N) of G, H. But later on Euclid uses the same phrases about the *new* equimultiples with reference to the original magnitudes, reciting that "there have been taken, of A, C, equimultiples K, L and of B, D, other, *chance*, equimultiples M, N"; whereas M, N are not *any equimultiples whatever* of B, D, but are *any* equimultiples of the *particular* multiples (G, H) which have been taken of B, D respectively, though *these latter* have been taken at random. Simson would, in the first place, add ἃ ἔτυχεν in the passages where *any* equimultiples E, F are taken of A, C and *any* equimultiples K, L are taken of E, F, because the words are "wholly necessary" and, in the second place, would leave them out where M, N are called ἄλλα, ἃ ἔτυχεν, ἰσάκις πολλαπλάσια of B, D, because it is not true that of B, D have been taken "any equimultiples whatever (ἃ ἔτυχε), M, N." Simson adds: "And it is strange that neither Mr Briggs, who did right to leave out these words in one place of Prop. 13 of this book, nor Dr Gregory, who changed them into the word 'some' in three places, and left them out in a fourth of that same Prop. 13, did not also leave them out in this place of Prop. 4 and in the second of the two places where they occur in Prop. 17 of this book, in neither of which they can stand consistent with truth: And in none of all these places, even in those which they corrected in their Latin translation, have they cancelled the words ἃ ἔτυχε in the Greek text, as they ought to have done. The same words ἃ ἔτυχε are found in four places of Prop. 11 of this book, in the first and last of which they are necessary, but in the second and third, though they are true, they are quite superfluous; as they likewise are in the second of the two places in which they are found in the 12th prop. and in the like places of Prop. 22, 23 of this book; but are wanting in the last place of Prop. 23, as also in Prop. 25, Book XI."

As will be seen, Simson's emendations amount to alterations of the text so considerable as to suggest doubt whether we should be justified in making them in the absence of MS. authority. The phrase "equimultiples of A, C and other, chance, equimultiples of B, D" recurs so constantly as to suggest that it was for Euclid a quasi-stereotyped phrase, and that it is equally genuine wherever it occurs. Is it then absolutely necessary to insert ἃ ἔτυχε in places where it does not occur, and to leave it out in the places where Simson holds it to be wrong? I think the text can be defended as it stands. In the first place to say "take equimultiples of A, C" is a fair enough way of saying take *any* equimultiples *whatever* of A, C. The other difficulty is greater, but may, I think, be only due to the adoption of *any whatever* as the translation of ἃ ἔτυχε. As a matter of fact, the words only mean *chance* equimultiples, equimultiples which are the result of random selection. Is it not justifiable to describe the product of two chance numbers, numbers selected at random, as being a "*chance* number," since it is the result of two random selections? I think so, and I have translated ἃ ἔτυχε accordingly as implying, in the case in question, "other equimultiples whatever they may happen to be."

To this proposition Theon added the following:

"Since then it was proved that, if K is in excess of M, L is also in excess of N, if it is equal, (the other is) equal, and if less, less,

it is clear also that,

if M is in excess of K, N is also in excess of L, if it is equal, (the other is) equal, and if less, less;

and for this reason,

as G is to E, so also is H to F.

PORISM. From this it is manifest that, if four magnitudes be proportional, they will also be proportional inversely."

Simson rightly pointed out that the demonstration of what Theon intended to prove, viz. that, if E, G, F, H be proportionals, they are proportional inversely, i.e. G is to E as H is to F, does not in the least depend upon this 4th proposition or the proof of it; for, when it is said that, "if K exceeds M, L also exceeds N etc.," this is not proved from the fact that E, G, F, H are proportionals (which is the conclusion of Prop. 4), but from the fact that A, B, C, D are proportionals.

The proposition that, if A, B, C, D are proportionals, they are also proportionals inversely is not given by Euclid, but Simson supplies the proof in his Prop. B. The fact is really obvious at once from the 5th definition of Book V. (cf. p. 127 above), and Euclid probably omitted the proposition as unnecessary.

Simson added, in place of Theon's corollary, the following:

"Likewise, if the first has the same ratio to the second which the third has to the fourth, then also any equimultiples whatever of the first and third have the same ratio to the second and fourth: And, in like manner, the first and the third have the same ratio to any equimultiples whatever of the second and fourth."

The proof, of course, follows exactly the method of Euclid's proposition itself, with the only difference that, instead of one of the two pairs of equimultiples, the magnitudes themselves are taken. In other words, the conclusion that

$$ma \text{ is to } nb \text{ as } mc \text{ is to } nd$$

is equally true when either m or n is equal to unity.

As De Morgan says, Simson's corollary is only necessary to those who will not admit M into the list M, $2M$, $3M$ etc.; the exclusion is grammatical and nothing else. The same may be said of Simson's Prop. A to the effect that, "If the first of four magnitudes has to the second the same ratio which the third has to the fourth : then, if the first be greater than the second, the third is also greater than the fourth; and if equal, equal; if less, less." This is needless to those who believe *once* A to be a proper component of the list of multiples, in spite of *multus* signifying many.

Proposition 5.

If a magnitude be the same multiple of a magnitude that a part subtracted is of a part subtracted, the remainder will also be the same multiple of the remainder that the whole is of the whole.

5   For let the magnitude AB be the same multiple of the magnitude CD that the part AE subtracted is of the part CF subtracted ;

I say that the remainder EB is also the same multiple of the remainder FD that the whole AB is of the whole CD.

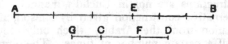

10   For, whatever multiple AE is of CF, let EB be made that multiple of CG.

Then, since AE is the same multiple of CF that EB is of GC,

therefore AE is the same multiple of CF that AB is of GF.

[v. 1]

15   But, by the assumption, AE is the same multiple of CF that AB is of CD.

Therefore AB is the same multiple of each of the magnitudes GF, CD ;

therefore GF is equal to CD.

20   Let CF be subtracted from each ;

therefore the remainder GC is equal to the remainder FD.

And, since AE is the same multiple of CF that EB is of GC,

and GC is equal to DF,

25 therefore AE is the same multiple of CF that EB is of FD.

But, by hypothesis,

AE is the same multiple of CF that AB is of CD;

therefore EB is the same multiple of FD that AB is of CD.

That is, the remainder EB will be the same multiple of 30 the remainder FD that the whole AB is of the whole CD.

Therefore etc.

Q. E. D.

10. **let EB be made that multiple of CG,** τοσαυταπλάσιον γεγονέτω καὶ τὸ EB τοῦ ΓΗ. From this way of stating the construction one might suppose that CG was given and EB had to be found equal to a certain multiple of it. But in fact EB is what is given and CG has to be found, i.e. CG has to be constructed as a certain *sub*multiple of EB.

This proposition corresponds to v. 1, with subtraction taking the place of addition. It proves the formula

$$ma - mb = m(a - b).$$

Euclid's construction assumes that, if AE is any multiple of CF, and EB is any other magnitude, a fourth straight line can be found such that EB is the same multiple of it that AE is of CF, or in other words that, given any magnitude, we can divide it into any number of equal parts. This is however not proved, even of straight lines, much less other magnitudes, until vi. 9. Peletarius had already seen this objection to the construction. The difficulty is not got over by regarding it merely as a *hypothetical* construction; for hypothetical constructions are not in Euclid's manner. The remedy is to substitute the alternative construction given by Simson, after Peletarius and Campanus' translation from the Arabic, which only requires us to add a magnitude to itself a certain number of times. The demonstration follows Euclid's line exactly.

"Take AG the same multiple of FD that AE is of CF;

therefore AE is the same multiple of CF that EG is of CD.

[v. 1]

But AE, by hypothesis, is the same multiple of CF that AB is of CD; therefore EG is the same multiple of CD that AB is of CD;

wherefore EG is equal to AB.

Take from them the common magnitude AE; the remainder AG is equal to the remainder EB.

Wherefore, since AE is the same multiple of CF that AG is of FD, and since AG is equal to EB,

therefore AE is the same multiple of CF that EB is of FD.

But AE is the same multiple of CF that AB is of CD;

therefore EB is the same multiple of FD that AB is of CD."

Q. E. D.

Euclid's proof amounts to this.

Suppose a magnitude x taken such that

$$ma - mb = mx, \text{ say.}$$

Add mb to each side, whence (by v. 1)

$$ma = m(x + b).$$

Therefore $\quad a = x + b, \text{ or } x = a - b,$

so that $\quad ma - mb = m(a - b).$

Simson's proof, on the other hand, argues thus.

Take $x = m(a - b)$, the same multiple of $(a - b)$ that mb is of b.

Then, by addition of mb to both sides, we have [v. 1]

$$x + mb = ma,$$

or $\quad x = ma - mb.$

That is, $\quad ma - mb = m(a - b).$

PROPOSITION 6.

If two magnitudes be equimultiples of two magnitudes, and any magnitudes subtracted from them be equimultiples of the same, the remainders also are either equal to the same or equimultiples of them.

For let two magnitudes AB, CD be equimultiples of two magnitudes E, F, and let AG, CH subtracted from them be equimultiples of the same two E, F; I say that the remainders also, GB, HD, are either equal to E, F or equimultiples of them.

For, first, let GB be equal to E; I say that HD is also equal to F.

For let CK be made equal to F.

Since AG is the same multiple of E that CH is of F, while GB is equal to E and KC to F, therefore AB is the same multiple of E that KH is of F.

[v. 2]

But, by hypothesis, AB is the same multiple of E that CD is of F; therefore KH is the same multiple of F that CD is of F.

Since then each of the magnitudes KH, CD is the same multiple of F, therefore KH is equal to CD.

Let *CH* be subtracted from each;

therefore the remainder *KC* is equal to the remainder *HD*.

But *F* is equal to *KC*;

therefore *HD* is also equal to *F*.

Hence, if *GB* is equal to *E*, *HD* is also equal to *F*.

Similarly we can prove that, even if *GB* be a multiple of *E*, *HD* is also the same multiple of *F*.

Therefore etc.

Q. E. D.

This proposition corresponds to v. 2, with subtraction taking the place of addition. It asserts namely that, if n is less than m, $ma - na$ is the same multiple of a that $mb - nb$ is of b. The enunciation distinguishes the cases in which $m - n$ is equal to 1 and greater than 1 respectively.

Simson observes that, while only the first case (the simpler one) is proved in the Greek, both are given in the Latin translation from the Arabic; and he supplies accordingly the proof of the second case, which Euclid leaves to the reader. The fact is that it is exactly the same as the other except that, in the construction, *CK* is made the same multiple of *F* that *GB* is of *E*, and at the end, when it has been proved that *KC* is equal to *HD*, instead of concluding that *HD* is equal to *F*, we have to say "Because *GB* is the same multiple of *E* that *KC* is of *F*, and *KC* is equal to *HD*, therefore *HD* is the same multiple of *F* that *GB* is of *E*."

Proposition 7.

Equal magnitudes have to the same the same ratio, as also has the same to equal magnitudes.

Let *A*, *B* be equal magnitudes and *C* any other, chance, magnitude;

I say that each of the magnitudes *A*, *B* has the same ratio to *C*, and *C* has the same ratio to each of the magnitudes *A*, *B*.

For let equimultiples *D*, *E* of *A*, *B* be taken, and of *C* another, chance, multiple *F*.

Then, since *D* is the same multiple of *A* that *E* is of *B*, while *A* is equal to *B*,

therefore *D* is equal to *E*.

But *F* is another, chance, magnitude.

If therefore D is in excess of F, E is also in excess of F, if equal to it, equal; and, if less, less.

And D, E are equimultiples of A, B,
while F is another, chance, multiple of C;

therefore, as A is to C, so is B to C.   [v. Def. 5]

I say next that C also has the same ratio to each of the magnitudes A, B.

For, with the same construction, we can prove similarly that D is equal to E;
and F is some other magnitude.

If therefore F is in excess of D, it is also in excess of E, if equal, equal; and, if less, less.

And F is a multiple of C, while D, E are other, chance, equimultiples of A, B;

therefore, as C is to A, so is C to B.   [v. Def. 5]

Therefore etc.

PORISM. From this it is manifest that, if any magnitudes are proportional, they will also be proportional inversely.

Q. E. D.

In this proposition there is a similar use of ὃ ἔτυχεν to that which has been discussed under Prop. 4. *Any* multiple F of C is taken and then, four lines lower down, we are told that "F is another, chance, magnitude." It is of course not *any magnitude whatever*, and Simson leaves out the sentence, but this time without calling attention to it.

Of the Porism to this proposition Heiberg says that it is properly put here in the best MS.; for, as August had already observed, if it was in its right place where Theon put it (at the end of v. 4), the second part of the proof of this proposition would be unnecessary. But the truth is that the Porism is no more in place here. The most that the proposition proves is that, if A, B are equal, and C any other magnitude, then two conclusions are simultaneously established, (1) that A is to C as B is to C and (2) that C is to A as C is to B. The second conclusion is not established from the first conclusion (as it ought to be in order to justify the inference in the Porism), but from a hypothesis on which the first conclusion itself depends; and moreover it is not a proportion in its general form, i.e. between *four* magnitudes, that is in question, but only the particular case in which the consequents are equal.

Aristotle tacitly assumes *inversion* (combined with the solution of the problem of Eucl. VI. 11) in *Meteorologica* III. 5, 376 a 14—16.

PROPOSITION 8.

Of unequal magnitudes, the greater has to the same a greater ratio than the less has; and the same has to the less a greater ratio than it has to the greater.

Let AB, C be unequal magnitudes, and let AB be greater; let D be another, chance, magnitude;

I say that AB has to D a greater ratio than C has to D, and D has to C a greater ratio than it has to AB.

For, since AB is greater than C, let BE be made equal to C;

then the less of the magnitudes AE, EB, if multiplied, will sometime be greater than D.   [v. Def. 4]

[*Case* 1.]

First, let AE be less than EB;

let AE be multiplied, and let FG be a multiple of it which is greater than D;

then, whatever multiple FG is of AE, let GH be made the same multiple of EB and K of C;

and let L be taken double of D, M triple of it, and successive multiples increasing by one, until what is taken is a multiple of D and the first that is greater than K. Let it be taken, and let it be N which is quadruple of D and the first multiple of it that is greater than K.

Then, since K is less than N first,

therefore K is not less than M.

And, since FG is the same multiple of AE that GH is of EB,

therefore FG is the same multiple of AE that FH is of AB.

[v. 1]

But FG is the same multiple of AE that K is of C;

therefore FH is the same multiple of AB that K is of C;

therefore FH, K are equimultiples of AB, C.

Again, since GH is the same multiple of EB that K is of C,

and EB is equal to C,

therefore GH is equal to K.

But K is not less than M;

therefore neither is GH less than M.

And FG is greater than D;

therefore the whole FH is greater than D, M together.

But D, M together are equal to N, inasmuch as M is triple of D, and M, D together are quadruple of D, while N is also quadruple of D; whence M, D together are equal to N.

But FH is greater than M, D;

therefore FH is in excess of N,

while K is not in excess of N.

And FH, K are equimultiples of AB, C, while N is another, chance, multiple of D;

therefore AB has to D a greater ratio than C has to D.

[v. Def. 7]

I say next, that D also has to C a greater ratio than D has to AB.

For, with the same construction, we can prove similarly that N is in excess of K, while N is not in excess of FH.

And N is a multiple of D,

while FH, K are other, chance, equimultiples of AB, C;

therefore D has to C a greater ratio than D has to AB.

[v. Def. 7]

[*Case* 2.]

Again, let AE be greater than EB.

Then the less, EB, if multiplied, will sometime be greater than D.　　　　　　[v. Def. 4]

Let it be multiplied, and let GH be a multiple of EB and greater than D;

and, whatever multiple GH is of EB, let FG be made the same multiple of AE, and K of C.

Then we can prove similarly that FH, K are equimultiples of AB, C;

and, similarly, let N be taken a multiple of D but the first that is greater than FG,

so that FG is again not less than M.

But GH is greater than D;

therefore the whole FH is in excess of D, M, that is, of N.

Now K is not in excess of N, inasmuch as FG also, which is greater than GH, that is, than K, is not in excess of N.

And in the same manner, by following the above argument, we complete the demonstration.

Therefore etc.

Q. E. D.

The two separate cases found in the Greek text of the demonstration can practically be compressed into one. Also the expositor of the two cases makes them differ more than they need. It is necessary in each case to select the *smaller* of the two segments AE, EB of AB with a view to taking a multiple of it which is greater than D; in the first case therefore AE is taken, in the second EB. But, while in the first case successive multiples of D are taken in order to find the first multiple that is greater than GH (or K), in the second case the multiple is taken which is the first that is greater than FG. This difference is not necessary; the first multiple of D that is greater than GH would equally serve in the second case. Lastly, the use of the magnitude K might have been dispensed with in both cases; it is of no practical use and only lengthens the proofs. For these reasons Simson considers that Theon, or some other unskilful editor, has vitiated the proposition. This however seems an unsafe assumption; for, while it was not the habit of the great Greek geometers to discuss separately a number of different cases (e.g. in I. 7 and I. 35 Euclid proves one case and leaves the others to the reader), there are many exceptions to prove the rule, e.g. Eucl. III. 25 and 33; and we know that many fundamental propositions, afterwards proved generally, were first discovered in relation to particular cases and then generalised, so that Book v., presenting a comparatively new theory, might fairly be expected to exhibit more instances than the earlier books do of unnecessary subdivision. The use of the K is no more conclusive against the genuineness of the proofs.

Nevertheless Simson's version of the proof is certainly shorter, and moreover it takes account of the case in which AE is *equal* to EB, and of the case in which AE, EB are both greater than D (though these cases are scarcely worth separate mention).

"If the magnitude which is not the greater of the two AE, EB be (1) not less than D, take FG, GH the doubles of AE, EB.

But if that which is not the greater of the two AE, EB be (2) less than D, this magnitude can be multiplied so as to become greater than D whether it be AE or EB.

Let it be multiplied until it becomes greater than D, and let the other be multiplied as often; let FG be the multiple thus taken of AE and GH the same multiple of EB,

therefore FG and GH are each of them greater than D.

And, in every one of the cases, take L the double of D, M its triple and so on, till the multiple of D be that which first becomes greater than GH.

Let N be that multiple of D which is first greater than GH, and M the multiple of D which is next less than N.

Then, because N is the multiple of D which is the first that becomes greater than GH,

the next preceding multiple is not greater than GH;

that is, GH is not less than M.

And, since FG is the same multiple of AE that GH is of EB,

GH is the same multiple of EB that FH is of AB;　　　　[v. 1]

wherefore FH, GH are equimultiples of AB, EB.

And it was shown that GH was not less than M;

and, by the construction, FG is greater than D;

therefore the whole FH is greater than M, D together.

But M, D together are equal to N;

therefore FH is greater than N.

But GH is not greater than N;

and FH, GH are equimultiples of AB, BE,

and N is a multiple of D;

therefore AB has to D a greater ratio than BE (or C) has to D.　　[v. Def. 7]

Also D has to BE a greater ratio than it has to AB.

For, having made the same construction, it may be shown, in like manner, that N is greater than GH but that it is not greater than FH;

and N is a multiple of D,

and GH, FH are equimultiples of EB, AB;

Therefore D has to EB a greater ratio than it has to AB."　　[v. Def. 7]

The proof may perhaps be more readily grasped in the more symbolical form thus.

Take the mth equimultiples of C, and of the excess of AB over C (that is, of AE), such that each is greater than D;

and, of the multiples of D, let pD be the first that is greater than mC, and nD the next less multiple of D.

Then, since mC is not less than nD,

and, by the construction, $m(AE)$ is greater than D,

the sum of mC and $m(AE)$ is greater than the sum of nD and D.

That is, $m(AB)$ is greater than pD.

And, by the construction, mC is less than pD.

Therefore [v. Def. 7] AB has to D a greater ratio than C has to D.

Again, since pD is less than $m(AB)$,

and pD is greater than mC,

D has to C a greater ratio than D has to AB.

Proposition 9.

Magnitudes which have the same ratio to the same are equal to one another; and magnitudes to which the same has the same ratio are equal.

For let each of the magnitudes A, B have the same ratio to C;

I say that A is equal to B.

For, otherwise, each of the magnitudes A, B would not have had the same ratio to C;   [v. 8]
but it has;

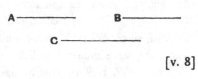

therefore A is equal to B.

Again, let C have the same ratio to each of the magnitudes A, B;

I say that A is equal to B.

For, otherwise, C would not have had the same ratio to each of the magnitudes A, B;   [v. 8]
but it has;

therefore A is equal to B.

Therefore etc.

Q. E. D.

If A is to C as B is to C,
or if C is to A as C is to B, then A is equal to B.

Simson gives a more explicit proof of this proposition which has the advantage of referring back to the fundamental 5th and 7th definitions, instead of quoting the results of previous propositions, which, as will be seen from the next note, may be, in the circumstances, unsafe.

"Let A, B have each of them the same ratio to C;

A is equal to B.

For, if they are not equal, one of them is greater than the other;
let A be the greater.

Then, by what was shown in the preceding proposition, there are some equimultiples of A and B, and some multiple of C, such that the multiple of A is greater than the multiple of C, but the multiple of B is not greater than that of C.

Let such multiples be taken, and let D, E be the equimultiples of A, B, and F the multiple of C, so that D may be greater than F, and E not greater than F.

But, because A is to C as B is to C,
and of A, B are taken equimultiples D, E, and of C is taken a multiple F, and D is greater than F,

E must also be greater than F.   [v. Def. 5]

But E is not greater than F: which is impossible.

Next, let C have the same ratio to each of the magnitudes A and B;

A is equal to B.

For, if not, one of them is greater than the other;
let A be the greater.

Therefore, as was shown in Prop. 8, there is some multiple F of C, and some equimultiples E and D of B and A, such that F is greater than E and not greater than D.

But, because C is to B as C is to A,

and F the multiple of the first is greater than E the multiple of the second,

F the multiple of the third is greater than D the multiple of the fourth.

[v. Def. 5]

But F is not greater than D: which is impossible.

Therefore A is equal to B."

PROPOSITION 10.

Of magnitudes which have a ratio to the same, that which has a greater ratio is greater; and that to which the same has a greater ratio is less.

For let A have to C a greater ratio than B has to C;
I say that A is greater than B.

A ——————————   B———————

C————————————

For, if not, A is either equal to B or less.

Now A is not equal to B;

for in that case each of the magnitudes A, B would have had the same ratio to C;   [v. 7]

but they have not;

therefore A is not equal to B.

Nor again is A less than B;

for in that case A would have had to C a less ratio than B has to C;   [v. 8]

but it has not;

therefore A is not less than B.

But it was proved not to be equal either;

therefore A is greater than B.

Again, let C have to B a greater ratio than C has to A;
I say that B is less than A.

For, if not, it is either equal or greater.

Now B is not equal to A;

for in that case C would have had the same ratio to each of the magnitudes A, B;   [v. 7]

but it has not;

therefore A is not equal to B.

Nor again is B greater than A;

for in that case C would have had to B a less ratio than it
has to A;   [v. 8]

but it has not;

therefore B is not greater than A.

But it was proved that it is not equal either;

therefore B is less than A.

Therefore etc.   Q. E. D.

No better example can, I think, be found of the acuteness which Simson
brought to bear in his critical examination of the *Elements*, and of his great
services to the study of Euclid, than is furnished by the admirable note on
this proposition where he points out a serious flaw in the proof as given in
the text.

For the first time Euclid is arguing about *greater* and *less* ratios, and it
will be found by an examination of the steps of the proof that he assumes
more with regard to the meaning of the terms than he is entitled to assume,
having regard to the fact that the definition of greater ratio (Def. 7) is all
that, as yet, he has to go upon. That we cannot argue, at present, about
greater and *less* as applied to *ratios* in the same way as about the same terms
in relation to *magnitudes* is indeed sufficiently indicated by the fact that Euclid
does not assume for ratios what is in Book I. an axiom, viz. that things which
are equal to the same thing are equal to one another; on the contrary, he
proves, in Prop. 11, that ratios which are the same with the same ratio are the
same with one another.

Let us now examine the steps of the proof in the text. First we are told
that

"A is greater than B.

For, if not, it is either equal to B or less than it.

Now A is not equal to B;

for in that case each of the two magnitudes A, B would have had the
same ratio to C: [v. 7]

but they have not:

therefore A is not equal to B."

As Simson remarks, the force of this reasoning is as follows.

If A has to C the same ratio as B has to C,

then—supposing any equimultiples of A, B to be taken and any multiple
of C—

by Def. 5, if the multiple of A be greater than the multiple of C, the multiple
of B is also greater than that of C.

But it follows from the hypothesis (that A has a greater ratio to C than B
has to C) that,

by Def. 7, there must be *some* equimultiples of A, B and *some* multiple of
C such that the multiple of A is greater than the multiple of C, but the
multiple of B is *not* greater than the same multiple of C.

And this directly contradicts the preceding deduction from the supposition
that A has to C the same ratio as B has to C;

therefore that supposition is impossible.

The proof now goes on thus:

"Nor again is A less than B;

for, in that case, A would have had to C a less ratio than B has to C;

[v. 8]

but it has not;

therefore A is not less than B."

It is here that the difficulty arises. As before, we must use Def. 7. "A would have had to C a less ratio than B has to C," or the equivalent statement that B would have had to C a greater ratio than A has to C, means that there would have been *some* equimultiples of B, A and *some* multiple of C such that

(1) the multiple of B is greater than the multiple of C, but

(2) the multiple of A is *not* greater than the multiple of C,

and it ought to have been proved that this can never happen if the hypothesis of the proposition is true, viz. that A has to C a greater ratio than B has to C: that is, it should have been proved that, in the latter case, the multiple of A is *always* greater than the multiple of C whenever the multiple of B is greater than the multiple of C (for, when this is demonstrated, it will be evident that B cannot have a greater ratio to C than A has to C). But this is not proved (cf. the remark of De Morgan quoted in the note on v. Def. 7, p. 130), and hence it is not proved that the above inference from the supposition that A is less than B is inconsistent with the hypothesis in the enunciation. The proof therefore fails.

Simson suggests that the proof is not Euclid's, but the work of some one who apparently "has been deceived in applying what is manifest, when understood of magnitudes, unto ratios, viz. that a magnitude cannot be both greater and less than another."

The proof substituted by Simson is satisfactory and simple.

"Let A have to C a greater ratio than B has to C;

A is greater than B.

For, because A has a greater ratio to C than B has to C, there are some equimultiples of A, B and some multiple of C such that

the multiple of A is greater than the multiple of C, but the multiple of B is *not* greater than it.

[v. Def. 7]

Let them be taken, and let D, E be equimultiples of A, B, and F a multiple of C, such that

D is greater than F,

but E is not greater than F.

Therefore D is greater than E.

And, because D and E are equimultiples of A and B, and D is greater than E,

therefore A is greater than B.   [Simson's 4th Ax.]

Next, let C have a greater ratio to B than it has to A;

B is less than A.

For there is some multiple F of C and some equimultiples E and D of B and A such that

F is greater than E but not greater than D.   [v. Def. 7]

Therefore E is less than D;

and, because E and D are equimultiples of B and A,

therefore B is less than A."

PROPOSITION 11.

Ratios which are the same with the same ratio are also the same with one another.

For, as A is to B, so let C be to D,
and, as C is to D, so let E be to F;
I say that, as A is to B, so is E to F.

```
A———        C——        E——
B——         D——        F—
G————————   H————————  K———
L————————   M————————  N———
```

For of A, C, E let equimultiples G, H, K be taken, and of B, D, F other, chance, equimultiples L, M, N.

Then since, as A is to B, so is C to D,
and of A, C equimultiples G, H have been taken,
and of B, D other, chance, equimultiples L, M,
therefore, if G is in excess of L, H is also in excess of M,
if equal, equal,
and if less, less.

Again, since, as C is to D, so is E to F,
and of C, E equimultiples H, K have been taken,
and of D, F other, chance, equimultiples M, N,
therefore, if H is in excess of M, K is also in excess of N,
if equal, equal,
and if less, less.

But we saw that, if H was in excess of M, G was also in excess of L; if equal, equal; and if less, less;
so that, in addition, if G is in excess of L, K is also in excess of N,
if equal, equal,
and if less, less.

And G, K are equimultiples of A, E,
while L, N are other, chance, equimultiples of B, F;
therefore, as A is to B, so is E to F.
Therefore etc.

Q. E. D.

Algebraically, if $\qquad a : b = c : d,$

and $\qquad\qquad\qquad c : d = e : f,$

then $\qquad\qquad\qquad a : b = e : f.$

The idiomatic use of the imperfect in quoting a result previously obtained is noteworthy. Instead of saying "But *it was proved that*, if H *is* in excess of M, G *is* also in excess of L," the Greek text has "But if H *was* in excess of M, G *was* also in excess of L," ἀλλὰ εἰ ὑπερεῖχε τὸ Θ τοῦ Μ, ὑπερεῖχε καὶ τὸ Η τοῦ Λ.

This proposition is tacitly used in combination with v. 16 and v. 24 in the geometrical passage in Aristotle, *Meteorologica* III. 5, 376 a 22—26.

PROPOSITION 12.

If any number of magnitudes be proportional, as one of the antecedents is to one of the consequents, so will all the antecedents be to all the consequents.

Let any number of magnitudes A, B, C, D, E, F be proportional, so that, as A is to B, so is C to D and E to F;

I say that, as A is to B, so are A, C, E to B, D, F.

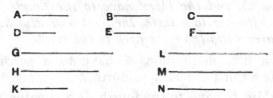

For of A, C, E let equimultiples G, H, K be taken, and of B, D, F other, chance, equimultiples L, M, N.

Then since, as A is to B, so is C to D, and E to F,
and of A, C, E equimultiples G, H, K have been taken,
and of B, D, F other, chance, equimultiples L, M, N,
therefore, if G is in excess of L, H is also in excess of M, and K of N,
if equal, equal,
and if less, less ;
so that, in addition,
if G is in excess of L, then G, H, K are in excess of L, M, N,
if equal, equal,
and if less, less.

Now G and G, H, K are equimultiples of A and A, C, E, since, if any number of magnitudes whatever are respectively equimultiples of any magnitudes equal in multitude, whatever multiple one of the magnitudes is of one, that multiple also will all be of all.   [v. 1]

For the same reason

L and L, M, N are also equimultiples of B and B, D, F;

therefore, as A is to B, so are A, C, E to B, D, F.

[v. Def. 5]

Therefore etc.

Q. E. D.

Algebraically, if $a : a' = b : b' = c : c'$ etc., each ratio is equal to the ratio $(a + b + c + \ldots) : (a' + b' + c' + \ldots)$.

This theorem is quoted by Aristotle, *Eth. Nic.* v. 7, 1131 b 14, in the shortened form "the whole is to the whole what each part is to each part (respectively)."

PROPOSITION 13.

If a first magnitude have to a second the same ratio as a third to a fourth, and the third have to the fourth a greater ratio than a fifth has to a sixth, the first will also have to the second a greater ratio than the fifth to the sixth.

For let a first magnitude A have to a second B the same ratio as a third C has to a fourth D,

and let the third C have to the fourth D a greater ratio than a fifth E has to a sixth F;

I say that the first A will also have to the second B a greater ratio than the fifth E to the sixth F.

A—————— C—————— M—————— G——————
B———— D———— N———— K——————
E——————
F——————
H——————
L——————

For, since there are some equimultiples of C, E,

and of D, F other, chance, equimultiples, such that the multiple of C is in excess of the multiple of D, while the multiple of E is not in excess of the multiple of F,

[v. Def. 7]

let them be taken,

and let G, H be equimultiples of C, E,

and K, L other, chance, equimultiples of D, F,

so that G is in excess of K, but H is not in excess of L;

and, whatever multiple G is of C, let M be also that multiple of A,

and, whatever multiple K is of D, let N be also that multiple of B.

Now, since, as A is to B, so is C to D,

and of A, C equimultiples M, G have been taken,

and of B, D other, chance, equimultiples N, K,

therefore, if M is in excess of N, G is also in excess of K,

if equal, equal,

and if less, less.

[v. Def. 5]

But G is in excess of K;

therefore M is also in excess of N.

But H is not in excess of L;

and M, H are equimultiples of A, E,

and N, L other, chance, equimultiples of B, F;

therefore A has to B a greater ratio than E has to F.

[v. Def. 7]

Therefore etc.

Q. E. D.

Algebraically, if　　　　$a : b = c : d$,

　　and　　　　　　　　$c : d > e : f$,

　　then　　　　　　　　$a : b > e : f$.

After the words "for, since" in the first line of the proof, Theon added "C has to D a greater ratio than E has to F," so that "there are some equimultiples" began, with him, the principal sentence.

The Greek text has, after "of D, F other, chance, equimultiples," "*and* the multiple of C is in excess of the multiple of D...." The meaning being "such that," I have substituted this for "and," after Simson.

The following will show the method of Euclid's proof.

Since　　　　　　　　　$c : d > e : f$,

there will be some equimultiples mc, me of c, e, and some equimultiples nd, nf of d, f, such that

$$mc > nd, \text{ while } me \not> nf.$$

But, since $\qquad\qquad a : b = c : d,$

therefore, according as $ma > = < nb, \quad mc > = < nd.$

And $mc > nd$;

therefore $ma > nb$, while (from above) $me \not> nf.$

Therefore $\qquad\qquad a : b > e : f.$

Simson adds as a corollary the following :

" If the first have a greater ratio to the second than the third has to the fourth, but the third the same ratio to the fourth which the fifth has to the sixth, it may be demonstrated in like manner that the first has a greater ratio to the second than the fifth has to the sixth."

This however scarcely seems to be worth separate statement, since it only amounts to changing the order of the two parts of the hypothesis.

Proposition 14.

If a first magnitude have to a second the same ratio as a third has to a fourth, and the first be greater than the third, the second will also be greater than the fourth; if equal, equal; and if less, less.

For let a first magnitude A have the same ratio to a second B as a third C has to a fourth D ; and let A be greater than C ;

I say that B is also greater than D.

```
A———————        C————
B————————        D————
```

For, since A is greater than C,

and B is another, chance, magnitude,

therefore A has to B a greater ratio than C has to B.   [v. 8]

But, as A is to B, so is C to D ;

therefore C has also to D a greater ratio than C has to B.

[v. 13]

But that to which the same has a greater ratio is less ;

[v. 10]

therefore D is less than B ;

so that B is greater than D.

Similarly we can prove that, if A be equal to C, B will also be equal to D ;

and, if A be less than C, B will also be less than D.

Therefore etc.

Q. E. D.

Algebraically, if $a : b = c : d,$

then, according as $a > = < c,$ $b > = < d.$

Simson adds the specific proof of the second and third parts of this proposition, which Euclid dismisses with "Similarly we can prove...."

"Secondly, if A be equal to C, B is equal to D; for A is to B as C, that is A, is to D;

therefore B is equal to D.   [v. 9]

Thirdly, if A be less than C, B shall be less than D.

For C is greater than A ;

and, because C is to D as A is to B,

D is greater than B, by the first case.

Wherefore B is less than D."

Aristotle, *Meteorol.* III. 5, 376 a 11—14, quotes the equivalent proposition that, if $a > b$, $c > d$.

PROPOSITION 15.

Parts have the same ratio as the same multiples of them taken in corresponding order.

For let AB be the same multiple of C that DE is of F;
I say that, as C is to F, so is AB to DE.

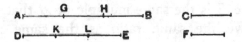

For, since AB is the same multiple of C that DE is of F, as many magnitudes as there are in AB equal to C, so many are there also in DE equal to F.

Let AB be divided into the magnitudes AG, GH, HB equal to C,

and DE into the magnitudes DK, KL, LE equal to F;

then the multitude of the magnitudes AG, GH, HB will be equal to the multitude of the magnitudes DK, KL, LE.

And, since AG, GH, HB are equal to one another,

and DK, KL, LE are also equal to one another,

therefore, as AG is to DK, so is GH to KL, and HB to LE.

[v. 7]

Therefore, as one of the antecedents is to one of the consequents, so will all the antecedents be to all the consequents ;   [v. 12]

therefore, as AG is to DK, so is AB to DE.

But AG is equal to C and DK to F;
>   therefore, as C is to F, so is AB to DE.

Therefore etc.   Q. E. D.

Algebraically, $a : b = ma : mb$.

Proposition 16.

If four magnitudes be proportional, they will also be proportional alternately.

Let A, B, C, D be four proportional magnitudes,
so that, as A is to B, so is C to D;
I say that they will also be so alternately, that is, as A is
to C, so is B to D.

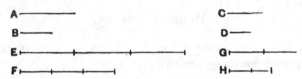

For of A, B let equimultiples E, F be taken,
and of C, D other, chance, equimultiples G, H.

Then, since E is the same multiple of A that F is of B,
and parts have the same ratio as the same multiples of
them,   [v. 15]
therefore, as A is to B, so is E to F.

But as A is to B, so is C to D;
therefore also, as C is to D, so is E to F.   [v. 11]

Again, since G, H are equimultiples of C, D,
therefore, as C is to D, so is G to H.   [v. 15]

But, as C is to D, so is E to F;
therefore also, as E is to F, so is G to H.   [v. 11]

But, if four magnitudes be proportional, and the first be
greater than the third,
>   the second will also be greater than the fourth;

if equal, equal;
and if less, less.   [v. 14]

Therefore, if E is in excess of G, F is also in excess of H,
if equal, equal,
and if less, less.

Now E, F are equimultiples of A, B,

and G, H other, chance, equimultiples of C, D;

therefore, as A is to C, so is B to D.　　　　[v. Def. 5]

Therefore etc.

Q. E. D.

3. "Let A, B, C, D be four proportional magnitudes, so that, as A is to B, so is C to D." In a number of expressions like this it is absolutely necessary, when translating into English, to interpolate words which are not in the Greek. Thus the Greek here is: Ἔστω τέσσαρα μεγέθη ἀνάλογον τὰ Α, Β, Γ, Δ, ὡς τὸ Α πρὸς τὸ Β, οὕτως τὸ Γ πρὸς τὸ Δ, literally "Let A, B, C, D be four proportional magnitudes, as A to B, so C to D." The same remark applies to the corresponding expressions in the next propositions, v. 17, 18, and to other forms of expression in v. 20—23 and later propositions: e.g. in v. 20 we have a phrase meaning literally "Let there be magnitudes...which taken two and two are in the same ratio, as A to B, so D to E," etc.: in v. 21 "(magnitudes)...which taken two and two are in the same ratio, and let the proportion of them be perturbed, as A to B, so E to F," etc. In all such cases (where the Greek is so terse as to be almost ungrammatical) I shall insert the words necessary in English, without further remark.

Algebraically, if　　　　$a : b = c : d,$

then　　　　$a : c = b : d.$

Taking equimultiples ma, mb of a, b, and equimultiples nc, nd of c, d, we have, by v. 15,

$$a : b = ma : mb,$$
$$c : d = nc : nd.$$

And, since　　　　$a : b = c : d,$

we have [v. 11]　　　　$ma : mb = nc : nd.$

Therefore [v. 14], according as $ma > = < nc$,　$mb > = < nd$, ·

so that　　　　$a : c = b : d.$

Aristotle tacitly uses the theorem in *Meteorologica* III. 5, 376 a 22—24.

The four magnitudes in this proposition must all be *of the same kind*, and Simson inserts "of the same kind" in the enunciation.

This is the first of the propositions of Eucl. v. which Smith and Bryant (*Euclid's Elements of Geometry*, 1901, pp. 298 sqq.) prove by means of vi. 1 so far as the only geometrical magnitudes in question are *straight lines* or *rectilineal areas*; and certainly the proofs are more easy to follow than Euclid's. The proof of this proposition is as follows.

To prove that, *If four magnitudes of the same kind* [straight lines or rectilineal areas] *be proportionals, they will be proportionals when taken alternately.*

Let P, Q, R, S be the four magnitudes of the same kind such that

$$P : Q = R : S;$$

then it is required to prove that

$$P : R = Q : S.$$

First, let all the magnitudes be areas.

Construct a rectangle $abcd$ equal to the area P, and to bc apply the rectangle $bcef$ equal to Q.

Also to ab, bf apply rectangles ag, bk equal to R, S respectively.

Then, since the rectangles *ac*, *be* have the same height, they are to one another as their bases.   [VI. 1]

Hence $P : Q = ab : bf.$

But $P : Q = R : S.$

Therefore $R : S = ab : bf,$ [v. 11]

i.e.   rect. *ag* : rect. *bk* = *ab* : *bf*.

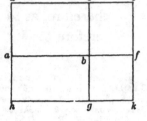

Hence (by the converse of VI. 1) the rectangles *ag*, *bk* have the same height, so that *k* is on the line *hg*.

Hence the rectangles *ac*, *ag* have the same height, namely *ab* ; also *be*, *bk* have the same height, namely *bf*.

Therefore   rect. *ac* : rect. *ag* = *bc* : *bg*,

and   rect. *be* : rect. *bk* = *bc* : *bg*.   [VI. 1]

Therefore   rect. *ac* : rect. *ag* = rect. *be* : rect. *bk*.   [v. 11]

That is, $P : R = Q : S.$

Secondly, let the magnitudes be straight lines *AB*, *BC*, *CD*, *DE*. Construct the rectangles *Ab*, *Bc*, *Cd*, *De* with the same height.

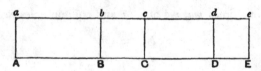

Then $Ab : Bc = AB : BC,$

and $Cd : De = CD : DE.$ [VI. 1]

But $AB : BC = CD : DE.$

Therefore $Ab : Bc = Cd : De.$ [v. 11]

Hence, by the first case,

$$Ab : Cd = Bc : De,$$

and, since these rectangles have the same height,

$$AB : CD = BC : DE.$$

PROPOSITION 17.

If magnitudes be proportional componendo, *they will also be proportional* separando.

Let *AB*, *BE*, *CD*, *DF* be magnitudes proportional *componendo*, so that, as *AB* is to *BE*, so is *CD* to *DF* ; I say that they will also be proportional *separando*, that is, as *AE* is to *EB*, so is *CF* to *DF*.

For of *AE*, *EB*, *CF*, *FD* let equimultiples *GH*, *HK*, *LM*, *MN* be taken,

and of *EB*, *FD* other, chance, equimultiples, *KO*, *NP*,

Then, since GH is the same multiple of AE that HK is of EB,
therefore GH is the same multiple of AE that GK is of AB.

[v. 1]

But GH is the same multiple of AE that LM is of CF;
therefore GK is the same multiple of AB that LM is of CF.

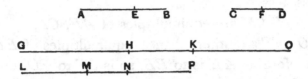

Again, since LM is the same multiple of CF that MN is of FD,
therefore LM is the same multiple of CF that LN is of CD.

[v. 1]

But LM was the same multiple of CF that GK is of AB;
therefore GK is the same multiple of AB that LN is of CD.

Therefore GK, LN are equimultiples of AB, CD.

Again, since HK is the same multiple of EB that MN is of FD,
and KO is also the same multiple of EB that NP is of FD,
therefore the sum HO is also the same multiple of EB that MP is of FD.

[v. 2]

And, since, as AB is to BE, so is CD to DF,
and of AB, CD equimultiples GK, LN have been taken,
and of EB, FD equimultiples HO, MP,
therefore, if GK is in excess of HO, LN is also in excess of MP,
if equal, equal,
and if less, less.

Let GK be in excess of HO;
then, if HK be subtracted from each,

GH is also in excess of KO.

But we saw that, if GK was in excess of HO, LN was also in excess of MP;

therefore LN is also in excess of MP,

and, if MN be subtracted from each,

LM is also in excess of NP;

so that, if GH is in excess of KO, LM is also in excess of NP.

Similarly we can prove that,

if GH be equal to KO, LM will also be equal to NP,

and if less, less.

And GH, LM are equimultiples of AE, CF,

while KO, NP are other, chance, equimultiples of EB, FD;

therefore, as AE is to EB, so is CF to FD.

Therefore etc.

Q. E. D.

Algebraically, if $\qquad a : b = c : d,$

then $\qquad (a - b) : b = (c - d) : d$

I have already noted the somewhat strange use of the participles of συγκεῖσθαι and διαιρεῖσθαι to convey the sense of the technical σύνθεσις and διαίρεσις λόγου, or what we denote by *componendo* and *separando*. ἐὰν συγκείμενα μεγέθη ἀνάλογον ᾖ, καὶ διαιρεθέντα ἀνάλογον ἔσται is, literally, "if magnitudes compounded be proportional, they will also be proportional separated," by which is meant "if one magnitude made up of two parts is to one of its parts as another magnitude made up of two parts is to one of its parts, the remainder of the first whole is to the part of it first taken as the remainder of the second whole is to the part of it first taken." In the algebraical formula above a, c are the wholes and b, $a - b$ and d, $c - d$ are the parts and remainders respectively. The formula might also be stated thus:

If $\qquad a + b : b = c + d : d,$

then $\qquad a : b = c : d,$

in which case $a + b$, $c + d$ are the wholes and b, a and d, c the parts and remainders respectively. Looking at the last formula, we observe that "separated," διαιρεθέντα, is used with reference not to the magnitudes a, b, c, d but to the *compounded* magnitudes $a + b$, b, $c + d$, d.

As the proof is somewhat long, it will be useful to give a conspectus of it in the more symbolical form. To avoid minuses, we will take for the hypothesis the form

$a + b$ is to b as $c + d$ is to d.

Take any equimultiples of the four magnitudes a, b, c, d, viz.

ma, mb, mc, md,

and any other equimultiples of the consequents, viz.

nb and nd.

Then, by v. 1, $m(a + b)$, $m(c + d)$ are equimultiples of $a + b$, $c + d$,

and, by v. 2, $(m + n)b$, $(m + n)d$ are equimultiples of b, d.

Therefore, by Def. 5, since $a + b$ is to b as $c + d$ is to d,

according as $m(a + b) > = < (m + n)b$, $\quad m(c + d) > = < (m + n)d$.

Subtract from $m(a+b)$, $(m+n)b$ the common part mb, and from $m(c+d)$, $(m+n)d$ the common part md; and we have,

according as $ma > = < nb$, $mc > = < nd$.

But ma, mc are any equimultiples of a, c, and nb, nd any equimultiples of b, d,

therefore, by v. Def. 5,

a is to b as c is to d.

Smith and Bryant's proof follows, *mutatis mutandis*, their alternative proof of the next proposition (see pp. 173—4 below).

Proposition 18.

If magnitudes be proportional separando, *they will also be proportional* componendo.

Let AE, EB, CF, FD be magnitudes proportional *separando*, so that, as AE is to EB, so is CF to FD; I say that they will also be proportional *componendo*, that is, as AB is to BE, so is CD to FD.

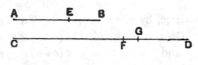

For, if CD be not to DF as AB to BE, then, as AB is to BE, so will CD be either to some magnitude less than DF or to a greater.

First, let it be in that ratio to a less magnitude DG.

Then, since, as AB is to BE, so is CD to DG, they are magnitudes proportional *componendo*;

so that they will also be proportional *separando*.   [v. 17]

Therefore, as AE is to EB, so is CG to GD.

But also, by hypothesis,

as AE is to EB, so is CF to FD.

Therefore also, as CG is to GD, so is CF to FD.   [v. 11]

But the first CG is greater than the third CF;

therefore the second GD is also greater than the fourth FD.   [v. 14]

But it is also less: which is impossible.

Therefore, as AB is to BE, so is not CD to a less magnitude than FD.

Similarly we can prove that neither is it in that ratio to a greater;

it is therefore in that ratio to *FD* itself.

Therefore etc.

Q. E. D.

Algebraically, if $a : b = c : d,$
then $(a + b) : b = (c + d) : d.$

In the enunciation of this proposition there is the same special use of διῃρημένα and συντεθέντα as there was of συγκείμενα and διαιρεθέντα in the last enunciation. Practically, as the algebraical form shows, διῃρημένα might have been left out.

The following is the method of proof employed by Euclid.

Given that $a : b = c : d,$
suppose, if possible, that

$$(a + b) : b = (c + d) : (d \pm x).$$

Therefore, *separando* [v. 17],

$$a : b = (c \mp x) : (d \pm x),$$

whence, by v. 11, $(c \mp x) : (d \pm x) = c : d.$

But $(c - x) < c,$ while $(d + x) > d,$
and $(c + x) > c,$ while $(d - x) < d,$

which relations respectively contradict v. 14.

Simson pointed out (as Saccheri before him saw) that Euclid's demonstration is not legitimate, because it assumes without proof that *to any three magnitudes, two of which, at least, are of the same kind, there exists a fourth proportional.* Clavius and, according to him, other editors made this an axiom. But it is far from axiomatic; it is not till VI. 12 that Euclid shows, by construction, that it is true even in the particular case where the three given magnitudes are all straight lines.

In order to remove the defect it is necessary either (1) to prove beforehand the proposition thus assumed by Euclid or (2) to prove v. 18 independently of it.

Saccheri ingeniously proposed that the assumed proposition should be proved, for areas and straight lines, by means of Euclid VI. 1, 2 and 12. As he says, there was nothing to prevent Euclid from interposing these propositions immediately after v. 17 and then proving v. 18 by means of them. VI. 12 enables us to *construct* the fourth proportional when the three given magnitudes are straight lines; and VI. 12 depends only on VI. 1 and 2. "Now," says Saccheri, "when we have once found the means of constructing a straight line which is a fourth proportional to three given straight lines, we obviously have the solution of the general problem 'To construct a straight line which shall have to a given straight line the same ratio which two *polygons* have (to one another).'" For it is sufficient to transform the polygons into two triangles of equal height and then to construct a straight line which shall be a fourth proportional to the bases of the triangles and the given straight line.

The method of Saccheri is, as will be seen, similar to that adopted by

Smith and Bryant (*loc. cit.*) in proving the theorems of Euclid v. 16, 17, 18, 22, so far as straight lines and rectilineal areas are concerned, by means of VI. 1.

De Morgan gives a sketch of a general proof of the assumed proposition that, B being any magnitude, and P and Q two magnitudes of the same kind, there does exist a magnitude A which is to B in the same ratio as P to Q.

"The right to reason upon any aliquot part of any magnitude is assumed; though, in truth, aliquot parts obtained by continual bisection would suffice: and it is taken as previously proved that the tests of greater and of less ratio are never both presented in any one scale of relation as compared with another" (see note on v. Def. 7 *ad fin.*).

"(1)　If M be to B in a greater ratio than P to Q, so is every magnitude greater than M, and so are *some less magnitudes*; and if M be to B in a less ratio than P to Q, so is every magnitude less than M, and so are *some greater magnitudes*. Part of this is in every system: the rest is proved thus. If M be to B in a greater ratio than P to Q, say, for instance, we find that $15M$ lies between $22B$ and $23B$, while $15P$ lies before $22Q$. Let $15M$ exceed $22B$ by Z; then, if N be less than M by anything less than the 15th part of Z, $15N$ is between $22B$ and $23B$: or N, less than M, is in a greater ratio to B than P to Q. And similarly for the other case.

(2)　M can certainly be taken so small as to be in a less ratio to B than P to Q, and so large as to be in a greater; and since we can never pass from the greater ratio back again to the smaller by increasing M, it follows that, while we pass from the first designated value to the second, we come upon an intermediate magnitude A such that every smaller is in a less ratio to B than P to Q, and every greater in a greater ratio. Now A cannot be in a less ratio to B than P to Q, for then some greater magnitudes would also be in a less ratio; nor in a greater ratio, for then some less magnitudes would be in a greater ratio; therefore A is in the same ratio to B as P to Q. The previously proved proposition above mentioned shows the three alternatives to be the only ones."

Alternative proofs of V. 18.

Simson bases his alternative on v. 5, 6. As the 18th proposition is the converse of the 17th, and the latter is proved by means of v. 1 and 2, of which v. 5 and 6 are converses, the proof of v. 18 by v. 5 and 6 would be natural; and Simson holds that Euclid must have proved v. 18 in this way because "the 5th and 6th do not enter into the demonstration of any proposition in this book as we have it, nor can they be of any use in any proposition of the Elements," and "the 5th and 6th have undoubtedly been put into the 5th book for the sake of some propositions in it, as all the other propositions about equimultiples have been."

Simson's proof is however, as it seems to me, intolerably long and difficult to follow unless it be put in the symbolical form as follows.

Suppose that a is to b as c is to d;

it is required to prove that $a + b$ is to b as $c + d$ is to d.

Take any equimultiples of the last four magnitudes, say

$$m(a+b), \quad mb, \quad m(c+d), \quad md,$$

and any equimultiples of b, d, as

$$nb, \quad nd.$$

Clearly, if nb is greater than mb,

                               nd is greater than md;

if equal, equal; and if less, less.

I. Suppose nb not greater than mb, so that nd is also not greater than md.

Now $m(a+b)$ is greater than mb:

therefore $m(a+b)$ is greater than nb.

Similarly $m(c+d)$ is greater than nd.

II. Suppose nb greater than mb.

Since $m(a+b)$, mb, $m(c+d)$, md are equimultiples of $(a+b)$, b, $(c+d)$, d,

                    ma is the same multiple of a that $m(a+b)$ is of $(a+b)$,

and mc is the same multiple of c that $m(c+d)$ is of $(c+d)$,

so that ma, mc are equimultiples of a, c.   [v. 5]

Again nb, nd are equimultiples of b, d,

          and so are mb, md;

therefore $(n-m)b$, $(n-m)d$ are equimultiples of b, d and, whether $n-m$ is equal to unity or to any other integer [v. 6], it follows, by Def. 5, that, since a, b, c, d are proportionals,

if ma is greater than $(n-m)b$,

then mc is greater than $(n-m)d$;

if equal, equal; and if less, less.

(1) If now $m(a+b)$ is greater than nb, subtracting mb from each, we have

                    ma is greater than $(n-m)b$;

therefore mc is greater than $(n-m)d$,

and, if we add md to each,

                    $m(c+d)$ is greater than nd.

(2) Similarly it may be proved that,

if $m(a+b)$ is equal to nb,

then $m(c+d)$ is equal to nd,

and (3) that, if $m(a+b)$ is less than nb,

then $m(c+d)$ is less than nd.

But (under I. above) it was proved that, in the case where nb is not greater than mb,

                    $m(a+b)$ is always greater than nb,

and $m(c+d)$ is always greater than nd.

Hence, whatever be the values of m and n, $m(c+d)$ is always greater than, equal to, or less than nd according as $m(a+b)$ is greater than, equal to, or less than nb.

Therefore, by Def. 5,

                    $a+b$ is to b as $c+d$ is to d.

Todhunter gives the following short demonstration from Austin (*Examination of the first six books of Euclid's Elements*).

"Let AE be to EB as CF is to FD:

                    AB shall be to BE as CD is to DF.

For, because AE is to EB as CF is to FD,
therefore, alternately,

$$AE \text{ is to } CF \text{ as } EB \text{ is to } FD. \qquad \text{[v. 16]}$$

And, as one of the antecedents is to its consequent, so is the sum of the antecedents to the sum of the consequents: [v. 12]

therefore, as EB is to FD, so are AE, EB together to CF, FD together;

that is, AB is to CD as EB is to FD.

Therefore, alternately,

$$AB \text{ is to } BE \text{ as } CD \text{ is to } FD.\text{''}$$

The objection to this proof is that it is only valid in the case where the proposition v. 16 used in it is valid, i.e. where all four magnitudes are of the same kind.

Smith and Bryant's proof avails where all four magnitudes are straight lines, where all four magnitudes are rectilineal areas, or where one antecedent and its consequent are straight lines and the other antecedent and its consequent rectilineal areas.

Suppose that $A : B = C : D$.

First, let all the magnitudes be areas.

Construct a rectangle *abcd* equal to A, and to *bc* apply the rectangle *bcef* equal to B.

Also to *ab*, *bf* apply the rectangles *ag*, *bk* equal to C, D respectively.

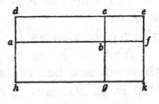

Then, since the rectangles *ac*, *be* have equal heights *bc*, they are to one another as their bases. [vi. 1]

Hence $ab : bf$ = rect. ac : rect. be

$$= A : B$$
$$= C : D$$
$$= \text{rect. } ag : \text{rect. } bk.$$

Therefore [vi. 1, converse] the rectangles *ag*, *bk* have the same height, so that *k* is on the straight line *hg*.

Hence $A + B : B$ = rect. ae : rect. be

$$= af : bf$$
$$= \text{rect. } ak : \text{rect. } bk$$
$$= C + D : D.$$

Secondly, let the magnitudes A, B be straight lines and the magnitudes C, D areas.

Let *ab*, *bf* be equal to the straight lines A, B, and to *ab*, *bf* apply the rectangles *ag*, *bk* equal to C, D respectively.

Then, as before, the rectangles *ag*, *bk* have the same height.

Now $A + B : B = af : bf$

$$= \text{rect. } ak : \text{rect. } bk$$
$$= C + D : D.$$

Thirdly, let all the magnitudes be straight lines.

Apply to the straight lines C, D rectangles P, Q having the same height.

Then $\qquad P : Q = C : D.$ [vi. 1]

Hence, by the second case,

$$A + B : B = P + Q : Q.$$

Also $\qquad P + Q : Q = C + D : D.$

Therefore $\qquad A + B : B = C + D : D.$

PROPOSITION 19.

If, as a whole is to a whole, so is a part subtracted to a part subtracted, the remainder will also be to the remainder as whole to whole.

For, as the whole AB is to the whole CD, so let the part AE subtracted be to the part CF subtracted;

I say that the remainder EB will also be to the remainder FD as the whole AB to the whole CD.

For since, as AB is to CD, so is AE to CF,

alternately also, as BA is to AE, so is DC to CF. [v. 16]

And, since the magnitudes are proportional *componendo*, they will also be proportional *separando*, [v. 17]

that is, as BE is to EA, so is DF to CF,

and, alternately,

as BE is to DF, so is EA to FC. [v. 16]

But, as AE is to CF, so by hypothesis is the whole AB to the whole CD.

Therefore also the remainder EB will be to the remainder FD as the whole AB is to the whole CD. [v. 11]

Therefore etc.

[PORISM. From this it is manifest that, if magnitudes be proportional *componendo*, they will also be proportional *convertendo*.]

Q. E. D.

Algebraically, if $a : b = c : d$ (where $c < a$ and $d < b$), then

$$(a - c) : (b - d) = a : b.$$

The "Porism" at the end of this proposition is led up to by a few lines which Heiberg brackets because it is not Euclid's habit to explain a Porism, and indeed a Porism, from its very nature, should not need any explanation, being a sort of by-product appearing without effort or trouble, ἀπραγματεύτως (Proclus, p. 303, 6). But Heiberg thinks that Simson does wrong in finding fault with the argument leading to the "Porism," and that it does contain the true demonstration of *conversion* of a ratio. In this it appears to me that Heiberg is clearly mistaken, the supposed proof on the basis of Prop. 19 being no more correct than the similar attempt to prove the *inversion* of a ratio from Prop. 4. The words are: "And since it was proved that, as AB is to CD, so is EB to FD,

alternately also, as AB is to BE, so is CD to FD :

therefore magnitudes when compounded are proportional.

But it was proved that, as BA is to AE, so is DC to CF, and this is *convertendo*."

It will be seen that this amounts to proving *from the hypothesis* $a : b = c : d$ that the following transformations are simultaneously true, viz. :

$$a : a - c = b : b - d,$$

and $$a : c = b : d.$$

The former is not proved from the latter as it ought to be if it were intended to prove *conversion*.

The inevitable conclusion is that both the "Porism" and the argument leading up to it are interpolations, though no doubt made, as Heiberg says, before Theon's time.

The *conversion* of ratios does not depend upon v. 19 at all but, as Simson shows in his Proposition E (containing a proof already given by Clavius), on Props. 17 and 18. Prop. E is as follows.

If four magnitudes be proportionals, they are also proportionals by conversion, that is, the first is to its excess above the second as the third is to its excess above the fourth.

Let AB be to BE as CD to DF :
then BA is to AE as DC to CF.

Because AB is to BE as CD to DF, by division [*separando*],

AE is to EB as CF to FD,   [v. 17]

and, by inversion,

BE is to EA as DF to FC.

[Simson's Prop. B directly obtained from v. Def. 5]

Wherefore, by composition [*componendo*],

BA is to AE as DC to CF.   [v. 18]

Proposition 20.

If there be three magnitudes, and others equal to them in multitude, which taken two and two are in the same ratio, and if ex aequali *the first be greater than the third, the fourth will also be greater than the sixth; if equal, equal; and, if less, less.*

Let there be three magnitudes A, B, C, and others D, E, F equal to them in multitude, which taken two and two are in the same ratio, so that,

as A is to B, so is D to E,

and   as B is to C, so is E to F;

and let A be greater than C *ex aequali*;

I say that D will also be greater than F; if A is equal to C, equal; and, if less, less.

A ——————   D ———————
B ———   E ——
C —————   F ———————

For, since A is greater than C,

and B is some other magnitude,

and the greater has to the same a greater ratio than the less has,   [v. 8]

therefore A has to B a greater ratio than C has to B.

But, as A is to B, so is D to E,

and, as C is to B, inversely, so is F to E;

therefore D has also to E a greater ratio than F has to E. [v. 13]

But, of magnitudes which have a ratio to the same, that which has a greater ratio is greater;   [v. 10]

therefore D is greater than F.

Similarly we can prove that, if A be equal to C, D will also be equal to F; and if less, less.

Therefore etc.

Q. E. D.

Though, as already remarked, Euclid has not yet given us any definition of *compounded ratios*, Props. 20—23 contain an important part of the theory of such ratios. The term "compounded ratio" is not used, but the propositions connect themselves with the definitions of *ex aequali* in its two forms, the ordinary form defined in Def. 17 and that called *perturbed proportion* in Def. 18. The compounded ratios dealt with in these propositions are those compounded of successive ratios in which the consequent of one is the antecedent of the next, or the antecedent of one is the consequent of the next.

Prop. 22 states the fundamental proposition about the ratio *ex aequali* in its ordinary form, to the effect that,

if a is to b as d is to e,

and b is to c as e is to f,

then a is to c as d is to f,

with the extension to any number of such ratios; Prop. 23 gives the corresponding theorem for the case of *perturbed proportion*, namely that,

if　　　　　　　　　　a is to b as e is to f,

and　　　　　　　　　b is to c as d is to e,

then　　　　　　　　 a is to c as d is to f.

Each depends on a preliminary proposition, Prop. 22 on Prop. 20 and Prop. 23 on Prop. 21. The course of the proof will be made most clear by using the algebraic notation.

The preliminary Prop. 20 asserts that,

if　　　　　　　　　　$a : b = d : e$,

and　　　　　　　　　$b : c = e : f$,

then, according as $a > = < c$,　$d > = < f$.

For, according as a is greater than, equal to, or less than c,

the ratio $a : b$ is greater than, equal to, or less than the ratio $c : b$,　[v. 8 or v. 7]

or (since　　　　　　 $d : e = a : b$,

and　　　　　　　　　$c : b = f : e$)

the ratio $d : e$ is greater than, equal to, or less than the ratio $f : e$,

[by aid of v. 13 and v. 11]

and therefore d is greater than, equal to, or less than f.　 [v. 10 or v. 9]

It is next proved in Prop. 22 that, by v. 4, the given proportions can be transformed into

$$ma : nb = md : ne,$$

and　　　　　　　　　$nb : pc = ne : pf$,

whence, by v. 20,

according as　ma is greater than, equal to, or less than pc,

md is greater than, equal to, or less than pf,

so that, by Def. 5,

$$a : c = d : f.$$

Prop. 23 depends on Prop. 21 in the same way as Prop. 22 on Prop. 20, but the transformation of the ratios in Prop. 23 is to the following :

(1)　　　　　　　　　$ma : mb = ne : nf$

(by a double application of v. 15 and by v. 11),

(2)　　　　　　　　　$mb : nc = md : ne$

(by v. 4, or equivalent steps),

and Prop. 21 is then used.

Simson makes the proof of Prop. 20 slightly more explicit, but the main difference from the text is in the addition of the two other cases which Euclid dismisses with "Similarly we can prove." These cases are :

"Secondly, let A be equal to C; then shall D be equal to F.

Because A and C are equal to one another,

A is to B as C is to B.　　　　　　　　　　　　[v. 7]

But　　　　　　　　A is to B as D is to E,

and　　　　　　　　C is to B as F is to E,

wherefore　　　　 D is to E as F to E ;　　　　　　　[v. 11]

and therefore D is equal to F.　　　　　　　　　　　[v. 9]

Next, let A be less than C; then shall D be less than F.

For C is greater than A,

and, as was shown in the first case,

C is to B as F to E,

and, in like manner,

B is to A as E to D;

therefore F is greater than D, by the first case; and therefore D is less than F."

PROPOSITION 21.

If there be three magnitudes, and others equal to them in multitude, which taken two and two together are in the same ratio, and the proportion of them be perturbed, then, if ex aequali the first magnitude is greater than the third, the fourth will also be greater than the sixth; if equal, equal; and if less, less.

Let there be three magnitudes A, B, C, and others D, E, F equal to them in multitude, which taken two and two are in the same ratio, and let the proportion of them be perturbed, so that,

as A is to B, so is E to F,

and,   as B is to C, so is D to E,

and let A be greater than C *ex aequali*;

I say that D will also be greater than F; if A is equal to C, equal; and if less, less.

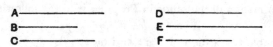

For, since A is greater than C,

and B is some other magnitude,

therefore A has to B a greater ratio than C has to B.   [v. 8]

But, as A is to B, so is E to F,

and, as C is to B, inversely, so is E to D.

Therefore also E has to F a greater ratio than E has to D.

[v. 13]

But that to which the same has a greater ratio is less;

[v. 10]

therefore F is less than D;

therefore D is greater than F.

Similarly we can prove that,

if A be equal to C, D will also be equal to F;

and if less, less.

Therefore etc.   Q. E. D.

Algebraically, if $a:b=e:f,$

and $b:c=d:e,$

then, according as $a>=<c$, $d>=<f.$

Simson's alterations correspond to those which he makes in Prop. 20.   After the first case he proceeds thus.

"Secondly, let A be equal to C; then shall D be equal to F.

Because A and C are equal,

A is to B as C is to B.   [v. 7]

But A is to B as E is to F,

and C is to B as E is to D:

wherefore E is to F as E to D,   [v. 11]

and therefore D is equal to F.   [v. 9]

Next, let A be less than C; then shall D be less than F.

For C is greater than A,

and, as was shown,

C is to B as E to D,

and, in like manner,

B is to A as F to E;

therefore F is greater than D, by the first case,

and therefore D is less than F."

The proof may be shown thus.

According as $a>=<c$, $a:b>=<c:b.$

But $a:b=e:f,$ and, by inversion, $c:b=e:d.$

Therefore, according as $a>=<c$, $e:f>=<e:d,$

and therefore $d>=<f.$

PROPOSITION 22.

If there be any number of magnitudes whatever, and others equal to them in multitude, which taken two and two together are in the same ratio, they will also be in the same ratio ex aequali.

Let there be any number of magnitudes A, B, C, and others D, E, F equal to them in multitude, which taken two and two together are in the same ratio, so that,

as A is to B, so is D to E,

and,   as B is to C, so is E to F;

I say that they will also be in the same ratio *ex aequali*,

$<$ that is, as A is to C, so is D to $F>$.

For of A, D let equimultiples G, H be taken,
and of B, E other, chance, equimultiples K, L ;
and, further, of C, F other, chance, equimultiples M, N.

Then, since, as A is to B, so is D to E,
and of A, D equimultiples G, H have been taken,
and of B, E other, chance, equimultiples K, L,

therefore, as G is to K, so is H to L.   [v. 4]

For the same reason also,

as K is to M, so is L to N.

Since, then, there are three magnitudes G, K, M, and others H, L, N equal to them in multitude, which taken two and two together are in the same ratio,

therefore, *ex aequali*, if G is in excess of M, H is also in excess of N ;

if equal, equal ; and if less, less.   [v. 20]

And G, H are equimultiples of A, D,

and M, N other, chance, equimultiples of C, F.

Therefore, as A is to C, so is D to F.   [v. Def. 5]

Therefore etc.

Q. E. D.

Euclid enunciates this proposition as true of *any number of magnitudes whatever* forming two sets connected in the manner described, but his proof is confined to the case where each set consists of three magnitudes only. The extension to any number of magnitudes is, however, easy, as shown by Simson.

"Next let there be four magnitudes A, B, C, D, and other four E, F, G, H, which two and two have the same ratio, viz. :

as A is to B, so is E to F,
and   as B is to C, so is F to G,
and   as C is to D, so is G to H ;
A shall be to D as E to H.

Because A, B, C are three magnitudes, and E, F, G other three, which taken two and two have the same ratio,

by the foregoing case,

A is to C as E to G.

But C is to D as G is to H;
wherefore again, by the first case,

A is to D as E to H.

And so on, whatever be the number of magnitudes."

PROPOSITION 23.

*If there be three magnitudes, and others equal to them in
multitude, which taken two and two together are in the same
ratio, and the proportion of them be perturbed, they will also
be in the same ratio ex aequali.*

Let there be three magnitudes A, B, C, and others equal
to them in multitude, which, taken two and two together, are
in the same proportion, namely D, E, F; and let the proportion of them be perturbed, so that,

as A is to B, so is E to F,

and,   as B is to C, so is D to E;
I say that, as A is to C, so is D to F.

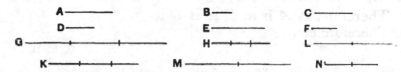

Of A, B, D let equimultiples G, H, K be taken,
and of C, E, F other, chance, equimultiples L, M, N.

Then, since G, H are equimultiples of A, B,
and parts have the same ratio as the same multiples of
them,   [v. 15]

therefore, as A is to B, so is G to H.

For the same reason also,

as E is to F, so is M to N.

And, as A is to B, so is E to F;
therefore also, as G is to H, so is M to N.   [v. 11]

Next, since, as B is to C, so is D to E,
alternately, also, as B is to D, so is C to E.   [v. 16]

And, since H, K are equimultiples of B, D,
and parts have the same ratio as their equimultiples,

therefore, as B is to D, so is H to K.   [v. 15]

But, as B is to D, so is C to E;

therefore also, as H is to K, so is C to E.   [v. 11]

Again, since L, M are equimultiples of C, E,

therefore, as C is to E, so is L to M.   [v. 15]

But, as C is to E, so is H to K;

therefore also, as H is to K, so is L to M,   [v. 11]

and, alternately, as H is to L, so is K to M.   [v. 16]

But it was also proved that,

as G is to H, so is M to N.

Since, then, there are three magnitudes G, H, L, and others equal to them in multitude K, M, N, which taken two and two together are in the same ratio,

and the proportion of them is perturbed,

therefore, *ex aequali*, if G is in excess of L, K is also in excess of N;

if equal, equal; and if less, less.   [v. 21]

And G, K are equimultiples of A, D,

and L, N of C, F.

Therefore, as A is to C, so is D to F.

Therefore etc.

Q. E. D.

There is an important difference between the version given by Simson of one part of the proof of this proposition and that found in the Greek text of Heiberg. Peyrard's MS. has the version given by Heiberg, but Simson's version has the authority of other MSS. The Basel *editio princeps* gives both versions (Simson's being the first). After it has been proved by means of v. 15 and v. 11 that,

as G is to H, so is M to N,

or, with the notation used in the note on Prop. 20,

$$ma : mb = ne : nf,$$

it has to be proved further that,

as H is to L, so is K to M,

or $$mb : nc = md : ne,$$

and it is clear that the latter result may be directly inferred from v. 4. The reading translated by Simson makes this inference:

"And because, as B is to C, so is D to E,

and H, K are equimultiples of B, D,

and L, M of C, E,

therefore, as H is to L, so is K to M."   [v. 4]

The version in Heiberg's text is not only much longer (it adopts the roundabout method of using each of three Propositions v. 11, 15, 16 twice over), but it is open to the objection that it uses v. 16 which is only applicable if the four magnitudes are *of the same kind*; whereas v. 23, the proposition now in question, is not subject to this restriction.

Simson rightly observes that in the last step of the proof it should be stated that "*G*, *K* are *any* equimultiples *whatever* of *A*, *D*, and *L*, *N any whatever* of *C*, *F*."

He also gives the extension of the proposition to any number of magnitudes, enunciating it thus:

"If there be any number of magnitudes, and as many others, which, taken two and two, in a cross order, have the same ratio; the first shall have to the last of the first magnitudes the same ratio which the first of the others has to the last";

and adding to the proof as follows:

"Next, let there be four magnitudes *A*, *B*, *C*, *D*, and other four *E*, *F*, *G*, *H*, which, taken two and two in a cross order, have the same ratio, viz.:

$$A \text{ to } B \text{ as } G \text{ to } H,$$
$$B \text{ to } C \text{ as } F \text{ to } G,$$
and $$C \text{ to } D \text{ as } E \text{ to } F;$$

then *A* is to *D* as *E* to *H*.

A	B	C	D
E	F	G	H

Because *A*, *B*, *C* are three magnitudes, and *F*, *G*, *H* other three which, taken two and two in a cross order, have the same ratio,

by the first case,　　*A* is to *C* as *F* to *H*.

But　　*C* is to *D* as *E* is to *F*;

wherefore again, by the first case,

A is to *D* as *E* to *H*.

And so on, whatever be the number of magnitudes."

PROPOSITION 24.

If a first magnitude have to a second the same ratio as a third has to a fourth, and also a fifth have to the second the same ratio as a sixth to the fourth, the first and fifth added together will have to the second the same ratio as the third and sixth have to the fourth.

Let a first magnitude *AB* have to a second *C* the same ratio as a third *DE* has to a fourth *F*; and let also a fifth *BG* have to the second *C* the same ratio as a sixth *EH* has to the fourth *F*;

I say that the first and fifth added together, *AG*, will have to the second *C* the same ratio as the third and sixth, *DH*, has to the fourth *F*.

For since, as BG is to C, so is EH to F,
inversely, as C is to BG, so is F to EH.

Since, then, as AB is to C, so is DE to F,
and, as C is to BG, so is F to EH,
therefore, *ex aequali,* as AB is to BG, so is DE to EH. [v. 22]

And, since the magnitudes are proportional *separando,* they
will also be proportional *componendo* ;   [v. 18]
therefore, as AG is to GB, so is DH to HE.

But also, as BG is to C, so is EH to F ;
therefore, *ex aequali,* as AG is to C, so is DH to F.   [v. 22]

Therefore etc.   Q. E. D.

Algebraically, if $a : c = d : f,$
and $b : c = e : f,$
then $(a + b) : c = (d + e) : f.$

This proposition is of the same character as those which precede the
propositions relating to compounded ratios ; but it could not be placed earlier
than it is because v. 22 is used in the proof of it.

Inverting the second proportion to
$$c : b = f : e,$$
it follows, by v. 22, that $a : b = d : e,$
whence, by v. 18, $(a + b) : b = (d + e) : e,$
and from this and the second of the two given proportions we obtain, by a
fresh application of v. 22,
$$(a + b) : c = (d + e) : f.$$

The first use of v. 22 is important as showing that the opposite process to
compounding ratios, or what we should now call *division* of one ratio by
another, does not require any new and separate propositions.

Aristotle tacitly uses v. 24 in combination with v. 11 and v. 16, *Meteorologica*
III. 5, 376 a 22—26.

Simson adds two corollaries, one of which (Cor. 2) notes the extension to
any number of magnitudes.

"The proposition holds true of two ranks of magnitudes whatever be their
number, of which each of the first rank has to the second magnitude the same
ratio that the corresponding one of the second rank has to a fourth magnitude ;
as is manifest."

Simson's Cor. 1 states the corresponding proposition to the above with
separando taking the place of *componendo,* viz., that corresponding to the
algebraical form
$$(a - b) : c = (d - e) : f.$$

"Cor. 1. If the same hypothesis be made as in the proposition, the
excess of the first and fifth shall be to the second as the excess of the third
and sixth to the fourth. The demonstration of this is the same with that of
the proposition if division be used instead of composition." That is, we use
v. 17 instead of v. 18, and conclude that
$$(a - b) : b = (d - e) : e.$$

PROPOSITION 25.

If four magnitudes be proportional, the greatest and the least are greater than the remaining two.

Let the four magnitudes AB, CD, E, F be proportional so that, as AB is to CD, so is E to F, and let AB be the greatest of them and F the least;

I say that AB, F are greater than CD, E.

For let AG be made equal to E, and CH equal to F.

Since, as AB is to CD, so is E to F,

and E is equal to AG, and F to CH,

therefore, as AB is to CD, so is AG to CH.

And since, as the whole AB is to the whole CD, so is the part AG subtracted to the part CH subtracted,

the remainder GB will also be to the remainder HD as the whole AB is to the whole CD.   [v. 19]

But AB is greater than CD;

therefore GB is also greater than HD.

And, since AG is equal to E, and CH to F,

therefore AG, F are equal to CH, E.

And if, GB, HD being unequal, and GB greater, AG, F be added to GB and CH, E be added to HD,

it follows that AB, F are greater than CD, E.

Therefore etc.

Q. E. D.

Algebraically, if $a : b = c : d$,

and a is the greatest of the four magnitudes and d the least,

$$a + d > b + c.$$

Simson is right in inserting a word in the setting-out, "let AB be the greatest of them and *<consequently>* F the least." This follows from the particular case, really included in Def. 5, which Simson makes the subject of his proposition A, the case namely where the equimultiples taken are *once* the several magnitudes.

The proof is as follows.

Since $a : b = c : d$,

$$a - c : b - d = a : b.$$ [v. 19]

But $a > b$; therefore $(a - c) > (b - d)$.   [v. 16 and 14]

Add to each $(c + d)$;

therefore $(a + d) > (b + c)$.

There is an important particular case of this proposition, which is, however, not mentioned here, viz. the case where $b = c$. The result shows, in this case, that *the arithmetic mean between two magnitudes is greater than their geometric mean*. The truth of this is proved for straight lines in VI. 27 by "geometrical algebra," and the theorem forms the διορισμός for equations of the second degree.

Simson adds at the end of Book v. four propositions, F, G, H, K, which, however, do not seem to be of sufficient practical use to justify their inclusion here. But he adds at the end of his notes to the Book the following paragraph which deserves quotation word for word.

"The 5th book being thus corrected, I most readily agree to what the learned Dr Barrow says, 'that there is nothing in the whole body of the elements of a more subtile invention, nothing more solidly established, and more accurately handled than the doctrine of proportionals.' And there is some ground to hope that geometers will think that this could not have been said with as good reason, since Theon's time till the present."

Simson's claim herein will readily be admitted by all readers who are competent to form a judgment upon his criticisms and elucidations of Book v.

BOOK VI.

INTRODUCTORY NOTE.

The theory of proportions has been established in Book v. in a perfectly general form applicable to all kinds of magnitudes (although the representation of magnitudes by straight lines gives it a *geometrical* appearance); it is now necessary to apply the theory to the particular case of *geometrical* investigation. The only thing still required in order that this may be done is a proof of the existence of such a magnitude as bears to any given finite magnitude any given finite ratio; and this proof is supplied, so far as regards the subject matter of geometry, by vi. 12 which shows how to construct a fourth proportional to three given straight lines.

A few remarks on the enormous usefulness of the theory of proportions to geometry will not be out of place. We have already in Books i. and ii. made acquaintance with one important part of what has been well called geometrical algebra, the method, namely, of *application of areas*. We have seen that this method, working by the representation of products of two quantities as rectangles, enables us to solve some particular quadratic equations. But the limitations of such a method are obvious. So long as general quantities are represented by *straight lines* only, we cannot, if our geometry is *plane*, deal with products of more than two such quantities; and, even by the use of three dimensions, we cannot work with products of more than three quantities, since no geometrical meaning could be attached to such a product. This limitation disappears so soon as we can represent any general quantity, corresponding to what we denote by a letter in algebra, by a *ratio*; and this we can do because, on the general theory of proportion established in Book v., a ratio may be a ratio of two incommensurable quantities as well as of commensurables. Ratios can be *compounded ad infinitum*, and the division of one ratio by another is equally easy, since it is the same thing as compounding the first ratio with the inverse of the second. Thus e.g. it is seen at once that the *coefficients* in a quadratic of the most general form can be represented by ratios between straight lines, and the solution by means of Books i. and ii. of problems corresponding to quadratic equations with particular coefficients can now be extended to cover *any* quadratic with real roots. As indicated, we can perform, by composition of ratios, the operation corresponding to multiplying algebraical quantities, and this to any extent. We can divide quantities by compounding a ratio with the inverse of the ratio representing the divisor. For the addition and subtraction of quantities we have only to use the geometrical equivalent of bringing to a common denominator, which is effected by means of the fourth proportional.

DEFINITIONS.

1. **Similar rectilineal figures** are such as have their angles severally equal and the sides about the equal angles proportional.

[2. **Reciprocally related** figures. *See note.*]

3. A straight line is said to have been **cut in extreme and mean ratio** when, as the whole line is to the greater segment, so is the greater to the less.

4. The **height** of any figure is the perpendicular drawn from the vertex to the base.

DEFINITION 1.

Ὅμοια σχήματα εὐθύγραμμά ἐστιν, ὅσα τάς τε γωνίας ἴσας ἔχει κατὰ μίαν καὶ τὰς περὶ τὰς ἴσας γωνίας πλευρὰς ἀνάλογον.

This definition is quoted by Aristotle, *Anal. post.* II. 17, 99 a 13, where he says that *similarity* (τὸ ὅμοιον) in the case of figures "consists, let us say (ἴσως), in their having their sides proportional and their angles equal." The use of the word ἴσως may suggest that, in Aristotle's time, this definition had not quite established itself in the text-books (Heiberg, *Mathematisches zu Aristoteles*, p. 9).

It was pointed out in Van Swinden's *Elements of Geometry* (Jacobi's edition, 1834, pp. 114—5) that Euclid omits to state an essential part of the definition, namely that "the corresponding sides must be opposite to equal angles," which is necessary in order that the corresponding sides may follow in the same order in both figures.

At the same time the definition states more than is absolutely necessary, for it is true to say that *two polygons are similar when, if the sides and angles are taken in the same order, the angles are equal and the sides about the equal angles are proportional, omitting*

(1) *three consecutive angles,*

or (2) *two consecutive angles and the side common to them,*

or (3) *two consecutive sides and the angle included by them,*

and making no assumption with regard to the omitted sides and angles.

Austin objected to this definition on the ground that it is not obvious that the properties (1) of having their angles respectively equal and (2) of having the sides about the equal angles proportional can *co-exist* in two figures; but, a definition not being concerned to prove the *existence* of the thing defined, the objection falls to the ground. We are properly left to satisfy ourselves as to the existence of similar figures in the course of the exposition in Book VI., where we learn how to construct on any given straight line a rectilineal figure similar to a given one (VI. 18).

DEFINITION 2.

The Greek text gives here a definition of *reciprocally related figures* (ἀντιπεπονθότα σχήματα). "[Two] figures are *reciprocally related* when there are in each of the two figures antecedent and consequent ratios" (Ἀντιπεπονθότα δὲ σχήματά ἐστιν, ὅταν ἐν ἑκατέρῳ τῶν σχημάτων ἡγούμενοί τε καὶ ἑπόμενοι λόγοι ὦσιν). No intelligible meaning can be attached to "antecedent and consequent ratios" here; the sense would require rather "an antecedent and a consequent of (two equal) ratios in each figure." Hence Candalla and Peyrard read λόγων ὅροι ("terms of ratios") instead of λόγοι. Camerer reads λόγων without ὅροι. But the objection to the definition lies deeper. It is never used; when we come, in VI. 14, 15, XI. 34 etc. to parallelograms, triangles etc. having the property indicated, they are not called "reciprocal" parallelograms etc., but parallelograms etc. "*the sides of which are* reciprocally proportional," ὧν ἀντιπεπόνθασιν αἱ πλευραί. Hence Simson appears to be right in condemning the definition; it may have been interpolated from Heron, who has it.

Simson proposes in his note to substitute the following definition. "Two magnitudes are said to be reciprocally proportional to two others when one of the first is to one of the other magnitudes as the remaining one of the last two is to the remaining one of the first." This definition requires that the magnitudes shall be all of the same kind.

DEFINITION 3.

Ἄκρον καὶ μέσον λόγον εὐθεῖα τετμῆσθαι λέγεται, ὅταν ᾖ ὡς ἡ ὅλη πρὸς τὸ μεῖζον τμῆμα, οὕτως τὸ μεῖζον πρὸς τὸ ἔλαττον.

DEFINITION 4.

Ὕψος ἐστὶ παντὸς σχήματος ἡ ἀπὸ τῆς κορυφῆς ἐπὶ τὴν βάσιν κάθετος ἀγομένη.

The definition of "height" is not found in Campanus and is perhaps rightly suspected, since it does not apply in terms to parallelograms, parallelepipeds, cylinders and prisms, though it is used in the *Elements* with reference to these latter figures. Aristotle does not appear to know altitude (ὕψος) in the mathematical sense; he uses κάθετος of triangles (*Meteorologica* III. 3, 373 a 11). The term is however readily understood, and scarcely requires definition.

[DEFINITION 5.

Λόγος ἐκ λόγων συγκεῖσθαι λέγεται, ὅταν αἱ τῶν λόγων πηλικότητες ἐφ' ἑαυτὰς πολλαπλασιασθεῖσαι ποιῶσί τινα.

"A ratio is said to be compounded of ratios when the sizes (πηλικότητες) of the ratios multiplied together make some (? ratio, or size)."]

As already remarked (pp. 116, 132), it is beyond doubt that this definition of ratio is interpolated. It has little MS. authority. The best MS. (P) only has it in the margin; it is omitted altogether in Campanus' translation from the

Arabic; and the other MSS. which contain it do not agree in the position which they give to it. There is no reference to the definition in the place where compound ratio is mentioned for the first time (VI. 23), nor anywhere else in Euclid; neither is it ever referred to by the other great geometers, Archimedes, Apollonius and the rest. It appears to be only twice mentioned at all, (1) in the passage of Eutocius referred to above (p. 116) and (2) by Theon in his commentary on Ptolemy's σύνταξις. Moreover the content of the definition is in itself suspicious. It speaks of the "sizes of ratios being multiplied together (literally, into themselves)," an operation unknown to geometry. There is no wonder that Eutocius, and apparently Theon also, in their efforts to explain it, had to give the word πηλικότης a meaning which has no application except in the case of such ratios as can be expressed by numbers (Eutocius e.g. making it the "number by which the ratio is called"). Nor is it surprising that Wallis should have found it necessary to substitute for the "quantitas" of Commandinus a different translation, "quantuplicity," which he said was represented by the "*exponent of the ratio*" (rationis exponens), what Peletarius had described as "denominatio ipsae proportionis" and Clavius as "denominator." The fact is that the definition is ungeometrical and useless, as was already seen by Savile, in whose view it was one of the two blemishes in the body of geometry (the other being of course Postulate 5).

It is right to add that Hultsch (art. "Eukleides" in Pauly-Wissowa's *Real-Encyclopädie der classischen Altertumswissenschaft*) thought the definition genuine. His grounds are (1) that it stood in the παλαιὰ ἔκδοσις represented by P (though P has it in the margin only) and (2) that some explanation on the subject must have been given by way of preparation for VI. 23, while there is nothing in the definition which is *inconsistent* with the mode of statement of VI. 23. If the definition is after all genuine, I should be inclined to regard it as a mere survival from earlier textbooks, like the first of the two alternative definitions of a solid angle (XI. Def. 11); for its form seems to suit the old theory of proportion, applicable to commensurable magnitudes only, better than the generalised theory of Eudoxus.

BOOK VI. PROPOSITIONS.

PROPOSITION 1.

Triangles and parallelograms which are under the same height are to one another as their bases.

Let *ABC*, *ACD* be triangles and *EC*, *CF* parallelograms under the same height;

5 I say that, as the base *BC* is to the base *CD*, so is the triangle *ABC* to the triangle *ACD*, and the parallelogram *EC* to the parallelogram *CF*.

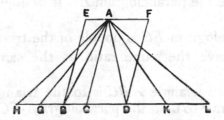

For let *BD* be produced in both directions to the points *H*, *L* and let [any number of straight lines] *BG*, *GH* be
10 made equal to the base *BC*, and any number of straight lines *DK*, *KL* equal to the base *CD*;

let *AG*, *AH*, *AK*, *AL* be joined.

Then, since *CB*, *BG*, *GH* are equal to one another,

the triangles *ABC*, *AGB*, *AHG* are also equal to one
15 another.   [I. 38]

Therefore, whatever multiple the base *HC* is of the base *BC*, that multiple also is the triangle *AHC* of the triangle *ABC*.

For the same reason,

20 whatever multiple the base *LC* is of the base *CD*, that multiple also is the triangle *ALC* of the triangle *ACD*;

and, if the base *HC* is equal to the base *CL*, the triangle *AHC* is also equal to the triangle *ACL*,   [I. 38]

if the base HC is in excess of the base CL, the triangle AHC
25 is also in excess of the triangle ACL,

and, if less, less.

Thus, there being four magnitudes, two bases BC, CD
and two triangles ABC, ACD,

equimultiples have been taken of the base BC and the
30 triangle ABC, namely the base HC and the triangle AHC,

and of the base CD and the triangle ADC other, chance, equimultiples, namely the base LC and the triangle ALC;

and it has been proved that,

if the base HC is in excess of the base CL, the triangle AHC
35 is also in excess of the triangle ALC;

if equal, equal; and, if less, less.

Therefore, as the base BC is to the base CD, so is the
triangle ABC to the triangle ACD.   [v. Def. 5]

Next, since the parallelogram EC is double of the triangle
40 ABC,   [I. 41]

and the parallelogram FC is double of the triangle ACD,

while parts have the same ratio as the same multiples of
them,   [v. 15]

therefore, as the triangle ABC is to the triangle ACD, so is
45 the parallelogram EC to the parallelogram FC.

Since, then, it was proved that, as the base BC is to CD,
so is the triangle ABC to the triangle ACD,

and, as the triangle ABC is to the triangle ACD, so is the
parallelogram EC to the parallelogram CF,

50 therefore also, as the base BC is to the base CD, so is the
parallelogram EC to the parallelogram FC.   [v. 11]

Therefore etc.

Q. E. D.

4. **Under the same height.** The Greek text has "under the same height AC," with
a figure in which the side AC common to the two triangles is perpendicular to the base and
is therefore itself the "height." But, even if the two triangles are placed contiguously so as
to have a common side AC, it is quite gratuitous to require it to be perpendicular to the base.
Theon, on this occasion making an improvement, altered to "which are (ὄντα) under the
same height, (namely) the perpendicular drawn from A to BD." I have ventured to alter so
far as to omit "AC" and to draw the figure in the usual way.

14. **ABC, AGB, AHG.** Euclid, indifferent to exact order, writes "AHG, AGB, ABC."

46. **Since then it was proved that, as the base BC is to CD, so is the triangle
ABC to the triangle ACD.** Here again words have to be supplied in translating the
extremely terse Greek ἐπεὶ οὖν ἐδείχθη, ὡς μὲν ἡ βάσις ΒΓ πρὸς τὴν ΓΔ, οὕτως τὸ ΑΒΓ
τρίγωνον πρὸς τὸ ΑΓΔ τρίγωνον, literally "since was proved, as the base BC to CD, so the
triangle ABC to the triangle ACD." Cf. note on v. 16, p. 165.

The proof assumes—what is however an obvious deduction from I. 38—that, of triangles or parallelograms on *unequal* bases and between the same parallels, the greater is that which has the greater base.

It is of course not necessary that the two given triangles should have a common side, as in the figure; the proof is just as easy if they have not. The proposition being equally true of triangles and parallelograms of *equal* heights, Simson states this fact in a corollary thus:

"From this it is plain that triangles and parallelograms that have equal altitudes are to one another as their bases.

Let the figures be so placed as to have their bases in the same straight line; and, if we draw perpendiculars from the vertices of the triangles to the bases, the straight line which joins the vertices is parallel to that in which their bases are, because the perpendiculars are both equal and parallel to one another [I. 33]. Then, if the same construction be made as in the proposition, the demonstration will be the same."

The object of placing the bases in one straight line is to get the triangles and parallelograms *within the same parallels*. Cf. Proclus' remark on I. 38 (p. 405, 17) that having the same height is the same thing as being in the same parallels.

Rectangles, or right-angled triangles, which have one of the sides about the right angle of the same length can be placed so that the equal sides coincide and the others are in a straight line. If then we call the common side the base, the rectangles or the right-angled triangles are to one another as their heights, by VI. 1. Now, instead of each right-angled triangle or rectangle, we can take any other triangle or parallelogram respectively with an equal base and between the same parallels. Thus

Triangles and parallelograms having equal bases are to one another as their heights.

Legendre and those authors of modern text-books who follow him in basing their treatment of proportion on the algebraical definition are obliged to divide their proofs of propositions like this into two parts, the first of which proves the particular theorem in the case where the magnitudes are commensurable, and the second extends it to the case where they are incommensurable.

Legendre (*Éléments de Géométrie*, III. 3) uses for this extension a rigorous method by *reductio ad absurdum* similar to that used by Archimedes in his treatise *On the equilibrium of planes*, I. 7. The following is Legendre's proof of the extension of VI. 1 to incommensurable parallelograms and bases.

The proposition having been proved for commensurable bases, let there be two rectangles *ABCD*, *AEFD* as in the figure, on bases *AB*, *AE* which are incommensurable with one another.

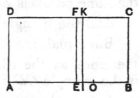

To prove that   rect. *ABCD* : rect. *AEFD* = *AB* : *AE*.

For, if not, let   rect. *ABCD* : rect. *AEFD* = *AB* : *AO*, ... (1)

where *AO* is (for instance) greater than *AE*.

Divide *AB* into equal parts each of which is less than *EO*, and mark off on *AO* lengths equal to one of the parts; then there will be at least one point of division between *E* and *O*.

Let it be *I*, and draw *IK* parallel to *EF*.

Then the rectangles $ABCD$, $AIKD$ are in the ratio of the bases AB, AI, since the latter are commensurable.

Therefore, inverting the proportion,

$$\text{rect. } AIKD : \text{rect. } ABCD = AI : AB \ldots \ldots \ldots \ldots \ldots (2).$$

From this and (1), *ex aequali*,

$$\text{rect. } AIKD : \text{rect. } AEFD = AI : AO.$$

But $AO > AI$; therefore rect. $AEFD >$ rect. $AIKD$.

But this is impossible, for the rectangle $AEFD$ is less than the rectangle $AIKD$.

Similarly an impossibility can be proved if $AO < AE$.

Therefore   rect. $ABCD$: rect. $AEFD = AB : AE$.

Some modern American and German text-books adopt the less rigorous method of appealing to the theory of *limits*.

PROPOSITION 2.

If a straight line be drawn parallel to one of the sides of a triangle, it will cut the sides of the triangle proportionally; and, if the sides of the triangle be cut proportionally, the line joining the points of section will be parallel to the remaining side of the triangle.

For let DE be drawn parallel to BC, one of the sides of the triangle ABC;

I say that, as BD is to DA, so is CE to EA.

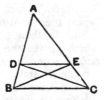

For let BE, CD be joined.

Therefore the triangle BDE is equal to the triangle CDE;

for they are on the same base DE and in the same parallels DE, BC.   [I. 38]

And the triangle ADE is another area.

But equals have the same ratio to the same;   [V. 7]

therefore, as the triangle BDE is to the triangle ADE, so is the triangle CDE to the triangle ADE.

But, as the triangle BDE is to ADE, so is BD to DA;

for, being under the same height, the perpendicular drawn from E to AB, they are to one another as their bases.   [VI. 1]

For the same reason also,

as the triangle CDE is to ADE, so is CE to EA.

Therefore also, as BD is to DA, so is CE to EA.   [V. 11]

Again, let the sides AB, AC of the triangle ABC be cut proportionally, so that, as BD is to DA, so is CE to EA ; and let DE be joined.

I say that DE is parallel to BC.

For, with the same construction,

since, as BD is to DA, so is CE to EA,

but, as BD is to DA, so is the triangle BDE to the triangle ADE,

and, as CE is to EA, so is the triangle CDE to the triangle ADE,   [vi. 1]

therefore also,

as the triangle BDE is to the triangle ADE, so is the triangle CDE to the triangle ADE.   [v. 11]

Therefore each of the triangles BDE, CDE has the same ratio to ADE.

Therefore the triangle BDE is equal to the triangle CDE;
                                                                                           [v. 9]

and they are on the same base DE.

But equal triangles which are on the same base are also in the same parallels.   [I. 39]

Therefore DE is parallel to BC.

Therefore etc.

Q. E. D.

Euclid evidently did not think it worth while to distinguish in the enunciation, or in the figure, the cases in which the parallel to the base cuts the other two sides produced (a) beyond the point in which they intersect, (b) in the other direction. Simson gives the three figures and inserts words in the enunciation, reading "it shall cut the other sides, *or those sides produced*, proportionally" and "if the sides, *or the sides produced*, be cut proportionally."

Todhunter observes that the second part of the enunciation ought to make it clear which segments in the proportion correspond to which. Thus e.g., if AD were double of DB, and CE double of EA, the sides would be cut proportionally, but DE would not be parallel to BC. The omission could be supplied by saying "and if the sides of the triangle be cut proportionally *so that the segments adjacent to the third side are corresponding terms in the proportion.*"

PROPOSITION 3.

If an angle of a triangle be bisected and the straight line cutting the angle cut the base also, the segments of the base will have the same ratio as the remaining sides of the triangle; and, if the segments of the base have the same ratio as the

*remaining sides of the triangle, the straight line joined from
the vertex to the point of section will bisect the angle of the
triangle.*

Let ABC be a triangle, and let the angle BAC be bisected
by the straight line AD;

I say that, as BD is to CD, so
is BA to AC.

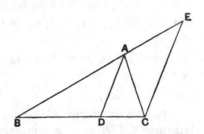

For let CE be drawn through
C parallel to DA, and let BA
be carried through and meet it
at E.

Then, since the straight line
AC falls upon the parallels AD,
EC,

     the angle ACE is equal to the angle CAD.   [I. 29]

But the angle CAD is by hypothesis equal to the angle
BAD;

therefore the angle BAD is also equal to the angle ACE.

Again, since the straight line BAE falls upon the parallels
AD, EC,

     the exterior angle BAD is equal to the interior angle
AEC.   [I. 29]

But the angle ACE was also proved equal to the angle
BAD;

     therefore the angle ACE is also equal to the angle AEC,

so that the side AE is also equal to the side AC.   [I. 6]

And, since AD has been drawn parallel to EC, one of
the sides of the triangle BCE,

therefore, proportionally, as BD is to DC, so is BA to AE.

                                                       [VI. 2]
But AE is equal to AC;

therefore, as BD is to DC, so is BA to AC.

Again, let BA be to AC as BD to DC, and let AD be
joined;

I say that the angle BAC has been bisected by the straight
line AD.

For, with the same construction,

since, as BD is to DC, so is BA to AC,

and also, as BD is to DC, so is BA to AE: for AD has been drawn parallel to EC, one of the sides of the triangle BCE: [VI. 2]

therefore also, as BA is to AC, so is BA to AE.   [V. 11]

Therefore AC is equal to AE,   [V. 9]
so that the angle AEC is also equal to the angle ACE. [I. 5]

But the angle AEC is equal to the exterior angle BAD, [I. 29]

and the angle ACE is equal to the alternate angle CAD; [id.]

therefore the angle BAD is also equal to the angle CAD.

Therefore the angle BAC has been bisected by the straight line AD.

Therefore etc.

Q. E. D.

The demonstration assumes that CE *will* meet BA produced in some point E. This is proved in the same way as it is proved in VI. 4 that BA, ED will meet if produced. The angles ABD, BDA in the figure of VI. 3 are together less than two right angles, and the angle BDA is equal to the angle BCE, since DA, CE are parallel. Therefore the angles ABC, BCE are together less than two right angles; and BA, CE must meet, by I. Post. 5.

The corresponding proposition about the segments into which BC is divided *externally* by the bisector of the *external angle* at A when that bisector meets BC produced (i.e. when the sides AB, AC are not equal) is important. Simson gives it as a separate proposition, A, noting the fact that Pappus assumes the result without proof (Pappus, VII. p. 730, 24).

The best plan is however, as De Morgan says, to combine Props. 3 and A in one proposition, which may be enunciated thus: *If an angle of a triangle be bisected internally or externally by a straight line which cuts the opposite side or the opposite side produced, the segments of that side will have the same ratio as the other sides of the triangle; and, if a side of a triangle be divided internally or externally so that its segments have the same ratio as the other sides of the triangle, the straight line drawn from the point of section to the angular point which is opposite to the first mentioned side will bisect the interior or exterior angle at that angular point.*

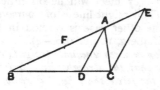 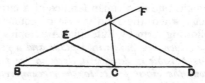

Let AC be the smaller of the two sides AB, AC, so that the bisector AD of the exterior angle at A may meet BC produced beyond C. Draw CE through C parallel to DA, meeting BA in E.

Then, if FAC is the exterior angle bisected by AD in the case of external bisection, and if a point F is taken on AB in the figure of VI. 3, the proof of

VI. 3 can be used almost word for the other case. We have only to speak of the angle "*FAC*" for the angle "*BAC*," and of the angle "*FAD*" for the angle "*BAD*" wherever they occur, to say "let *BA*, or *BA* produced, meet *CE* in *E*," and to substitute "*BA* or *BA* produced" for "*BAE*" lower down.

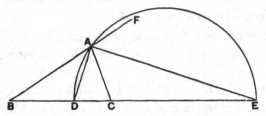

If *AD*, *AE* be the internal and external bisectors of the angle *A* in a triangle of which the sides *AB*, *AC* are unequal, *AC* being the smaller, and if *AD*, *AE* meet *BC* and *BC* produced in *D*, *E* respectively,

the ratios of *BD* to *DC* and of *BE* to *EC* are alike equal to the ratio of *BA* to *AC*.

Therefore   *BE* is to *EC* as *BD* to *DC*,

that is, *BE* is to *EC* as the difference between *BE* and *ED* is to the difference between *ED* and *EC*,

whence *BE*, *ED*, *EC* are in *harmonic progression*, or *DE* is a *harmonic mean* between *BE* and *EC*, or again *B*, *D*, *C*, *E* is a *harmonic range*.

Since the angle *DAC* is half of the angle *BAC*,

and the angle *CAE* half of the angle *CAF*,

while the angles *BAC*, *CAF* are equal to two right angles,

the angle *DAE* is a right angle.

Hence the circle described on *DE* as diameter passes through *A*.

Now, if the ratio of *BA* to *AC* is given, and if *BC* is given, the points *D*, *E* on *BC* and *BC* produced are given, and therefore so is the circle on *D*, *E* as diameter. Hence *the locus of a point such that its distances from two given points are in a given ratio (not being a ratio of equality) is a circle*.

This locus was discussed by Apollonius in his *Plane Loci*, Book II., as we know from Pappus (VII. p. 666), who says that the book contained the theorem that, if from two given points straight lines inflected to another point are in a given ratio, the point in which they meet will lie on either a straight line or a circumference of a circle. The straight line is of course the locus when the ratio is one of equality. The other case is quoted in the following form by Eutocius (Apollonius, ed. Heiberg, II. pp. 180—4).

Given two points in a plane and a proportion between unequal straight lines, it is possible to describe a circle in the plane so that the straight lines inflected from the given points to the circumference of the circle shall have a ratio the same as the given one.

Apollonius' construction, as given by Eutocius, is remarkable because he makes no use of either of the points *D*, *E*. He finds *O*, the centre of the required circle, and the length of its radius directly from the data *BC* and the given ratio which we will call *h : k*. But the construction was not discovered by Apollonius; it belongs to a much earlier date, since it appears in exactly the same form in Aristotle, *Meteorologica* III. 5, 376 a 3 sqq. The analysis leading up to the construction is, as usual, not given either by Aristotle or Eutocius. We are told to take three straight lines x, CO (a length measured along BC produced beyond C, where B, C are the points at which the greater and smaller of the inflected lines respectively terminate), and r, such that, if $h : k$ be the given ratio and $h > k$,

$$k : h = h : k + x, \quad\dots\dots\dots\dots\dots\dots\dots\dots\dots\dots(a)$$

$$x : BC = k : CO = h : r. \quad\dots\dots\dots\dots\dots\dots\dots(\beta)$$

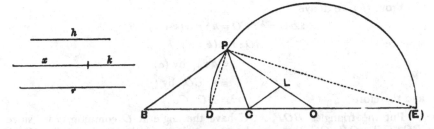

This determines the position of O, and the length of r, the radius of the required circle. The circle is then drawn, *any* point P is taken on it and joined to B, C respectively, and it is proved that

$$PB : PC = h : k.$$

We may conjecture that the analysis proceeded somewhat as follows.

It would be seen that B, C are "conjugate points" with reference to the circle on DE as diameter. (Cf. Apollonius, *Conics*, I. 36, where it is proved, in terms, for a circle as well as for an ellipse and a hyperbola, that, if the polar of B meets the diameter DE in C, then $EC : CD = EB : BD$.)

If O be the middle point of DE, and therefore the centre of the circle, D, E may be eliminated, as in the *Conics*, I. 37, thus.

Since $EC : CD = EB : BD$,

it follows that $EC + CD : EC \sim CD = EB + BD : EB \sim BD$,

or $2OD : 2OC = 2OB : 2OD$,

that is, $BO . OC = OD^2 = r^2$, say.

If therefore P be any point on the circle with centre O and radius r,

$$BO : OP = OP : OC,$$

so that BOP, POC are similar triangles.

In addition, $h : k = BD : DC = BE : EC$

$$= BD + BE : DE = BO : r.$$

Hence we require that

$$BO : r = r : OC = BP : PC = h : k. \quad\dots\dots\dots\dots (\delta)$$

Therefore, alternately,

$$k : CO = h : r,$$

which is the second relation in (β) above.

Now assume a length x such that each of the last ratios is equal to $x : BC$, as in (β).

Then $\qquad x : BC = k : CO = h : r.$

Therefore $\qquad x + k : BO = h : r,$

and, alternately, $\qquad x + k : h = BO : r$

$$= h : k, \text{ from } (\delta) \text{ above};$$

and this is the relation (a) which remained to be found.

Apollonius' proof of the construction is given by Eutocius, who begins by saying that it is manifest that r is a mean proportional between BO and OC. This is seen as follows:

From (β) we derive

$$x : BC = k : CO = h : r = (k + x) : BO,$$

whence $\qquad BO : r = (k + x) : h$

$$= h : k, \text{ by } (a),$$

$$= r : CO, \text{ by } (\beta),$$

and therefore $\qquad r^2 = BO . CO.$

But the triangles BOP, POC have the angle at O common, and, since $BO : OP = OP : OC$, the triangles are similar and the angles OPC, OBP are equal.

[Up to this point Aristotle's proof is exactly the same; from this point it diverges slightly.]

If now CL be drawn parallel to BP meeting OP in L, the angles BPC LCP are equal also.

Therefore the triangles BPC, PCL are similar, and

$$BP : PC = PC : CL,$$

whence $\qquad BP^2 : PC^2 = BP : CL$

$$= BO : OC, \text{ by parallels,}$$

$$= BO^2 : OP^2 \text{ (since } BO : OP = OP : OC).$$

Therefore $\qquad BP : PC = BO : OP$

$$= h : k \text{ (for } OP = r).$$

[Aristotle infers this more directly from the similar triangles POB, COP. Since these triangles are similar,

$$OP : CP = OB : BP,$$

whence $\qquad BP : PC = BO : OP$

$$= h : k.]$$

Apollonius proves lastly, by *reductio ad absurdum*, that the last equation cannot be true with reference to any point P which is not on the circle so described.

PROPOSITION 4.

In equiangular triangles the sides about the equal angles are proportional, and those are corresponding sides which subtend the equal angles.

Let *ABC*, *DCE* be equiangular triangles having the angle *ABC* equal to the angle *DCE*, the angle *BAC* to the angle *CDE*, and further the angle *ACB* to the angle *CED*;

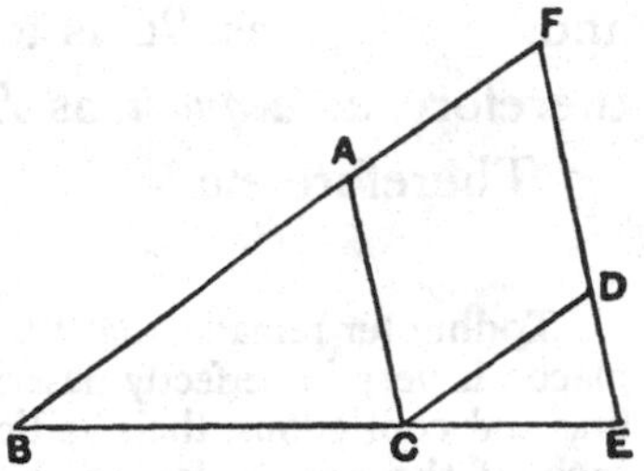

I say that in the triangles *ABC*, *DCE* the sides about the equal angles are proportional, and those are corresponding sides which subtend the equal angles.

For let *BC* be placed in a straight line with *CE*.

Then, since the angles *ABC*, *ACB* are less than two right angles,   [I. 17]

and the angle *ACB* is equal to the angle *DEC*,

therefore the angles *ABC*, *DEC* are less than two right angles;

therefore *BA*, *ED*, when produced, will meet.   [I. Post. 5]

Let them be produced and meet at *F*.

Now, since the angle *DCE* is equal to the angle *ABC*,

BF is parallel to *CD*.   [I. 28]

Again, since the angle *ACB* is equal to the angle *DEC*,

AC is parallel to *FE*.   [I. 28]

Therefore *FACD* is a parallelogram;

therefore *FA* is equal to *DC*, and *AC* to *FD*.   [I. 34]

And, since *AC* has been drawn parallel to *FE*, one side of the triangle *FBE*,

therefore, as *BA* is to *AF*, so is *BC* to *CE*.   [VI. 2]

But *AF* is equal to *CD*;

therefore, as *BA* is to *CD*, so is *BC* to *CE*,

and alternately, as *AB* is to *BC*, so is *DC* to *CE*.   [V. 16]

Again, since *CD* is parallel to *BF*,

therefore, as *BC* is to *CE*, so is *FD* to *DE*.   [VI. 2]

But *FD* is equal to *AC*;

therefore, as *BC* is to *CE*, so is *AC* to *DE*,

and alternately, as *BC* is to *CA*, so is *CE* to *ED*.   [V. 16]

Since then it was proved that,

as AB is to BC, so is DC to CE,

and,   as BC is to CA, so is CE to ED;

therefore, *ex aequali*, as BA is to AC, so is CD to DE. [v. 22]

Therefore etc.

Q. E. D.

Todhunter remarks that "the manner in which the two triangles are to be placed is very imperfectly described; their bases are to be in the same straight line and contiguous, their vertices are to be on the same side of the base, and each of the two angles which have a common vertex is to be equal to the remote angle of the other triangle." But surely Euclid's description is sufficient, except for not saying that B and D must be on the same side of BCE.

VI. 4 can be immediately deduced from VI. 2 if we superpose one triangle on the other three times in succession, so that each angle successively coincides with its equal, the triangles being similarly situated, e.g. if (A, B, C and D, E, F being the equal angles respectively) we apply the angle DEF to the angle ABC so that D lies on AB (produced if necessary) and F on BC (produced if necessary). De Morgan prefers this method. "Abandon," he says, "the peculiar mode of construction by which Euclid proves two cases at once; make an angle coincide with its equal, and suppose this process repeated three times, one for each angle."

PROPOSITION 5.

If two triangles have their sides proportional, the triangles will be equiangular and will have those angles equal which the corresponding sides subtend.

Let ABC, DEF be two triangles having their sides proportional, so that,

as AB is to BC, so is DE to EF,

as BC is to CA, so is EF to FD,

and further, as BA is to AC, so is ED to DF;

I say that the triangle ABC is equiangular with the triangle DEF, and they will have those angles equal which the corresponding sides subtend, namely the angle ABC to the angle DEF, the angle BCA to the angle EFD, and further the angle BAC to the angle EDF.

For on the straight line EF, and at the points E, F on it, let there be constructed the angle FEG equal to the angle ABC, and the angle EFG equal to the angle ACB;   [I. 23] therefore the remaining angle at A is equal to the remaining angle at G.   [I. 32]

Therefore the triangle ABC is equiangular with the triangle GEF.

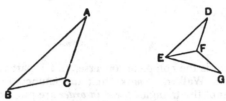

Therefore in the triangles ABC, GEF the sides about the equal angles are proportional, and those are corresponding sides which subtend the equal angles;   [VI. 4]

therefore, as AB is to BC, so is GE to EF.

But, as AB is to BC, so by hypothesis is DE to EF;

therefore, as DE is to EF, so is GE to EF.   [V. 11]

Therefore each of the straight lines DE, GE has the same ratio to EF;

therefore DE is equal to GE.   [V. 9]

For the same reason

DF is also equal to GF.

Since then DE is equal to EG,

and EF is common,

the two sides DE, EF are equal to the two sides GE, EF;

and the base DF is equal to the base FG;

therefore the angle DEF is equal to the angle GEF,   [I. 8]

and the triangle DEF is equal to the triangle GEF,

and the remaining angles are equal to the remaining angles, namely those which the equal sides subtend.   [I. 4]

Therefore the angle DFE is also equal to the angle GFE,

and the angle EDF to the angle EGF.

And, since the angle FED is equal to the angle GEF,

while the angle GEF is equal to the angle ABC,

therefore the angle ABC is also equal to the angle DEF.

For the same reason

the angle ACB is also equal to the angle DFE,
and further, the angle at A to the angle at D;

therefore the triangle ABC is equiangular with the triangle DEF.

Therefore etc.

Q. E. D.

This proposition is the complete converse, VI. 6 a partial converse, of VI. 4.

Todhunter, after Walker, remarks that the enunciation should make it clear that the sides of the triangles *taken in order* are proportional. It is quite possible that there should be two triangles ABC, DEF such that

AB is to BC as DE to EF,
and
BC is to CA as DF is to ED (instead of EF to FD),
so that
AB is to AC as DF to EF
(*ex aequali* in *perturbed proportion*);

in this case the sides of the triangles are proportional, but not in the same order, and the triangles are not necessarily equiangular to one another. For a numerical illustration we may suppose the sides of one triangle to be 3, 4 and 5 feet respectively, and those of another to be 12, 15 and 20 feet respectively.

In VI. 5 there is the same *apparent* avoidance of indirect demonstration which has been noticed on I. 48.

PROPOSITION 6.

If two triangles have one angle equal to one angle and the sides about the equal angles proportional, the triangles will be equiangular and will have those angles equal which the corresponding sides subtend.

Let ABC, DEF be two triangles having one angle BAC equal to one angle EDF and the sides about the equal angles proportional, so that,

as BA is to AC, so is ED to DF;

I say that the triangle ABC is equiangular with the triangle DEF, and will have the angle ABC equal to the angle DEF, and the angle ACB to the angle DFE.

For on the straight line DF, and at the points D, F on it, let there be constructed the angle FDG equal to either of the angles BAC, EDF, and the angle DFG equal to the angle ACB;   [I. 23]

therefore the remaining angle at B is equal to the remaining angle at G.   [I. 32]

Therefore the triangle ABC is equiangular with the triangle DGF.

Therefore, proportionally, as BA is to AC, so is GD to DF.   [VI. 4]

But, by hypothesis, as BA is to AC, so also is ED to DF; therefore also, as ED is to DF, so is GD to DF.   [V. 11]

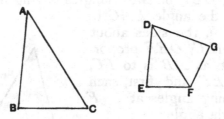

Therefore ED is equal to DG;   [V. 9]
and DF is common;

therefore the two sides ED, DF are equal to the two sides GD, DF; and the angle EDF is equal to the angle GDF;

therefore the base EF is equal to the base GF,

and the triangle DEF is equal to the triangle DGF,

and the remaining angles will be equal to the remaining angles, namely those which the equal sides subtend.   [I. 4]

Therefore the angle DFG is equal to the angle DFE,

and the angle DGF to the angle DEF.

But the angle DFG is equal to the angle ACB; therefore the angle ACB is also equal to the angle DFE.

And, by hypothesis, the angle BAC is also equal to the angle EDF;

therefore the remaining angle at B is also equal to the remaining angle at E;   [I. 32]

therefore the triangle ABC is equiangular with the triangle DEF.

Therefore etc.

Q. E. D.

PROPOSITION 7.

If two triangles have one angle equal to one angle, the sides about other angles proportional, and the remaining angles either both less or both not less than a right angle, the triangles will be equiangular and will have those angles equal, the sides about which are proportional.

Let *ABC*, *DEF* be two triangles having one angle equal to one angle, the angle *BAC* to the angle *EDF*, the sides about other angles *ABC*, *DEF* proportional, so that, as *AB* is to *BC*, so is *DE* to *EF*, and, first, each of the remaining angles at *C*, *F* less than a right angle ;

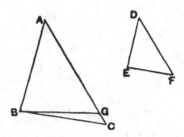

I say that the triangle *ABC* is equiangular with the triangle *DEF*, the angle *ABC* will be equal to the angle *DEF*, and the remaining angle, namely the angle at *C*, equal to the remaining angle, the angle at *F*.

For, if the angle *ABC* is unequal to the angle *DEF*, one of them is greater.

Let the angle *ABC* be greater ;

and on the straight line *AB*, and at the point *B* on it, let the angle *ABG* be constructed equal to the angle *DEF*.   [I. 23]

Then, since the angle *A* is equal to *D*,

and the angle *ABG* to the angle *DEF*,

therefore the remaining angle *AGB* is equal to the remaining angle *DFE*.   [I. 32]

Therefore the triangle *ABG* is equiangular with the triangle *DEF*.

Therefore, as *AB* is to *BG*, so is *DE* to *EF*.   [VI. 4]

But, as *DE* is to *EF*, so by hypothesis is *AB* to *BC* ;

therefore *AB* has the same ratio to each of the straight lines *BC*, *BG* ;   [V. 11]

therefore *BC* is equal to *BG*,   [V. 9]

so that the angle at *C* is also equal to the angle *BGC*.   [I. 5]

But, by hypothesis, the angle at C is less than a right angle;

therefore the angle BGC is also less than a right angle;

so that the angle AGB adjacent to it is greater than a right angle.   [I. 13]

And it was proved equal to the angle at F;

therefore the angle at F is also greater than a right angle.

But it is by hypothesis less than a right angle: which is absurd.

Therefore the angle ABC is not unequal to the angle DEF;

therefore it is equal to it.

But the angle at A is also equal to the angle at D;

therefore the remaining angle at C is equal to the remaining angle at F.   [I. 32]

Therefore the triangle ABC is equiangular with the triangle DEF.

But, again, let each of the angles at C, F be supposed not less than a right angle;

I say again that, in this case too, the triangle ABC is equiangular with the triangle DEF.

For, with the same construction, we can prove similarly that

BC is equal to BG;

so that the angle at C is also equal to the angle BGC.   [I. 5]

But the angle at C is not less than a right angle;

therefore neither is the angle BGC less than a right angle.

Thus in the triangle BGC the two angles are not less than two right angles: which is impossible.   [I. 17]

Therefore, once more, the angle ABC is not unequal to the angle DEF;

therefore it is equal to it.

But the angle at A is also equal to the angle at D;

therefore the remaining angle at C is equal to the remaining angle at F.   [I. 32]

Therefore the triangle ABC is equiangular with the triangle DEF.

Therefore etc.

Q. E. D.

Todhunter points out, after Walker, that some more words are necessary to make the enunciation precise: "If two triangles have one angle equal to one angle, the sides about other angles proportional *<so that the sides subtending the equal angles are homologous>*...."

This proposition is the extension to similar triangles of the *ambiguous case* already mentioned as omitted by Euclid in relation to *equality* of triangles in all respects (cf. note following I. 26, Vol. 1. p. 306). The enunciation of VI. 7 has suggested the ordinary method of enunciating the *ambiguous case* where *equality* and not similarity is in question. Cf. Todhunter's note on I. 26.

Another possible way of presenting this proposition is given by Todhunter. The essential theorem to prove is:

If two triangles have two sides of the one proportional to two sides of the other, and the angles opposite to one pair of corresponding sides equal, the angles which are opposite to the other pair of corresponding sides shall either be equal or be together equal to two right angles.

For the angles included by the proportional sides must be either equal or unequal.

If they are equal, then, since the triangles have two angles of the one equal to two angles of the other, respectively, they are equiangular to one another.

We have therefore only to consider the case in which the angles included by the proportional sides are unequal.

The proof is, except at the end, like that of VI. 7.

Let the triangles ABC, DEF have the angle at A equal to the angle at D;

let AB be to BC as DE to EF,

but let the angle ABC be *not* equal to the angle DEF.

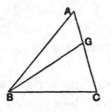 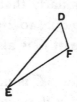

The angles ACB, DFE shall be together equal to two right angles.

For one of the angles ABC, DEF must be the greater.

Let ABC be the greater; and make the angle ABG equal to the angle DEF.

Then we prove, as in VI. 7, that the triangles ABG, DEF are equiangular, whence

AB is to BG as DE is to EF.

But AB is to BC as DE is to EF, by hypothesis.

Therefore BG is equal to BC,

and the angle BGC is equal to the angle BCA.

Now, since the triangles *ABG*, *DEF* are equiangular,

the angle *BGA* is equal to the angle *EFD*.

Add to them respectively the equal angles *BGC*, *BCA*; therefore the angles *BCA*, *EFD* are together equal to the angles *BGA*, *BGC*, i.e. to two right angles.

It follows therefore that the angles *BCA*, *EFD* must be either equal or supplementary.

But (1), if each of them is less than a right angle, they cannot be supplementary, and they must therefore be equal;

(2) if each of them is greater than a right angle, they cannot be supplementary and must therefore be equal;

(3) if one of them is a right angle, they are supplementary and also equal.

Simson distinguishes the last case (3) in his enunciation: "then, if each of the remaining angles be either less or not less than a right angle, *or if one of them be a right angle....*"

The change is right, on the principle of restricting the conditions to the minimum necessary to enable the conclusion to be inferred. Simson adds a separate proof of the case in which one of the remaining angles is a right angle.

"Lastly, let one of the angles at *C*, *F*, viz. the angle at *C*, be a right angle; in this case likewise the triangle *ABC* is equiangular to the triangle *DEF*.

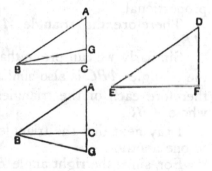

For, if they be not equiangular, make, at the point *B* of the straight line *AB*, the angle *ABG* equal to the angle *DEF*; then it may be proved, as in the first case, that *BG* is equal to *BC*.

But the angle *BCG* is a right angle;

therefore the angle *BGC* is also a right angle;

whence two of the angles of the triangle *BGC* are together not less than two right angles: which is impossible.

Therefore the triangle *ABC* is equiangular to the triangle *DEF*."

PROPOSITION 8.

If in a right-angled triangle a perpendicular be drawn from the right angle to the base, the triangles adjoining the perpendicular are similar both to the whole and to one another.

Let *ABC* be a right-angled triangle having the angle *BAC* right, and let *AD* be drawn from *A* perpendicular to *BC*;

I say that each of the triangles *ABD*, *ADC* is similar to the whole *ABC* and, further, they are similar to one another.

For, since the angle BAC is equal to the angle ADB, for each is right,

and the angle at B is common to the two triangles ABC and ABD,

therefore the remaining angle ACB is equal to the remaining angle BAD;   [I. 32]

therefore the triangle ABC is equiangular with the triangle ABD.

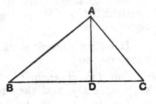

Therefore, as BC which subtends the right angle in the triangle ABC is to BA which subtends the right angle in the triangle ABD, so is AB itself which subtends the angle at C in the triangle ABC to BD which subtends the equal angle BAD in the triangle ABD, and so also is AC to AD which subtends the angle at B common to the two triangles.

[VI. 4]

Therefore the triangle ABC is both equiangular to the triangle ABD and has the sides about the equal angles proportional.

Therefore the triangle ABC is similar to the triangle ABD.   [VI. Def. 1]

Similarly we can prove that

the triangle ABC is also similar to the triangle ADC;

therefore each of the triangles ABD, ADC is similar to the whole ABC.

I say next that the triangles ABD, ADC are also similar to one another.

For, since the right angle BDA is equal to the right angle ADC,

and moreover the angle BAD was also proved equal to the angle at C,

therefore the remaining angle at B is also equal to the remaining angle DAC;   [I. 32]

therefore the triangle ABD is equiangular with the triangle ADC.

Therefore, as BD which subtends the angle BAD in the triangle ABD is to DA which subtends the angle at C in the triangle ADC equal to the angle BAD, so is AD itself which subtends the angle at B in the triangle ABD to DC which subtends the angle DAC in the triangle ADC equal

to the angle at B, and so also is BA to AC, these sides subtending the right angles;   [VI. 4]
therefore the triangle ABD is similar to the triangle ADC.
                                                        [VI. Def. 1]

Therefore etc.

PORISM. From this it is clear that, if in a right-angled triangle a perpendicular be drawn from the right angle to the base, the straight line so drawn is a mean proportional between the segments of the base.   Q. E. D.

Simson remarks on this proposition: "It seems plain that some editor has changed the demonstration that Euclid gave of this proposition: For, after he has demonstrated that the triangles are equiangular to one another, he particularly shows that their sides about the equal angles are proportionals, as if this had not been done in the demonstration of prop. 4 of this book: this superfluous part is not found in the translation from the Arabic, and is now left out."

This seems a little hypercritical, for the "particular showing" that the sides about the equal angles are proportionals is really nothing more than a somewhat full citation of VI. 4. Moreover to shorten his proof still more, Simson says, after proving that each of the triangles ABD, ADC is similar to the whole triangle ABC, "And the triangles ABD, ADC being both equiangular and similar to ABC are equiangular and similar to one another," thus assuming a particular case of VI. 21, which might well be proved here, as Euclid proves it, with somewhat more detail.

We observe that, here as generally, Euclid seems to disdain to give the reader such small help as might be afforded by arranging the letters used to denote the triangles so as to show the corresponding angular points in the same order for each pair of triangles; A is the first letter throughout, and the other two for each triangle are in the order of the figure from left to right. It may be in compensation for this that he states at such length which side corresponds to which when he comes to the proportions.

In the Greek texts there is an addition to the Porism inserted after "(Being) what it was required to prove," viz. "and further that between the base and any one of the segments the side adjacent to the segment is a mean proportional." Heiberg concludes that these words are an interpolation (1) because they come after the words ὅπερ ἔδει δεῖξαι which as a rule follow the Porism, (2) they are absent from the best Theonine MSS., though P and Campanus have them without the ὅπερ ἔδει δεῖξαι. Heiberg's view seems to be confirmed by the fact noted by Austin, that, whereas the first part of the Porism is quoted later in VI. 13, in the lemma before X. 33 and in the lemma after XIII. 13, the second part is *proved* in the former lemma, and elsewhere, as also in Pappus (III. p. 72, 9—23).

PROPOSITION 9.

From a given straight line to cut off a prescribed part.

Let AB be the given straight line;
thus it is required to cut off from AB a prescribed part.

Let the third part be that prescribed.

5   Let a straight line AC be drawn through from A containing with AB any angle ;

let a point D be taken at random on AC, and let DE, EC be made equal to AD.   [I. 3]

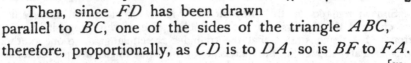

10   Let BC be joined, and through D let DF be drawn parallel to it.   [I. 31]

Then, since FD has been drawn parallel to BC, one of the sides of the triangle ABC,

therefore, proportionally, as CD is to DA, so is BF to FA.

[VI. 2]

15   But CD is double of DA ;

therefore BF is also double of FA ;

therefore BA is triple of AF.

Therefore from the given straight line AB the prescribed third part AF has been cut off.

Q. E. F.

6. **any angle.** The expression here and in the two following propositions is τυχοῦσα γωνία, corresponding exactly to τυχὸν σημεῖον which I have translated as "a point (taken) *at random*"; but "an angle (taken) at random" would not be so appropriate where it is a question, not of *taking* any angle at all, but of drawing a straight line casually so as to make any angle with another straight line.

Simson observes that "this is demonstrated in a particular case, viz. that in which the third part of a straight line is required to be cut off; which is not at all like Euclid's manner. Besides, the author of that demonstration, from four magnitudes being proportionals, concludes that the third of them is the same multiple of the fourth which the first is of the second ; now this is nowhere demonstrated in the 5th book, as we now have it ; but the editor assumes it from the confused notion which the vulgar have of proportionals."

The truth of the assumption referred to is proved by Simson in his proposition D given above (p. 128); hence he is able to supply a general and legitimate proof of the present proposition.

"Let AB be the given straight line; it is required to cut off any part from it.

From the point A draw a straight line AC making any angle with AB; in AC take any point D, and take AC the same multiple of AD that AB is of the part which is to be cut off from it;

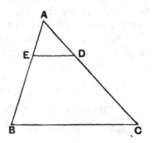

join BC, and draw DE parallel to it :

then AE is the part required to be cut off.

Because *ED* is parallel to one of the sides of the triangle *ABC*, viz. to *BC*,

　　　　　as *CD* is to *DA*, so is *BE* to *EA*,　　　　　[VI. 2]

and, *componendo*,

　　　　　CA is to *AD*, as *BA* to *AE*.　　　　　[V. 18]

But *CA* is a multiple of *AD*;

therefore *BA* is the same multiple of *AE*.　　　　　[Prop. D]

Whatever part therefore *AD* is of *AC*, *AE* is the same part of *AB*;

wherefore from the straight line *AB* the part required is cut off."

The use of Simson's Prop. D can be avoided, as noted by Camerer after Baermann, in the following way.　We first prove, as above, that

　　　　　CA is to *AD* as *BA* is to *AE*.

Then we infer that, alternately,

　　　　　CA is to *BA* as *AD* to *AE*.　　　　　[V. 16]

But *AD* is to *AE* as *n . AD* to *n . AE*

(where *n* is the number of times that *AD* is contained in *AC*);　　　　　[V. 15]

　　　　　whence *AC* is to *AB* as *n . AD* is to *n . AE*.　　　　　[V. 11]

In this proportion the first term is equal to the third; therefore [V. 14] the second is equal to the fourth,

　　　　　so that *AB* is equal to *n* times *AE*.

Prop. 9 is of course only a particular case of Prop. 10.

PROPOSITION 10.

To cut a given uncut straight line similarly to a given cut straight line.

Let *AB* be the given uncut straight line, and *AC* the straight line cut at the points *D*, *E*; and let them be so placed as to contain any angle; let *CB* be joined, and through *D*, *E* let *DF*, *EG* be drawn parallel to *BC*, and through *D* let *DHK* be drawn parallel to *AB*.　　　[I. 31]

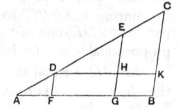

Therefore each of the figures *FH*, *HB* is a parallelogram;

therefore *DH* is equal to *FG* and *HK* to *GB*.　　　　　[I. 34]

Now, since the straight line *HE* has been drawn parallel to *KC*, one of the sides of the triangle *DKC*,

therefore, proportionally, as *CE* is to *ED*, so is *KH* to *HD*.

　　　　　[VI. 2]

But KH is equal to BG, and HD to GF;
therefore, as CE is to ED, so is BG to GF.

Again, since FD has been drawn parallel to GE, one of the sides of the triangle AGE,
therefore, proportionally, as ED is to DA, so is GF to FA.

[VI. 2]

But it was also proved that,
 as CE is to ED, so is BG to GF;
therefore, as CE is to ED, so is BG to GF,
 and, as ED is to DA, so is GF to FA.

Therefore the given uncut straight line AB has been cut similarly to the given cut straight line AC.

Q. E. F.

PROPOSITION 11.

To two given straight lines to find a third proportional.

Let BA, AC be the two given straight lines, and let them be placed so as to contain any angle;
thus it is required to find a third proportional to BA, AC.

For let them be produced to the points D, E, and let BD be made equal to AC;   [I. 3]
let BC be joined, and through D let DE be drawn parallel to it.   [I. 31]

Since, then, BC has been drawn parallel to DE, one of the sides of the triangle ADE,
proportionally, as AB is to BD, so is AC to CE. [VI. 2]

But BD is equal to AC;
therefore, as AB is to AC, so is AC to CE.

Therefore to two given straight lines AB, AC a third proportional to them, CE, has been found.

Q. E. F.

1. **to find.** The Greek word, here and in the next two propositions, is προσευρεῖν, literally "to find *in addition*."

This proposition is again a particular case of the succeeding Prop. 12.

Given a ratio between straight lines, VI. 11 enables us to find the ratio which is its duplicate.

PROPOSITION 12.

To three given straight lines to find a fourth proportional.

Let *A*, *B*, *C* be the three given straight lines;
thus it is required to find a fourth proportional to *A*, *B*, *C*.

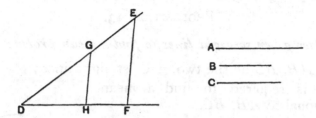

Let two straight lines *DE*, *DF* be set out containing any angle *EDF*;
let *DG* be made equal to *A*, *GE* equal to *B*, and further *DH* equal to *C*;
let *GH* be joined, and let *EF* be drawn through *E* parallel to it.   [I. 31]

Since, then, *GH* has been drawn parallel to *EF*, one of the sides of the triangle *DEF*,
therefore, as *DG* is to *GE*, so is *DH* to *HF*.   [VI. 2]

But *DG* is equal to *A*, *GE* to *B*, and *DH* to *C*;
therefore, as *A* is to *B*, so is *C* to *HF*.

Therefore to the three given straight lines *A*, *B*, *C* a fourth proportional *HF* has been found.

Q. E. F.

We have here the geometrical equivalent of the "rule of three."

It is of course immaterial whether, as in Euclid's proof, the first and second straight lines are measured on one of the lines forming the angle and the third on the other, or the first and third are measured on one and the second on the other.

If it should be desired that the first and the required fourth be measured on one of the lines, and the second and third on the other, we can use the following construction. Measure *DE* on one straight line equal to *A*, and on any other straight line making an angle with the first at the point *D* measure *DF* equal to *B*, and *DG* equal to *C*. Join *EF*, and through *G* draw *GH anti-parallel* to *EF*, i.e. make the angle *DGH* equal to the angle *DEF*; let *GH* meet *DE* (produced if necessary) in *H*.

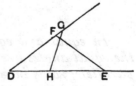

DH is then the fourth proportional.

For the triangles *EDF*, *GDH* are similar, and the sides about the equal angles are proportional, so that

$$DE \text{ is to } DF \text{ as } DG \text{ to } DH,$$

or   *A* is to *B* as *C* to *DH*.

PROPOSITION 13.

To two given straight lines to find a mean proportional.

Let *AB*, *BC* be the two given straight lines; thus it is required to find a mean proportional to *AB*, *BC*.

Let them be placed in a straight line, and let the semicircle *ADC* be described on *AC*;

let *BD* be drawn from the point *B* at right angles to the straight line *AC*, and let *AD*, *DC* be joined.

Since the angle *ADC* is an angle in a semicircle, it is right.   [III. 31]

And, since, in the right-angled triangle *ADC*, *DB* has been drawn from the right angle perpendicular to the base,

therefore *DB* is a mean proportional between the segments of the base, *AB*, *BC*.   [VI. 8, Por.]

Therefore to the two given straight lines *AB*, *BC* a mean proportional *DB* has been found.

Q. E. F.

This proposition, the Book VI. version of II. 14, is equivalent to the extraction of the square root. It further enables us, given a ratio between straight lines, to find the ratio which is its *sub-duplicate*, or the ratio of which it is duplicate.

PROPOSITION 14.

In equal and equiangular parallelograms the sides about the equal angles are reciprocally proportional; and equiangular parallelograms in which the sides about the equal angles are reciprocally proportional are equal.

Let *AB*, *BC* be equal and equiangular parallelograms having the angles at *B* equal, and let *DB*, *BE* be placed in a straight line ;

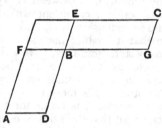

therefore *FB*, *BG* are also in a straight line.   [I. 14]

I say that, in *AB*, *BC*, the sides about the equal angles are reciprocally proportional, that is to say, that, as *DB* is to *BE*, so is *GB* to *BF*.

For let the parallelogram *FE* be completed.

Since, then, the parallelogram *AB* is equal to the parallelogram *BC*,

and *FE* is another area,

therefore, as *AB* is to *FE*, so is *BC* to *FE*.   [v. 7]

But, as *AB* is to *FE*, so is *DB* to *BE*,   [VI. 1]

and, as *BC* is to *FE*, so is *GB* to *BF*.   [*id.*]

therefore also, as *DB* is to *BE*, so is *GB* to *BF*.   [v. 11]

Therefore in the parallelograms *AB*, *BC* the sides about the equal angles are reciprocally proportional.

Next, let *GB* be to *BF* as *DB* to *BE* ;

I say that the parallelogram *AB* is equal to the parallelogram *BC*.

For since, as *DB* is to *BE*, so is *GB* to *BF*,

while, as *DB* is to *BE*, so is the parallelogram *AB* to the parallelogram *FE*,   [VI. 1]

and, as *GB* is to *BF*, so is the parallelogram *BC* to the parallelogram *FE*,   [VI. 1]

therefore also, as *AB* is to *FE*, so is *BC* to *FE* ;   [v. 11]

therefore the parallelogram *AB* is equal to the parallelogram *BC*.   [v. 9]

Therefore etc.

                              Q. E. D.

De Morgan says upon this proposition : "Owing to the disjointed manner in which Euclid treats compound ratio, this proposition is strangely out of place. It is a particular case of VI. 23, being that in which the ratio of the sides, compounded, gives a ratio of equality. The proper definition of four magnitudes being reciprocally proportional is that the ratio compounded of their ratios is that of equality."

It is true that VI. 14 is a particular case of VI. 23, but, if either is out of place, it is rather the latter that should be placed before VI. 14, since most of the propositions between VI. 15 and VI. 23 depend upon VI. 14 and 15. But it is perfectly consistent with Euclid's manner to give a particular case first and its extension later, and such an arrangement often has great advantages in that it enables the more difficult parts of a subject to be led up to more easily and gradually. Now, if De Morgan's view were here followed, we should, as it seems to me, be committing the mistake of explaining what is relatively easy to understand, viz. two ratios of which one is the inverse of the other, by a more complicated conception, that of compound ratio. In other words, it is easier for a learner to realise the relation indicated by the statement that the sides of equal and equiangular parallelograms are "reciprocally proportional" than to form a conception of parallelograms such that "the ratio compounded of the ratio of their sides is one of equality." For this reason I would adhere to Euclid's arrangement.

The conclusion that, since DB, BE are placed in a straight line, FB, BG are also in a straight line is referred to I. 14. The deduction is made clearer by the following steps.

The angle DBF is equal to the angle GBE;

add to each the angle FBE;

therefore the angles DBF, FBE are together equal to the angles GBE, FBE.
[C. N. 2]

But the angles DBF, FBE are together equal to two right angles,   [I. 13]

therefore the angles GBE, FBE are together equal to two right angles,
[C. N. 1]

and hence FB, BG are in one straight line.   [I. 14]

The result is also obvious from the converse of I. 15 given by Proclus (see note on I. 15).

The proposition VI. 14 contains a theorem and one partial converse of it; so also does VI. 15. To each proposition may be added the other partial converse, which may be enunciated as follows, the words in square brackets applying to the case of triangles (VI. 15).

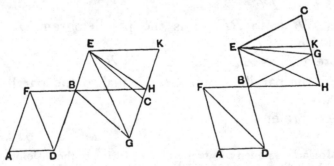

Equal parallelograms [triangles] which have the sides about one angle in each reciprocally proportional are equiangular [have the angles included by those sides either equal or supplementary.]

Let AB, BC be equal parallelograms, or let FBD, EBG be equal triangles, such that the sides about the angles at B are reciprocally proportional, i.e. such that

$$DB : BE = GB : BF.$$

We shall prove that the angles FBD, EBG are either equal or supplementary.

Place the figures so that DB, BE are in one straight line.

Then FB, BG are either in a straight line, or not in a straight line.

(1) If FB, BG are in a straight line, the figure of the proposition (with the diagonals FD, EG drawn) represents the facts, and

the angle FBD is equal to the angle EBG.   [I. 15]

(2) If FB, BG are not in a straight line,
produce FB to H so that BH may be equal to BG.

Join EH, and complete the parallelogram $EBHK$.

Now, since $DB : BE = GB : BF$
and $GB = HB$,
                                 $DB : BE = HB : BF$,

and therefore, by VI. 14 or 15,
the parallelograms AB, BK are equal, or the triangles FBD, EBH are equal.

But the parallelograms AB, BC are equal, and the triangles FBD, EBG are equal;

therefore the parallelograms BC, BK are equal, and the triangles EBH, EBG are equal.

Therefore these parallelograms or triangles are within the same parallels:
that is, G, C, H, K are in a straight line which is parallel to DE.   [I. 39]

Now, since BG, BH are equal,
the angles BGH, BHG are equal.

By parallels, it follows that

the angle EBG is equal to the angle DBH,

whence the angle EBG is supplementary to the angle FBD.

PROPOSITION 15.

In equal triangles which have one angle equal to one angle the sides about the equal angles are reciprocally proportional; and those triangles which have one angle equal to one angle, and in which the sides about the equal angles are reciprocally proportional, are equal.

Let ABC, ADE be equal triangles having one angle equal to one angle, namely the angle BAC to the angle DAE;

I say that in the triangles ABC, ADE the sides about the equal angles are reciprocally proportional, that is to say, that,

as CA is to AD, so is EA to AB.

For let them be placed so that CA is in a straight line with AD;

therefore EA is also in a straight line with AB. [I. 14]

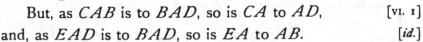

Let BD be joined.

Since then the triangle ABC is equal to the triangle ADE, and BAD is another area,

therefore, as the triangle CAB is to the triangle BAD, so is the triangle EAD to the triangle BAD. [V. 7]

But, as CAB is to BAD, so is CA to AD, [VI. 1]

and, as EAD is to BAD, so is EA to AB. [*id.*]

Therefore also, as CA is to AD, so is EA to AB. [V. 11]

Therefore in the triangles ABC, ADE the sides about the equal angles are reciprocally proportional.

Next, let the sides of the triangles ABC, ADE be reciprocally proportional, that is to say, let EA be to AB as CA to AD;

I say that the triangle ABC is equal to the triangle ADE.

For, if BD be again joined,

since, as CA is to AD, so is EA to AB,

while, as CA is to AD, so is the triangle ABC to the triangle BAD,

and, as EA is to AB, so is the triangle EAD to the triangle BAD, [VI. 1]

therefore, as the triangle ABC is to the triangle BAD, so is the triangle EAD to the triangle BAD. [V. 11]

Therefore each of the triangles ABC, EAD has the same ratio to BAD.

Therefore the triangle ABC is equal to the triangle EAD. [V. 9]

Therefore etc.

Q. E. D.

As indicated in the partial converse given in the last note, this proposition is equally true if the angle included by the two sides in one triangle is supplementary, instead of being equal, to the angle included by the two sides in the other.

Let *ABC*, *ADE* be two triangles such that the angles *BAC*, *DAE* are supplementary, and also

$$CA : AD = EA : AB.$$

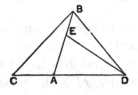

In this case we can place the triangles so that *CA* is in a straight line with *AD*, and *AB* lies along *AE* (since the angle *EAC*, being supplementary to the angle *EAD*, is equal to the angle *BAC*).

If we join *BD*, the proof given by Euclid applies to this case also.

It is true that VI. 15 can be immediately inferred from VI. 14, since a triangle is half of a parallelogram with the same base and height. But, Euclid's object being to give the student a grasp of *methods* rather than results, there seems to be no advantage in deducing one proposition from the other instead of using the same method on each.

PROPOSITION 16.

If four straight lines be proportional, the rectangle contained by the extremes is equal to the rectangle contained by the means; and, if the rectangle contained by the extremes be equal to the rectangle contained by the means, the four straight lines will be proportional.

Let the four straight lines *AB*, *CD*, *E*, *F* be proportional, so that, as *AB* is to *CD*, so is *E* to *F*;

I say that the rectangle contained by *AB*, *F* is equal to the rectangle contained by *CD*, *E*.

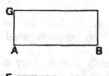

Let *AG*, *CH* be drawn from the points *A*, *C* at right angles to the straight lines *AB*, *CD*, and let *AG* be made equal to *F*, and *CH* equal to *E*.

Let the parallelograms *BG*, *DH* be completed.

Then since, as *AB* is to *CD*, so is *E* to *F*,

while *E* is equal to *CH*, and *F* to *AG*,

therefore, as *AB* is to *CD*, so is *CH* to *AG*.

Therefore in the parallelograms *BG*, *DH* the sides about the equal angles are reciprocally proportional.

But those equiangular parallelograms in which the sides about the equal angles are reciprocally proportional are equal;

[VI. 14]

therefore the parallelogram *BG* is equal to the parallelogram *DH*.

And *BG* is the rectangle *AB*, *F*, for *AG* is equal to *F*; and *DH* is the rectangle *CD*, *E*, for *E* is equal to *CH*; therefore the rectangle contained by *AB*, *F* is equal to the rectangle contained by *CD*, *E*.

Next, let the rectangle contained by *AB*, *F* be equal to the rectangle contained by *CD*, *E*; I say that the four straight lines will be proportional, so that, as *AB* is to *CD*, so is *E* to *F*.

For, with the same construction, since the rectangle *AB*, *F* is equal to the rectangle *CD*, *E*, and the rectangle *AB*, *F* is *BG*, for *AG* is equal to *F*, and the rectangle *CD*, *E* is *DH*, for *CH* is equal to *E*,

therefore *BG* is equal to *DH*.

And they are equiangular

But in equal and equiangular parallelograms the sides about the equal angles are reciprocally proportional.   [VI. 14]

Therefore, as *AB* is to *CD*, so is *CH* to *AG*.

But *CH* is equal to *E*, and *AG* to *F*; therefore, as *AB* is to *CD*, so is *E* to *F*.

Therefore etc.   Q. E. D.

This proposition is a particular case of VI. 14, but one which is on all accounts worth separate statement. It may also be enunciated in the following form:

Rectangles which have their bases reciprocally proportional to their heights are equal in area; and equal rectangles have their bases reciprocally proportional to their heights.

Since any parallelogram is equal to a rectangle of the same height and on the same base, and any triangle with the same height and on the same base is equal to half the parallelogram or rectangle, it follows that *Equal parallelograms or triangles have their bases reciprocally proportional to their heights and* vice versa.

The present place is suitable for giving certain important propositions, including those which Simson adds to Book VI. as Props. B, C and D, which are proved directly by means of VI. 16.

1. Proposition B is a particular case of the following theorem.

If a circle be circumscribed about a triangle ABC *and there be drawn through* A *any two straight lines either both within or both without the angle* BAC, *viz.*

AD *meeting* BC (*produced if necessary*) *in* D *and* AE *meeting the circle again in* E, *such that the angles* DAB, EAC *are equal, then the rectangle* AD, AE *is equal to the rectangle* BA, AC.

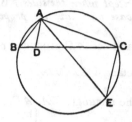
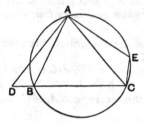

Join *CE*.

The angles *BAD*, *EAC* are equal, by hypothesis;

and the angles *ABD*, *AEC* are equal.   [III. 21, 22]

Therefore the triangles *ABD*, *AEC* are equiangular.

Hence   *BA* is to *AD* as *EA* is to *AC*,

and therefore the rectangle *BA*, *AC* is equal to the rectangle *AD*, *AE*.

[VI. 16]

There are now two particular cases to be considered.

(*a*) Suppose that *AD*, *AE* coincide;

ADE will then bisect the angle *BAC*.

(*b*) Suppose that *AD*, *AE* are in one straight line but that *D*, *E* are on opposite sides of *A*;

AD will then bisect the external angle at *A*.

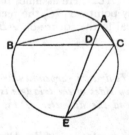
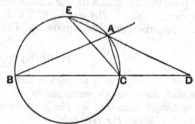

In the first case (*a*) we have

the rectangle *BA*, *AC* equal to the rectangle *EA*, *AD*;

and the rectangle *EA*, *AD* is equal to the rectangle *ED*, *DA* together with the square on *AD*,   [II. 3]

i.e. to the rectangle *BD*, *DC* together with the square on *AD*. [III. 35]

Therefore the rectangle *BA*, *AC* is equal to the rectangle *BD*, *DC* together with the square on *AD*.   [This is Simson's Prop. B]

In case (*b*) the rectangle *EA*, *AD* is equal to the excess of the rectangle *ED*, *DA* over the square on *AD*;

therefore the rectangle *BA*, *AC* is equal to the excess of the rectangle *BD*, *DC* over the square on *AD*.

The following converse of Simson's Prop. B may be given: *If a straight line AD be drawn from the vertex A of a triangle to meet the base, so that the square on AD together with the rectangle BD, DC is equal to the rectangle BA, AC, the line AD will bisect the angle BAC except when the sides AB, AC are equal, in which case every line drawn to the base will have the property mentioned.*

Let the circumscribed circle be drawn, and let AD produced meet it in E; join CE.

The rectangle BD, DC is equal to the rectangle ED, DA.   [III. 35]

Add to each the square on AD;

therefore the rectangle BA, AC is equal to the rectangle EA, AD.

[hyp. and II. 3]

Hence AB is to AD as AE to AC.   [VI. 16]

But the angle ABD is equal to the angle AEC.   [III. 21]

Therefore the angles BDA, ECA are either equal or supplementary.

[VI. 7 and note]

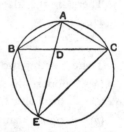

(*a*) If they are equal, the angles BAD, EAC are also equal, and AD bisects the angle BAC.

(*b*) If they are supplementary, the angle ADC must be equal to the angle ACE.

Therefore the angles BAD, ABD are together equal to the angles ACB, BCE, i.e. to the angles ACD, BAD.

Take away the common angle BAD, and

the angles ABD, ACD are equal, or

AB is equal to AC.

Euclid himself assumes, in Prop. 67 of the *Data*, the result of so much of this proposition as relates to the case where $BA = AC$. He assumes namely, without proof, that, if $BA = AC$, and if D be any point on BC, the rectangle BD, DC together with the square on AD is equal to the square on AB.

PROPOSITION C.

If from any angle of a triangle a straight line be drawn perpendicular to the opposite side, the rectangle contained by the other two sides of the triangle is equal to the rectangle contained by the perpendicular and the diameter of the circle circumscribed about the triangle.

Let ABC be a triangle and AD the perpendicular on AB. Draw the diameter AE of the circle circumscribed about the triangle ABC.

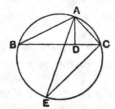
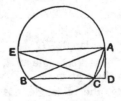

Then shall the rectangle BA, AC be equal to the rectangle EA, AD.
Join EC.

Since the right angle *BDA* is equal to the right angle *ECA* in a semicircle,　　　　　　　　　　　　　　　　　　　　　　　　[III. 31]

and the angles *ABD*, *AEC* in the same segment are equal,　　[III. 21]

the triangles *ABD*, *AEC* are equiangular.

Therefore,　　　　as *BA* is to *AD*, so is *EA* to *AC*,　　　[VI. 4]

whence the rectangle *BA*, *AC* is equal to the rectangle *EA*, *AD*.　　[VI. 16]

This result corresponds to the trigonometrical formula for *R*, the radius of the circumscribed circle,

$$R = \frac{abc}{4\Delta}.$$

PROPOSITION D.

This is the highly important lemma given by Ptolemy (ed. Heiberg, Vol. I, pp. 36—7) which is the basis of his calculation of the table of chords in the section of Book I. of the μεγάλη σύνταξις entitled "concerning the size of the straight lines [i.e. chords] in the circle" (περὶ τῆς πηλικότητος τῶν ἐν τῷ κύκλῳ εὐθειῶν).

The theorem may be enunciated thus.

The rectangle contained by the diagonals of any quadrilateral inscribed in a circle is equal to the sum of the rectangles contained by the pairs of opposite sides.

I shall give the proof in Ptolemy's words, with the addition only, in brackets, of two words applying to a second figure not given by Ptolemy.

"Let there be a circle with any quadrilateral *ABCD* inscribed in it, and let *AC*, *BD* be joined.

It is to be proved that the rectangle contained by *AC* and *BD* is equal to the sum of the rectangles *AB*, *DC* and *AD*, *BC*.

For let the angle *ABE* be made equal to the angle contained by *DB*, *BC*.

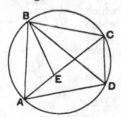 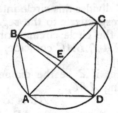

If then we add [or subtract] the angle *EBD*, the angle *ABD* will also be equal to the angle *EBC*.

But the angle *BDA* is also equal to the angle *BCE*,　　[III. 21]

for they subtend the same segment;

therefore the triangle *ABD* is equiangular with the triangle *EBC*.

Hence, proportionally,

as *BC* is to *CE*, so is *BD* to *DA*.　　　[VI. 4]

Therefore the rectangle *BC*, *AD* is equal to the rectangle *BD*, *CE*.

[VI. 16]

Again, since the angle *ABE* is equal to the angle *DBC*,

and the angle *BAE* is also equal to the angle *BDC*,　　[III. 21]

the triangle *ABE* is equiangular with the triangle *DBC*.

Therefore, proportionally,

as BA is to AE, so is BD to DC;   [VI. 4]

therefore the rectangle BA, DC is equal to the rectangle BD, AE.   [VI. 16]

But it was also proved that

the rectangle BC, AD is equal to the rectangle BD, CE;

therefore the rectangle AC, BD as a whole is equal to the sum of the rectangles AB, DC and AD, BC:

(being) what it was required to prove."

Another proof of this proposition, and of its converse, is indicated by Dr Lachlan (*Elements of Euclid*, pp. 273—4). It depends on two preliminary propositions.

(1) *If two circles be divided, by a chord in each, into segments which are similar respectively, the chords are proportional to the corresponding diameters.*

The proof is instantaneous if we join the ends of each chord to the centre of the circle which it divides, when we obtain two similar triangles.

(2) *If* D *be any point on the circle circumscribed about a triangle* ABC, *and* DX, DY, DZ *be perpendicular to the sides* BC, CA, AB *of the triangle respectively, then* X, Y, Z *lie in one straight line; and,* conversely, *if the feet of the perpendiculars from any point* D *on the sides of a triangle lie in one straight line,* D *lies on the circle circumscribed about the triangle.*

The proof depending on III. 21, 22 is well known.

Now suppose that D is *any* point in the plane of a triangle ABC, and that DX, DY, DZ are perpendicular to the sides BC, CA, AB respectively.

Join YZ, DA.

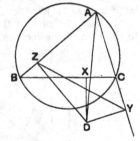

Then, since the angles at Y, Z are right, A, Y, D, Z lie on a circle of which DA is the diameter.

And YZ divides this circle into segments which are similar respectively to the segments into which BC divides the circle circumscribing ABC, since the angles ZAY, BAC coincide, and their supplements are equal.

Therefore, if d be the diameter of the circle circumscribing ABC,

BC is to d as YZ is to DA;

and therefore the rectangle AD, BC is equal to the rectangle d, YZ.

Similarly the rectangle BD, CA is equal to the rectangle d, ZX, and the rectangle CD, AB is equal to the rectangle d, XY.

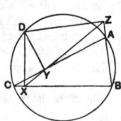

Hence, in a quadrilateral in general, the rectangle contained by the diagonals is less than the sum of the rectangles contained by the pairs of opposite sides.

Next, suppose that D lies on the circle circumscribed about ABC, but so that A, B, C, D follow each other on the circle in this order, as in the figure annexed.

Let DX, DY, DZ be perpendicular to BC, CA, AB respectively, so that X, Y, Z are in a straight line.

Then, since the rectangles AD, BC; BD, CA; CD, AB are equal to the rectangles d, YZ; d, ZX; d, XY respectively, and XZ is equal to the sum of

XY, *YZ*, so that the rectangle *d*, *XZ* is equal to the sum of the rectangles *d*, *XY* and *d*, *YZ*, it follows that

the rectangle *AC*, *BD* is equal to the sum of the rectangles *AD*, *BC* and *AB*, *CD*.

Conversely, if the latter statement is true, while we are supposed to know nothing about the position of *D*, it follows that

XZ must be equal to the sum of *XY*, *YZ*,

so that *X*, *Y*, *Z* must be in a straight line.

Hence, from the theorem (2) above, it follows that *D* must lie on the circle circumscribed about *ABC*, i.e. that *ABCD* is a quadrilateral about which a circle can be described.

All the above propositions can be proved on the basis of Book III. and without using Book VI., since it is possible by the aid of III. 21 and 35 alone to prove that *in equiangular triangles the rectangles contained by the non-corresponding sides about equal angles are equal to one another* (a result arrived at by combining VI. 4 and VI. 16). This is the method adopted by Casey, H. M. Taylor, and Lachlan; but I fail to see any particular advantage in it.

Lastly, the following proposition may be given which Playfair added as VI. E. It appears in the *Data* of Euclid, Prop. 93, and may be thus enunciated.

If the angle BAC *of a triangle* ABC *be bisected by the straight line* AD *meeting the circle circumscribed about the triangle in* D, *and if* BD *be joined, then*

the sum of BA, AC *is to* AD *as* BC *is to* BD.

Join *CD*. Then, since *AD* bisects the angle *BAC*, the subtended arcs *BD*, *DC*, and therefore the chords *BD*, *DC*, are equal.

(1) The result can now be easily deduced from Ptolemy's theorem.

For the rectangle *AD*, *BC* is equal to the sum of the rectangles *AB*, *DC* and *AC*, *BD*, i.e. (since *BD*, *CD* are equal) to the rectangle contained by *BA* + *AC* and *BD*.

Therefore the sum of *BA*, *AC* is to *AD* as *BC* is to *BD*.   [VI. 16]

(2) Euclid proves it differently in *Data*, Prop. 93. Let *AD* meet *BC* in *E*.

Then, since *AE* bisects the angle *BAC*,

$$BA \text{ is to } AC \text{ as } BE \text{ to } EC,$$ [VI. 3]

or, alternately,

$$AB \text{ is to } BE \text{ as } AC \text{ to } CE.$$ [V. 16]

Therefore also

$$BA + AC \text{ is to } BC \text{ as } AC \text{ to } CE.$$ [V. 12]

Again, since the angles *BAD*, *EAC* are equal, and the angles *ADB*, *ACE* are also equal,   [III. 21]

the triangles *ABD*, *AEC* are equiangular.

Therefore AC is to CE as AD to BD.   [VI. 4]

Hence $BA + AC$ is to BC as AD to BD,   [V. 11]
and, alternately,

         $BA + AC$ is to AD as BC is to BD.   [V. 16]

Euclid concludes that, if the circle ABC is given in magnitude, and the chord BC cuts off a segment of it containing a given angle (so that, by *Data* Prop. 87, BC and also BD are given in magnitude),

the ratio of $BA + AC$ to AD is given,

and further that (since, by similar triangles, BD is to DE as AC is to CE, while $BA + AC$ is to BC as AC is to CE),

the rectangle $(BA + AC)$, DE, being equal to the rectangle BC, BD, is also given.

PROPOSITION 17

If three straight lines be proportional, the rectangle contained by the extremes is equal to the square on the mean; and, if the rectangle contained by the extremes be equal to the square on the mean, the three straight lines will be proportional.

Let the three straight lines A, B, C be proportional, so that, as A is to B, so is B to C;
I say that the rectangle contained by A, C is equal to the square on B.

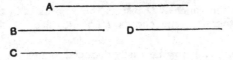

Let D be made equal to B.
Then, since, as A is to B, so is B to C,
and B is equal to D,
therefore, as A is to B, so is D to C.

But, if four straight lines be proportional, the rectangle contained by the extremes is equal to the rectangle contained by the means.   [VI. 16]

Therefore the rectangle A, C is equal to the rectangle B, D.

But the rectangle B, D is the square on B, for B is equal to D;
therefore the rectangle contained by A, C is equal to the square on B.

Next, let the rectangle A, C be equal to the square on B;
I say that, as A is to B, so is B to C.

For, with the same construction,
since the rectangle A, C is equal to the square on B,
while the square on B is the rectangle B, D, for B is equal
to D,
therefore the rectangle A, C is equal to the rectangle B, D.

But, if the rectangle contained by the extremes be equal
to that contained by the means, the four straight lines are
proportional.   [VI. 16]

Therefore, as A is to B, so is D to C.

But B is equal to D;

therefore, as A is to B, so is B to C.

Therefore etc.   Q. E. D.

VI. 17 is, of course, a particular case of VI. 16.

PROPOSITION 18.

*On a given straight line to describe a rectilineal figure
similar and similarly situated to a given rectilineal figure.*

Let AB be the given straight line and CE the given
rectilineal figure;
thus it is required to describe on the straight line AB a
rectilineal figure similar and similarly situated to the rectilineal figure CE.

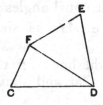

Let DF be joined, and on the straight line AB, and at
the points A, B on it, let the angle GAB be constructed
equal to the angle at C, and the angle ABG equal to the
angle CDF.   [I. 23]

Therefore the remaining angle CFD is equal to the angle
AGB;   [I. 32]

therefore the triangle FCD is equiangular with the triangle
GAB.

Therefore, proportionally, as FD is to GB, so is FC to
GA, and CD to AB.

Again, on the straight line BG, and at the points B, G on it, let the angle BGH be constructed equal to the angle DFE, and the angle GBH equal to the angle FDE.   [I. 23]

Therefore the remaining angle at E is equal to the remaining angle at H;   [I. 32]

therefore the triangle FDE is equiangular with the triangle GBH;

therefore, proportionally, as FD is to GB, so is FE to GH, and ED to HB.   [VI. 4]

But it was also proved that, as FD is to GB, so is FC to GA, and CD to AB;

therefore also, as FC is to AG, so is CD to AB, and FE to GH, and further ED to HB.

And, since the angle CFD is equal to the angle AGB, and the angle DFE to the angle BGH,

therefore the whole angle CFE is equal to the whole angle AGH.

For the same reason

the angle CDE is also equal to the angle ABH.

And the angle at C is also equal to the angle at A,

and the angle at E to the angle at H.

Therefore AH is equiangular with CE;

and they have the sides about their equal angles proportional;

therefore the rectilineal figure AH is similar to the rectilineal figure CE.   [VI. Def. 1]

Therefore on the given straight line AB the rectilineal figure AH has been described similar and similarly situated to the given rectilineal figure CE.

                                        Q. E. F.

Simson thinks the proof of this proposition has been vitiated, his grounds for this view being (1) that it is demonstrated only with reference to quadrilaterals, and does not show how it may be extended to figures of five or more sides, (2) that Euclid infers, from the fact of two triangles being equiangular, that a side of the one is to the corresponding side of the other as another side of the first is to the side corresponding to it in the other, i.e. he permutes, without mentioning the fact that he does so, the proportions obtained in VI. 4, whereas the proof of the very next proposition gives, in a similar case, the intermediate step of permutation. I think this is hypercriticism. As regards (2) it should be noted that the permuted form of the proportion is arrived at first in the *proof* of VI. 4; and the omission of the intermediate step of *alternando*, whether accidental or not, is of no importance. On the other hand, the use of this form of the proportion certainly simplifies the proof of the proposition, since it makes unnecessary the subsequent *ex aequali* steps of Simson's proof, their place being taken by the inference [v. 11] that ratios which are the same with a third ratio are the same with one another.

Nor is the first objection of any importance. We have only to take as the

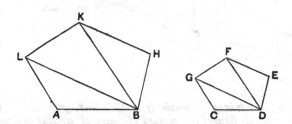

given polygon a polygon of five sides at least, as *CDEFG*, join one extremity of *CD*, say *D*, to each of the angular points other than *C* and *E*, and then use the same mode of construction as Euclid's for any number of successive triangles as *ABL*, *LBK*, etc., that may have to be made. Euclid's construction and proof for a quadrilateral are quite sufficient to show how to deal with the case of a figure of five or any greater number of sides.

Clavius has a construction which, given the power of moving a figure

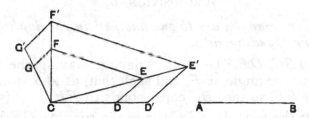

bodily from one position to any other, is easier. *CDEFG* being the given polygon, join *CE*, *CF*. Place *AB* on *CD* so that *A* falls on *C*, and let *B* fall on *D'*, which may either lie on *CD* or on *CD* produced.

Now draw *D'E'* parallel to *DE*, meeting *CE*, produced if necessary, in *E'*, *E'F'* parallel to *EF*, meeting *CF*, produced if necessary, in *F'*, and so on.

Let the parallel to the last side but one, *FG*, meet *CG*, produced if necessary, in *G'*.

Then *CD'E'F'G'* is similar and similarly situated to *CDEFG*, and it is constructed on *CD'*, a straight line equal to *AB*.

The proof of this is obvious.

A more general construction is indicated in the subjoined figure. If *CDEFG* be the given polygon, suppose its angular points all joined to any point *O* and the connecting straight lines produced both ways. Then, if *CD'*, a straight line equal to *AB*, be placed so that it is parallel to *CD*, and *C'*, *D'* lie respectively on *OC*, *OD* (this can of course be done by finding fourth proportionals), we have only to draw *D'E'*, *E'F'*, etc., parallel to the corresponding sides of the original polygon in the manner shown.

De Morgan would rearrange Props. 18 and 20 in the following manner. He would combine Prop. 18 and the first part of Prop. 20 into one, with the enunciation :

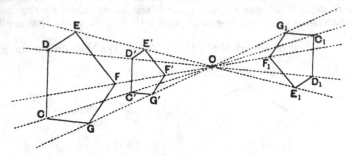

Pairs of similar triangles, similarly put together, give similar figures ; and every pair of similar figures is composed of pairs of similar triangles similarly put together.

He would then make the *problem* of VI. 18 an application of the first part. In form this would certainly appear to be an improvement; but, provided that the relation of the propositions is understood, the matter of form is perhaps not of great importance.

PROPOSITION 19.

Similar triangles are to one another in the duplicate ratio of the corresponding sides.

Let ABC, DEF be similar triangles having the angle at B equal to the angle at E, and such that, as AB is to BC, so
5 is DE to EF, so that BC corresponds to EF ;   [v. Def. 11]

I say that the triangle ABC has to the triangle DEF a ratio duplicate of that which BC has to EF.

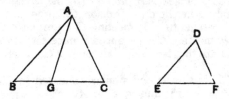

For let a third proportional BG be taken to BC, EF, so that, as BC is to EF, so is EF to BG ;   [VI. 11]
10 and let AG be joined.

Since then, as AB is to BC, so is DE to EF,

therefore, alternately, as AB is to DE, so is BC to EF. [v. 16]

But, as BC is to EF, so is EF to BG ;

therefore also, as AB is to DE, so is EF to BG.   [v. 11]

15   Therefore in the triangles ABG, DEF the sides about the equal angles are reciprocally proportional.

But those triangles which have one angle equal to one angle, and in which the sides about the equal angles are reciprocally proportional, are equal;   [VI. 15]

20 therefore the triangle ABG is equal to the triangle DEF.

Now since, as BC is to EF, so is EF to BG,

and, if three straight lines be proportional, the first has to the third a ratio duplicate of that which it has to the second,

[v. Def. 9]

therefore BC has to BG a ratio duplicate of that which CB
25 has to EF.

But, as CB is to BG, so is the triangle ABC to the triangle ABG ;   [VI. 1]

therefore the triangle ABC also has to the triangle ABG a ratio duplicate of that which BC has to EF.

30   But the triangle ABG is equal to the triangle DEF ;

therefore the triangle ABC also has to the triangle DEF a ratio duplicate of that which BC has to EF.

Therefore etc.

PORISM. From this it is manifest that, if three straight
35 lines be proportional, then, as the first is to the third, so is the figure described on the first to that which is similar and similarly described on the second.

Q. E. D.

4. and such that, as AB is to BC, so is DE to EF, literally "(triangles) having the angle at B equal to the angle at E, and (having), as AB to BC, so DE to EF."

Having combined Prop. 18 and the first part of Prop. 20 as just indicated, De Morgan would tack on to Prop. 19 the second part of Prop. 20, which asserts that, if similar polygons be divided into the same number of similar triangles, the triangles are "*homologous* to the wholes" (in the sense that the polygons have the same ratio as the corresponding triangles have), and that the polygons are to one another in the duplicate ratio of corresponding sides. This again, though no doubt an improvement of form, would necessitate the drawing over again of the figure of the altered Proposition 18 and a certain amount of repetition.

Agreeably to his suggestion that Prop. 23 should come before Prop. 14 which is a particular case of it, De Morgan would prove Prop. 19 for *parallelograms* by means of Prop. 23, and thence infer the truth of it for triangles or the halves of the parallelograms. He adds: "The method of Euclid is an elegant application of the *operation* requisite to compound equal ratios, by which the conception of the process is lost sight of." For the general reason given in the note on VI. 14 above, I think that Euclid showed the sounder discretion in the arrangement which he adopted. Moreover it is not easy to see how performing the actual operation of compounding two equal ratios can obscure the process, or the fact that two equal ratios are being compounded. On the definition of compounded ratios and duplicate ratio, De Morgan has himself acutely pointed out that "composition" is here used for the process of detecting the single alteration which produces the effect of two or more, the duplicate ratio being the result of compounding two equal ratios. The proof of VI. 19 does in fact exhibit the single alteration which produces the effect of two. And the *operation* was of the essence of the Greek geometry, because it was the manipulation of ratios in this manner, by simplification and transformation, that gave it so much power, as every one knows who has read, say, Archimedes or Apollonius. Hence the introduction of the necessary *operation*, as well as the theoretical proof, in this proposition seems to me to have been distinctly worth while, and, as it is somewhat simpler in this case than in the more general case of VI. 23, it was in accordance with the plan of enabling the difficulties of Book VI. to be more easily and gradually surmounted to give the simpler case first.

That Euclid wished to emphasise the importance of the *method* adopted, as well as of the result obtained, in VI. 19 seems to me clearly indicated by the Porism which follows the proposition. It is as if he should say: "I have shown you that similar triangles are to one another in the duplicate ratio of corresponding sides; but I have also shown you incidentally how it is possible to work conveniently with duplicate ratios, viz. by transforming them into simple ratios between straight lines. I shall have occasion to illustrate the use of this method in the proof of VI. 22."

The Porism to VI. 19 presents one difficulty. It will be observed that it speaks of the *figure* (εἶδος) described on the first straight line and of that which is similar and similarly described on the second. If "figure" could be regarded as loosely used for the figure *of the proposition*, i.e. for a *triangle*, there would be no difficulty. If on the other hand "the figure" means any rectilineal figure, i.e. any polygon, the Porism is not really established until the next proposition, VI. 20, has been proved, and therefore it is out of place here. Yet the correction τρίγωνον, *triangle*, for εἶδος, *figure*, is due to Theon alone; P and Campanus have "figure," and the reading of Philoponus and Psellus, τετράγωνον, *square*, partly supports εἶδος, since it can be reconciled with εἶδος but not with τρίγωνον. Again the second Porism to VI. 20, in which this Porism is reasserted for any rectilineal figure, and which is omitted by Campanus and only given by P in the margin, was probably interpolated by Theon. Heiberg concludes that Euclid wrote "figure" (εἶδος), and Theon, seeing the difficulty, changed the word into "triangle" here and added Por. 2 to VI. 20 in order to make the matter clear. If one may hazard a guess as to how Euclid made the slip, may it be that he first put it after VI 20 and then, observing that the expression of the duplicate ratio by a single ratio between two straight lines does not come in VI. 20 but in VI. 19, moved the Porism to the end of VI. 19 in order to make the connexion clearer, without noticing that, if this were done, εἶδος would need correction?

The following explanation at the end of the Porism is bracketed by Heiberg, viz. "Since it was proved that, as *CB* is to *BG*, so is the triangle

ABC to the triangle *ABG*, that is *DEF*." Such explanations in Porisms are not in Euclid's manner, and the words are not in Campanus, though they date from a time earlier than Theon.

<h2 style="text-align:center">PROPOSITION 20.</h2>

Similar polygons are divided into similar triangles, and into triangles equal in multitude and in the same ratio as the wholes, and the polygon has to the polygon a ratio duplicate of that which the corresponding side has to the corresponding
5 *side.*

Let *ABCDE*, *FGHKL* be similar polygons, and let *AB* correspond to *FG*;

I say that the polygons *ABCDE*, *FGHKL* are divided into similar triangles, and into triangles equal in multitude and in
10 the same ratio as the wholes, and the polygon *ABCDE* has to the polygon *FGHKL* a ratio duplicate of that which *AB* has to *FG*.

Let *BE*, *EC*, *GL*, *LH* be joined.

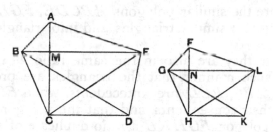

Now, since the polygon *ABCDE* is similar to the polygon
15 *FGHKL*,

the angle *BAE* is equal to the angle *GFL*;

and, as *BA* is to *AE*, so is *GF* to *FL*. [VI. Def. 1]

Since then *ABE*, *FGL* are two triangles having one angle equal to one angle and the sides about the equal angles
20 proportional,

therefore the triangle *ABE* is equiangular with the triangle *FGL*; [VI. 6]

so that it is also similar; [VI. 4 and Def. 1]

therefore the angle *ABE* is equal to the angle *FGL*.

25   But the whole angle ABC is also equal to the whole angle FGH because of the similarity of the polygons ;

therefore the remaining angle EBC is equal to the angle LGH.

And, since, because of the similarity of the triangles ABE,
30 FGL,

as EB is to BA, so is LG to GF,

and moreover also, because of the similarity of the polygons,

as AB is to BC, so is FG to GH,

therefore, *ex aequali*, as EB is to BC, so is LG to GH ; [v. 22]

35 that is, the sides about the equal angles EBC, LGH are proportional ;

therefore the triangle EBC is equiangular with the triangle LGH,   [VI. 6]

so that the triangle EBC is also similar to the triangle
40 LGH.   [VI. 4 and Def. 1]

For the same reason

the triangle ECD is also similar to the triangle LHK.

Therefore the similar polygons $ABCDE$, $FGHKL$ have been divided into similar triangles, and into triangles equal in
45 multitude.

I say that they are also in the same ratio as the wholes, that is, in such manner that the triangles are proportional, and ABE, EBC, ECD are antecedents, while FGL, LGH, LHK are their consequents, and that the polygon $ABCDE$
50 has to the polygon $FGHKL$ a ratio duplicate of that which the corresponding side has to the corresponding side, that is AB to FG.

For let AC, FH be joined.

Then since, because of the similarity of the polygons,
55 the angle ABC is equal to the angle FGH,

and, as AB is to BC, so is FG to GH,

the triangle ABC is equiangular with the triangle FGH ;
                                                      [VI. 6]

therefore the angle BAC is equal to the angle GFH,

and the angle BCA to the angle GHF.

60   And, since the angle BAM is equal to the angle GFN,

and the angle ABM is also equal to the angle FGN,

therefore the remaining angle AMB is also equal to the
remaining angle FNG;　　　　　　　　　　　　　　　　[I. 32]
therefore the triangle ABM is equiangular with the triangle
65 FGN.

Similarly we can prove that

the triangle BMC is also equiangular with the triangle GNH.

Therefore, proportionally, as AM is to MB, so is FN to
NG,

70 and,　　　　　　　　as BM is to MC, so is GN to NH;

so that, in addition, *ex aequali*,

　　　　　　　　as AM is to MC, so is FN to NH.

But, as AM is to MC, so is the triangle ABM to MBC,
and AME to EMC; for they are to one another as their
75 bases.　　　　　　　　　　　　　　　　　　　　　　[VI. 1]

Therefore also, as one of the antecedents is to one of the
consequents, so are all the antecedents to all the consequents;
　　　　　　　　　　　　　　　　　　　　　　　　[V. 12]
therefore, as the triangle AMB is to BMC, so is ABE to
CBE.

80　　　But, as AMB is to BMC, so is AM to MC;
therefore also, as AM is to MC, so is the triangle ABE to
the triangle EBC.

For the same reason also,

as FN is to NH, so is the triangle FGL to the triangle
85 GLH.

And, as AM is to MC, so is FN to NH;
therefore also, as the triangle ABE is to the triangle BEC,
so is the triangle FGL to the triangle GLH;
and, alternately, as the triangle ABE is to the triangle FGL,
90 so is the triangle BEC to the triangle GLH.

Similarly we can prove, if BD, GK be joined, that, as the
triangle BEC is to the triangle LGH, so also is the triangle
ECD to the triangle LHK.

And since, as the triangle ABE is to the triangle FGL,
95 so is EBC to LGH, and further ECD to LHK,

therefore also, as one of the antecedents is to one of the
consequents. so are all the antecedents to all the consequents;
　　　　　　　　　　　　　　　　　　　　　　　　[V. 12

therefore, as the triangle ABE is to the triangle FGL,
so is the polygon $ABCDE$ to the polygon $FGHKL$.

100   But the triangle ABE has to the triangle FGL a ratio duplicate of that which the corresponding side AB has to the corresponding side FG; for similar triangles are in the duplicate ratio of the corresponding sides.   [VI. 19]

Therefore the polygon $ABCDE$ also has to the polygon
105 $FGHKL$ a ratio duplicate of that which the corresponding side AB has to the corresponding side FG.

Therefore etc.

PORISM. Similarly also it can be proved in the case of quadrilaterals that they are in the duplicate ratio of the
110 corresponding sides. And it was also proved in the case of triangles; therefore also, generally, similar rectilineal figures are to one another in the duplicate ratio of the corresponding sides.

Q. E. D.

2. **in the same ratio as the wholes.** The same word ὁμόλογος is used which I have generally translated by "corresponding." But here it is followed by a dative, ὁμόλογα τοῖς ὅλοις "*homologous with* the wholes," instead of being used absolutely. The meaning can therefore here be nothing else but "in the same ratio with" or "proportional to the wholes"; and Euclid seems to recognise that he is making a special use of the word, because he explains it lower down (l. 46): "the triangles are homologous to the wholes, that is, in such manner that the triangles are proportional, and ABE, EBC, ECD are antecedents, while FGL, LGH, LHK are their consequents."

49. ἑπόμενα αὐτῶν, "*their* consequents," is a little awkward, but may be supposed to indicate which triangles correspond to which as consequent to antecedent.

An alternative proof of the second part of this proposition given after the Porisms is relegated by August and Heiberg to an Appendix as an interpolation. It is shorter than the proof in the text, and is the only one given by many editors, including Clavius, Billingsley, Barrow and Simson. It runs as follows:

"We will now also prove that the triangles are homologous in another and an easier manner.

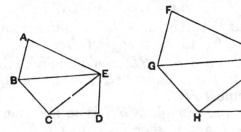

Again, let the polygons $ABCDE$, $FGHKL$ be set out, and let BE, EC, GL, LH be joined.

I say that, as the triangle ABE is to FGL, so is EBC to LGH and CDE to HKL.

For, since the triangle ABE is similar to the triangle FGL, the triangle ABE has to the triangle FGL a ratio duplicate of that which BE has to GL.

For the same reason also

the triangle BEC has to the triangle GLH a ratio duplicate of that which BE has to GL.

Therefore, as the triangle ABE is to the triangle FGL, so is BEC to GLH.

Again, since the triangle EBC is similar to the triangle LGH,

EBC has to LGH a ratio duplicate of that which the straight line CE has to HL.

For the same reason also

the triangle ECD has to the triangle LHK a ratio duplicate of that which CE has to HL.

Therefore, as the triangle EBC is to LGH, so is ECD to LHK.

But it was proved that,

as EBC is to LGH, so also is ABE to FGL.

Therefore also, as ABE is to FGL, so is BEC to GLH and ECD to LHK.

Q. E. D."

Now Euclid cannot fail to have noticed that the second part of his proposition could be proved in this way. It seems therefore that, in giving the other and longer method, he deliberately wished to avoid using the result of VI. 19, preferring to prove the first two parts of the theorem, as they can be proved, independently of any relation between the areas of similar triangles.

The first part of the Porism, stating that the theorem is true of *quadrilaterals*, would be superfluous but for the fact that technically, according to Book I. Def. 19, the term "polygon" (or figure of many sides, πολύπλευρον) used in the enunciation of the proposition is confined to rectilineal figures of *more than four* sides, so that a quadrilateral might seem to be excluded. The mention of the triangle in addition fills up the tale of "similar rectilineal figures."

The second Porism, Theon's interpolation, given in the text by the editors, but bracketed by Heiberg, is as follows:

"And, if we take O a third proportional to AB, FG, then BA has to O a ratio duplicate of that which AB has to FG.

But the polygon has also to the polygon, or the quadrilateral to the quadrilateral, a ratio duplicate of that which the corresponding side has to the corresponding side, that is AB to FG;

and this was proved in the case of triangles also;

so that it is also manifest generally that, if three straight lines be proportional, as the first is to the third, so will the figure described on the first be to the similar and similarly described figure on the second."

PROPOSITION 21.

Figures which are similar to the same rectilineal figure are also similar to one another.

For let each of the rectilineal figures A, B be similar to C; I say that A is also similar to B.

For, since A is similar to C,

it is equiangular with it and has the sides about the equal angles proportional. [VI. Def. 1]

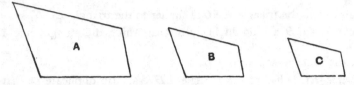

Again, since B is similar to C,

it is equiangular with it and has the sides about the equal angles proportional.

Therefore each of the figures A, B is equiangular with C and with C has the sides about the equal angles proportional;

therefore A is similar to B.

Q. E. D.

It will be observed that the text above omits a step which the editions generally have before the final inference "Therefore A is similar to B." The words omitted are "so that A is also equiangular with B and [with B] has the sides about the equal angles proportional." Heiberg follows P in leaving them out, conjecturing that they may be an addition of Theon's.

PROPOSITION 22.

If four straight lines be proportional, the rectilineal figures similar and similarly described upon them will also be proportional; and, if the rectilineal figures similar and similarly described upon them be proportional, the straight lines will themselves also be proportional.

Let the four straight lines AB, CD, EF, GH be proportional,

so that, as AB is to CD, so is EF to GH,

and let there be described on AB, CD the similar and similarly situated rectilineal figures KAB, LCD,

and on EF, GH the similar and similarly situated rectilineal figures MF, NH;

I say that, as KAB is to LCD, so is MF to NH.

For let there be taken a third proportional O to AB, CD, and a third proportional P to EF, GH. [VI. 11]

Then since, as AB is to CD, so is EF to GH,

and, as CD is to O, so is GH to P,

therefore, *ex aequali*, as AB is to O, so is EF to P.　　[v. 22]

But, as AB is to O, so is KAB to LCD,

and, as EF is to P, so is MF to NH;　　　[vi. 19, Por.]

therefore also, as KAB is to LCD, so is MF to NH.　　[v. 11]

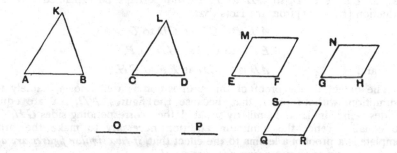

Next, let MF be to NH as KAB is to LCD;

I say also that, as AB is to CD, so is EF to GH.

For, if EF is not to GH as AB to CD,

let EF be to QR as AB to CD,　　　　　　　　　　[vi. 12]

and on QR let the rectilineal figure SR be described similar and similarly situated to either of the two MF, NH.　　[vi. 18]

Since then, as AB is to CD, so is EF to QR,

and there have been described on AB, CD the similar and similarly situated figures KAB, LCD,

and on EF, QR the similar and similarly situated figures MF, SR,

therefore, as KAB is to LCD, so is MF to SR.

But also, by hypothesis,

as KAB is to LCD, so is MF to NH;

therefore also, as MF is to SR, so is MF to NH.　　[v. 11]

Therefore MF has the same ratio to each of the figures NH, SR;

therefore NH is equal to SR.　　　　　　　　　　　　[v. 9]

But it is also similar and similarly situated to it;

therefore GH is equal to QR.

And, since, as AB is to CD, so is EF to QR,

while QR is equal to GH,

therefore, as AB is to CD, so is EF to GH.

Therefore etc.

Q. E. D.

The second assumption in the first step of the first part of the proof, viz. that, as CD is to O, so is GH to P, should perhaps be explained. It is a deduction [by v. 11] from the facts that

$$AB \text{ is to } CD \text{ as } CD \text{ to } O,$$

$$EF \text{ is to } GH \text{ as } GH \text{ to } P,$$

and AB is to CD as EF to GH.

The defect in the proof of this proposition is well known, namely the assumption, without proof, that, because the figures NH, SR are equal, besides being similar and similarly situated, their corresponding sides GH, QR are equal. Hence the minimum addition necessary to make the proof complete is a proof of a lemma to the effect that, *if two similar figures are also equal, any pair of corresponding sides are equal.*

To supply this lemma is one alternative; another is to prove, as a preliminary proposition, a much more general theorem, viz. that, *if the duplicate ratios of two ratios are equal, the two ratios are themselves equal.* When this is proved, the second part of VI. 22 is an immediate inference from it, and the effect is, of course, to substitute a new proof instead of supplementing Euclid's.

I. It is to be noticed that the lemma required as a minimum is very like what is needed to supplement VI. 28 and 29, in the proofs of which Euclid assumes that, *if two similar parallelograms are unequal, any side in the greater is greater than the corresponding side in the smaller.* Therefore, on the whole, it seems preferable to adopt the alternative of proving the simpler lemma which will serve to supplement all three proofs, viz. that, *if of two similar rectilineal figures the first is greater than, equal to, or less than, the second, any side of the first is greater than, equal to, or less than, the corresponding side of the second respectively.*

The case of *equality* of the figures is the case required for VI. 22; and the proof of it is given in the Greek text after the proposition. Since to give such a "lemma" after the proposition in which it is required is contrary to Euclid's manner, Heiberg concludes that it is an interpolation, though it is earlier than Theon. The lemma runs thus:

"But that, if rectilineal figures be equal and similar, their corresponding sides are equal to one another we will prove thus.

Let NH, SR be equal and similar rectilineal figures, and suppose that,

as HG is to GN, so is RQ to QS;

I say that RQ is equal to HG.

For, if they are unequal, one of them is greater;

let RQ be greater than HG.

Then, since, as RQ is to QS, so is HG to GN,
alternately also, as RQ is to HG, so is QS to GN;
and QR is greater than HG;

therefore QS is also greater than GN;

so that RS is also greater than HN*.

But it is also equal : which is impossible.

Therefore QR is not unequal to GH;

therefore it is equal to it."

[The step marked * is easy to see if it is remembered that it is only necessary to prove its truth in the case of *triangles* (since similar polygons are divisible into the same number of similar and similarly situated triangles having the same ratio to each other respectively as the polygons have). If the triangles be applied to each other so that the two corresponding sides of each, which are used in the question, and the angles included by them coincide, the truth of the inference is obvious.]

The lemma might also be arrived at by proving that, *if a ratio is greater than a ratio of equality, the ratio which is its duplicate is also greater than a ratio of equality; and if the ratio which is duplicate of another ratio is greater than a ratio of equality, the ratio of which it is the duplicate is also greater than a ratio of equality.* It is not difficult to prove this from the particular case of v. 25 in which the second magnitude is equal to the third, i.e. from the fact that in this case the sum of the extreme terms is greater than double the middle term.

II. We now come to the alternative which substitutes a new proof for the second part of the proposition, making the whole proposition an immediate inference from one to which it is practically equivalent, viz. that

(1) *If two ratios be equal, their duplicate ratios are equal, and* (2) *conversely, if the duplicate ratios of two ratios be equal, the ratios are equal.*

The proof of part (1) is after the manner of Euclid's own proof of the first part of VI. 22.

Let A be to B as C to D,

and let X be a third proportional to A, B, and Y a third proportional to C, D, so that

$$A \text{ is to } B \text{ as } B \text{ to } X,$$

and C is to D as D to Y;

whence A is to X in the duplicate ratio of A to B,

and C is to Y in the duplicate ratio of C to D.

Since A is to B as C is to D,

and B is to X as A is to B,

i.e. as C is to D,

i.e. as D is to Y,   [v. 11]

therefore, *ex aequali,* A is to X as C is to Y.

Part (2) is much more difficult and is the crux of the whole thing.

Most of the proofs depend on the assumption that, B being any magnitude and P and Q two magnitudes of the same kind, there does exist a magnitude A which is to B in the same ratio as P to Q. It is this same assumption which makes Euclid's proof of v. 18 illegitimate, since it is nowhere proved in Book v. Hence any proof of the proposition now in question which involves this assumption even in the case where B, P, Q are all straight lines should not properly be given as an addition to Book v.; it should at least be postponed until we have learnt, by means of VI. 12, giving the actual construction of a fourth proportional, that such a fourth proportional exists.

Two proofs which are given of the proposition depend upon the following lemma.

If A, B, C *be three magnitudes of one kind, and* D, E, F *three magnitudes of one kind, then, if*

the ratio of A to B *is greater than that of* D *to* E,

and   the ratio of B to C *greater than that of* E *to* F,

ex aequali,   the ratio of A to C *is greater than that of* D *to* F.

One proof of this does not depend upon the assumption referred to, and therefore, if this proof is used, the theorem can be added to Book v. The proof is that of Hauber (Camerer's Euclid, p. 358 of Vol. II.) and is reproduced by Mr H. M. Taylor. For brevity we will use symbols.

Take equimultiples mA, mD of A, D and nB, nE of B, E such that

$$mA > nB, \text{ but } mD \not> nE.$$

Also let pB, pE be equimultiples of B. E and qC, qF equimultiples of C, F such that

$$pB > qC, \text{ but } pE \not> qF.$$

Therefore, multiplying the first line by n and the second by n, we have

$$pmA > pnB, \; pmD \not> pnE,$$

and $npB > nqC, \; npE \not> nqF,$

whence $pmA > nqC, \; pmD \not> nqF.$

Now pmA, pmD are equimultiples of mA, mD,

and nqC, nqF equimultiples of qC, qF.

Therefore [v. 3] they are respectively equimultiples of A, D and of C, F.

Hence [v. Def. 7] $A : C > D : F.$

Another proof given by Clavius, though depending on the assumption referred to, is neat.

Take G such that

$$G : C = E : F.$$

Therefore $B : C > G : C,$ [v. 13]

and $B > G.$ [v. 10]

Therefore $A : G > A : B.$ [v. 8]

But	$A : B > D : E.$	
Therefore, *a fortiori*,	$A : G > D : E.$	
Suppose H taken such that		
	$H : G = D : E.$	
Therefore	$A > H.$	[v. 13, 10]
Hence	$A : C > H : C.$	[v. 8]
But	$H : G = D : E,$	
	$G : C = E : F.$	
Therefore, *ex aequali*,	$H : C = D : F.$	[v. 22]
Hence	$A : C > D : F.$	[v. 13]

Now we can prove that

Ratios of which equal ratios are duplicate are equal.

Suppose that	$A : B = B : C,$	
and	$D : E = E : F,$	
and further that	$A : C = D : F.$	

it is required to prove that

$$A : B = D : E.$$

For, if not, one of the ratios must be greater than the other.
Let $A : B$ be the greater.

Then, since	$A : B = B : C,$	
and	$D : E = E : F,$	
while	$A : B > D : E,$	
it follows that	$B : C > E : F.$	[v. 13]

Hence, by the lemma, *ex aequali*,

$$A : C > D : F,$$

which contradicts the hypothesis.

Thus the ratios $A : B$ and $D : E$ cannot be unequal; that is, they are equal.

Another proof, given by Dr Lachlan, also assumes the existence of a fourth proportional, but depends upon a simpler lemma to the effect that

It is impossible that two different ratios can have the same duplicate ratio.

For, if possible, let the ratio $A : B$ be duplicate both of $A : X$ and $A : Y$, so that

	$A : X = X : B,$	
and	$A : Y = Y : B.$	

Let X be greater than Y.

Then	$A : X < A : Y;$	[v. 8]
that is,	$X : B < Y : B,$	[v. 11, 13]
or	$X < Y.$	[v. 10]

But X is greater than Y: which is absurd, etc.

Hence	$X = Y.$

Now suppose that $A : B = B : C,$

                          $D : E = E : F,$

and $A : C = D : F.$

To prove that $A : B = D : E.$

If this is not so, suppose that

                          $A : B = D : Z.$

Since $A : C = D : F,$

therefore, inversely, $C : A = F : D.$

Therefore, *ex aequali*,

                          $C : B = F : Z,$ [v. 22]

or, inversely, $B : C = Z : F.$

Therefore $A : B = Z : F.$ [v. 11]

But $A : B = D : Z$, by hypothesis.

Therefore $D : Z = Z : F.$ [v. 11]

Also, by hypothesis, $D : E = E : F;$

whence, by the lemma, $E = Z.$

Therefore $A : B = D : E.$

De Morgan remarks that the best way of remedying the defect in Euclid is to insert the proposition (the lemma to the last proof) that *it is impossible that two different ratios can have the same duplicate ratio,* "which," he says, "immediately proves the second (or defective) case of the theorem." But this seems to be either too much or too little: too much, if we choose to make the *minimum* addition to Euclid (for that addition is a lemma which shall prove that, if a duplicate ratio is a ratio of equality, the ratio of which it is duplicate is also one of equality), and too little if the proof is to be altered in the more fundamental manner explained above.

I think that, if Euclid's attention had been drawn to the defect in his proof of VI. 22 and he had been asked to remedy it, he would have done so by supplying what I have called the minimum lemma and not by making the more fundamental alteration. This I infer from Prop. 24 of the *Data,* where he gives a theorem corresponding to the proposition that *ratios of which equal ratios are duplicate are equal.* The proposition in the *Data* is enunciated thus: *If three straight lines be proportional, and the first have to the third a given ratio, it will also have to the second a given ratio.*

A, B, C being the three straight lines, so that

$$A : B = B : C,$$

and $A : C$ being a given ratio, it is required to prove that $A : B$ is also a given ratio.

Euclid takes any straight line D, and first finds another, F, such that

$$D : F = A : C,$$

whence $D : F$ must be a given ratio, and, as D is given, F is therefore given.

Then he takes E a mean proportional between D, F, so that

$$D : E = E : F.$$

It follows [VI. 17] that

the rectangle D, F is equal to the square on E.

But D, F are both given;

therefore the square on E is given, so that E is also given.

[Observe that De Morgan's lemma is here assumed without proof. It may be proved (1) as it is by De Morgan, whose proof is that given above, p. 245, (2) in the manner of the "minimum lemma," pp. 242—3 above, or (3) as it is by Proclus on I. 46 (see note on that proposition).]

Hence the ratio $D : E$ is given.

Now, since $A : C = D : F$,

and $A : C =$ (square on A): (rect. A, C),

while $D : F =$ (square on D): (rect. D, F),   [VI. 1]

therefore (square on A) : (rect. A, C) = (square on D) : (rect. D, F).   [V. 11]

But, since $A : B = B : C$, (rect. A, C) = (sq. on B);   [VI. 17]

and (rect. D, F) = (sq. on E), from above;

therefore (square on A) : (square on B) = (sq. on D) : (sq. on E).

Therefore, says Euclid,

$$A : B = D : E,$$

that is, *he assumes the truth of* VI. 22 *for squares*.

Thus he deduces his proposition from VI. 22, instead of proving VI. 22 by means of it (or the corresponding proposition used by Mr Taylor and Dr Lachlan).

PROPOSITION 23.

Equiangular parallelograms have to one another the ratio compounded of the ratios of their sides.

Let AC, CF be equiangular parallelograms having the angle BCD equal to the angle ECG;

5 I say that the parallelogram AC has to the parallelogram CF the ratio compounded of the ratios of the sides.

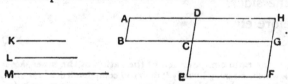

For let them be placed so that BC is in a straight line with CG;

therefore DC is also in a straight line with CE.

10   Let the parallelogram DG be completed;

let a straight line K be set out, and let it be contrived that,

as BC is to CG, so is K to L,

and,   as DC is to CE, so is L to M.   [VI. 12]

Then the ratios of K to L and of L to M are the same
15 as the ratios of the sides, namely of BC to CG and of DC
to CE.

But the ratio of K to M is compounded of the ratio of K
to L and of that of L to M;

so that K has also to M the ratio compounded of the ratios
20 of the sides.

Now since, as BC is to CG, so is the parallelogram AC
to the parallelogram CH,　　　　　　　　　　　　　　　　[VI. 1]

while, as BC is to CG, so is K to L,

therefore also, as K is to L, so is AC to CH.　　　　　[V. 11]

25　　Again, since, as DC is to CE, so is the parallelogram CH
to CF,　　　　　　　　　　　　　　　　　　　　　　　　[VI. 1]

while, as DC is to CE, so is L to M,

therefore also, as L is to M, so is the parallelogram CH to
the parallelogram CF.　　　　　　　　　　　　　　　　[V. 11]

30　　Since then it was proved that, as K is to L, so is the
parallelogram AC to the parallelogram CH,

and, as L is to M, so is the parallelogram CH to the
parallelogram CF,

therefore, *ex aequali*, as K is to M, so is AC to the parallelo-
35 gram CF.

But K has to M the ratio compounded of the ratios of
the sides;

therefore AC also has to CF the ratio compounded of the
ratios of the sides.

40　　Therefore etc.

Q. E. D.

1, 6, 19, 36. **the ratio compounded of the ratios of the sides,** λόγον τὸν συγκείμενον
ἐκ τῶν πλευρῶν which, meaning literally " the ratio compounded *of the sides*," is negligently
written here and commonly for λόγον τὸν συγκείμενον ἐκ τῶν τῶν πλευρῶν (sc. λόγων).
　11. **let it be contrived that, as BC is to CG, so is K to L.** The Greek phrase is
of the usual terse kind, untranslatable literally : καὶ γεγονέτω ὡς μὲν ἡ ΒΓ πρὸς τὴν ΓΗ,
οὕτως ἡ Κ πρὸς τὸ Λ, the words meaning "and let (there) be made, as BC to CG, so K to
L," where L is the straight line which has to be constructed.

The second definition of the *Data* says that *A ratio is said to be given if
we can find* (πορίσασθαι) [*another ratio that is*] *the same with it.* Accordingly
VI. 23 not only proves that equiangular parallelograms have to one another a
ratio which is compounded of two others, but shows that that ratio is "given"
when its component ratios are given, or that it can be represented as a simple
ratio between straight lines.

Just as VI. 23 exhibits the operation necessary for *compounding* two ratios, a proposition (8) of the *Data* indicates the operation by which we may divide one ratio by another. The proposition proves that *Things which have a given ratio to the same thing have also a given ratio to one another.* Euclid's procedure is of course to *compound* one ratio with the *inverse* of the other; but, when this is once done and the result of Prop. 8 obtained, he uses the result in the later propositions as a substitute for the method of composition. Thus he uses the *division* of ratios, instead of composition, in the propositions of the *Data* which deal with the same subject-matter as VI. 23. The effect is to represent the ratio of two equiangular parallelograms as a ratio between straight lines one of which is *one side of one of the parallelograms.* Prop. 56 of the *Data* shows us that, if we want to express the ratio of the parallelogram *AC* to the parallelogram *CF* in the figure

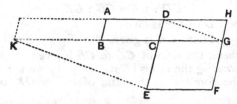

of VI. 23 in the form of a ratio in which, for example, the side *BC* is the antecedent term, the required ratio of the parallelograms is $BC : X$, where

$$DC : CE = CG : X,$$

or X is a fourth proportional to *DC* and the two sides of the parallelogram *CF*.

Measure *CK* along *CB*, produced if necessary, so that

$$DC : CE = CG : CK$$

(whence *CK* is equal to *X*).

[This may be simply done by joining *DG* and then drawing *EK* parallel to it meeting *CB* in *K*.]

Complete the parallelogram *AK*.

Then, since $DC : CE = CG : CK$,

the parallelograms *DK*, *CF* are equal.   [VI. 14]

Therefore $(AC) : (CF) = (AC) : (DK)$ [V. 7]

$$= BC : CK$$ [VI. 1]

$$= BC : X.$$

Prop. 68 of the *Data* uses the same construction to prove that, *If two equiangular parallelograms have to one another a given ratio, and one side have to one side a given ratio, the remaining side will also have to the remaining side a given ratio.*

I do not use the figure of the *Data* but, for convenience' sake, I adhere to the figure given above. Suppose that the ratio of the parallelograms is given, and also that of *CD* to *CE*.

Apply to *CD* the parallelogram *DK* equal to *CF* and such that *CK*, *CB* coincide in direction.   [I. 45]

Then the ratio of *AC* to *KD* is given, being equal to that of *AC* to *CF*.

And $(AC) : (KD) = CB : CK$;

therefore the ratio of *CB* to *CK* is given.

But, since $KD = CF$,

                        $CD : CE = CG : CK.$ [VI. 14]

Hence $CG : CK$ is a given ratio.

And $CB : CK$ was proved to be a given ratio.

Therefore the ratio of CB to CG is given.   [*Data*, Prop. 8]

Lastly we may refer to Prop. 70 of the *Data*, the first part of which proves what corresponds exactly to VI. 23, namely that, *If in two equiangular parallelograms the sides containing the equal angles have a given ratio to one another* [i.e. one side in one to one side in the other], *the parallelograms themselves will also have a given ratio to one another.* [Here the ratios of BC to CG and of CD to CE are given.]

The construction is the same as in the last case, and we have KD equal to CF, so that

                        $CD : CE = CG : CK.$ [VI. 14]

But the ratio of CD to CE is given;
therefore the ratio of CG to CK is given.

And, by hypothesis, the ratio of CG to CB is given.

Therefore, by *dividing* the ratios [*Data*, Prop. 8], we see that the ratio of CB to CK, and therefore [VI. 1] the ratio of AC to DK, or of AC to CF, is given.

Euclid extends these propositions to the case of two parallelograms which have *given* but not equal angles.

Pappus (VII. p. 928) exhibits the result of VI. 23 in a different way, which throws new light on compounded ratios. He proves, namely, that *a parallelogram is to an equiangular parallelogram as the rectangle contained by the adjacent sides of the first is to the rectangle contained by the adjacent sides of the second.*

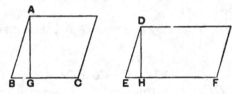

Let AC, DF be equiangular parallelograms on the bases BC, EF, and let the angles at B, E be equal.

Draw perpendiculars AG, DH to BC, EF respectively.

Since the angles at B, G are equal to those at E, H,

                        the triangles ABG, DEH are equiangular.

Therefore $BA : AG = ED : DH.$ [VI. 4]

     But $BA : AG = $ (rect. BA, BC) : (rect. AG, BC),

     and $ED : DH = $ (rect. ED, EF) : (rect. DH, EF).   [VI. 1]

Therefore [V. 11 and V. 16]

(rect. AB, BC) : (rect. DE, EF) = (rect. AG, BC) : (rect. DH, EF)
                                            $= (AC) : (DF)$.

Thus it is proved that the ratio compounded of the ratios $AB : DE$ and $BC : EF$ is equal to the ratio of the rectangle AB, BC to the rectangle DE, EF.

Since each parallelogram in the figure of the proposition can be divided into pairs of equal triangles, and all the triangles which are the halves of either parallelogram have two sides respectively equal and the angles included by them equal or supplementary, it can be at once deduced from VI. 23 (or it can be independently proved by the same method) that *triangles which have one angle of the one equal or supplementary to one angle of the other are in the ratio compounded of the ratios of the sides about the equal or supplementary angles.*   Cf. Pappus VII. pp. 894—6.

VI. 23 also shows that *rectangles, and therefore parallelograms or triangles, are to one another in the ratio compounded of the ratios of their bases and heights.*

The converse of VI. 23 is also true, as is easily proved by *reductio ad absurdum.*   More generally, *if two parallelograms or triangles are in the ratio compounded of the ratios of two adjacent sides, the angles included by those sides are either equal or supplementary.*

PROPOSITION 24.

In any parallelogram the parallelograms about the diameter are similar both to the whole and to one another.

Let $ABCD$ be a parallelogram, and AC its diameter, and let EG, HK be parallelograms about AC;

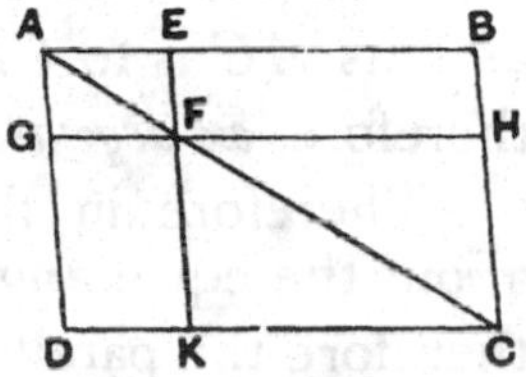

I say that each of the parallelograms EG, HK is similar both to the whole $ABCD$ and to the other.

For, since EF has been drawn parallel to BC, one of the sides of the triangle ABC,

proportionally, as BE is to EA, so is CF to FA.   [VI. 2]

Again, since FG has been drawn parallel to CD, one of the sides of the triangle ACD,

proportionally, as CF is to FA, so is DG to GA.   [VI. 2]

But it was proved that,

as CF is to FA, so also is BE to EA;

therefore also, as BE is to EA, so is DG to GA,

and therefore, *componendo,*

as BA is to AE, so is DA to AG,   [V. 18]

and, alternately,

as BA is to AD, so is EA to AG.   [V. 16]

Therefore in the parallelograms $ABCD$, EG, the sides about the common angle BAD are proportional.

And, since GF is parallel to DC,

the angle AFG is equal to the angle DCA;
and the angle DAC is common to the two triangles ADC, AGF;

therefore the triangle ADC is equiangular with the triangle AGF.

For the same reason

the triangle ACB is also equiangular with the triangle AFE,
and the whole parallelogram $ABCD$ is equiangular with the parallelogram EG.

Therefore, proportionally,

as AD is to DC, so is AG to GF,
as DC is to CA, so is GF to FA,
as AC is to CB, so is AF to FE,

and further, as CB is to BA, so is FE to EA.

And, since it was proved that,

as DC is to CA, so is GF to FA,
and, as AC is to CB, so is AF to FE,
therefore, *ex aequali*, as DC is to CB, so is GF to FE. [v. 22]

Therefore in the parallelograms $ABCD$, EG the sides about the equal angles are proportional;
therefore the parallelogram $ABCD$ is similar to the parallelogram EG.   [VI. Def. 1]

For the same reason

the parallelogram $ABCD$ is also similar to the parallelogram KH;

therefore each of the parallelograms EG, HK is similar to $ABCD$.

But figures similar to the same rectilineal figure are also similar to one another;   [VI. 21]

therefore the parallelogram EG is also similar to the parallelogram HK.

Therefore etc.

Q. E. D.

Simson was of opinion that this proof was made up by some unskilful editor out of two others, the first of which proved by parallels (VI. 2) that the sides about the common angle in the parallelograms are proportional, while the other used the similarity of triangles (VI. 4). It is of course true that, when we have proved by VI. 2 the fact that the sides about the common angle are proportional, we can infer the proportionality of the other sides directly from I. 34 combined with V. 7. But it does not seem to me unnatural that Euclid should (1) deliberately refrain from making any use of I. 34 and (2) determine beforehand that he would prove the sides proportional *in a definite order* beginning with the sides EA, AG and BA, AD about the common angle and then taking the remaining sides in the order indicated by the order of the letters A, G, F, E. Given that Euclid started the proof with such a fixed intention in his mind, the course taken presents no difficulty, nor is the proof unsystematic or unduly drawn out. And its genuineness seems to me supported by the fact that the proof, when once the first two sides about the common angle have been disposed of, follows closely the order and method of VI. 18. Moreover, it could readily be adapted to the more general case of two polygons having a common angle and the other corresponding sides respectively parallel.

The parallelograms in the proposition are of course similarly situated as well as similar; and those "about the diameter" may be "about" the diameter *produced* as well as about the diameter itself.

From the first part of the proof it follows that parallelograms which have one angle equal to one angle and the sides about those angles proportional are similar.

Prop. 26 is the converse of Prop. 24, and there seems to be no reason why they should be separated as they are in the text by the interposition of VI. 25. Campanus has VI. 24 and 26 as VI. 22 and 23 respectively, VI. 23 as VI. 24, and VI. 25 as we have it.

PROPOSITION 25.

To construct one and the same figure similar to a given rectilineal figure and equal to another given rectilineal figure.

Let ABC be the given rectilineal figure to which the figure to be constructed must be similar, and D that to which it must be equal;

thus it is required to construct one and the same figure similar to ABC and equal to D.

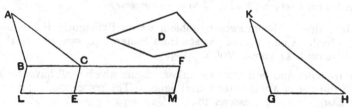

Let there be applied to BC the parallelogram BE equal to the triangle ABC [I. 44], and to CE the parallelogram CM equal to D in the angle FCE which is equal to the angle CBL.   [I. 45]

Therefore BC is in a straight line with CF, and LE with EM.

Now let GH be taken a mean proportional to BC, CF [VI. 13], and on GH let KGH be described similar and similarly situated to ABC.   [VI. 18]

Then, since, as BC is to GH, so is GH to CF,

and, if three straight lines be proportional, as the first is to the third, so is the figure on the first to the similar and similarly situated figure described on the second,   [VI. 19, Por.] therefore, as BC is to CF, so is the triangle ABC to the triangle KGH.

But, as BC is to CF, so also is the parallelogram BE to the parallelogram EF.   [VI. 1]

Therefore also, as the triangle ABC is to the triangle KGH, so is the parallelogram BE to the parallelogram EF; therefore, alternately, as the triangle ABC is to the parallelogram BE, so is the triangle KGH to the parallelogram EF.
                                                                      [V. 16]

But the triangle ABC is equal to the parallelogram BE; therefore the triangle KGH is also equal to the parallelogram EF.

But the parallelogram EF is equal to D; therefore KGH is also equal to D.

And KGH is also similar to ABC.

Therefore one and the same figure KGH has been constructed similar to the given rectilineal figure ABC and equal to the other given figure D.

                                                          Q. E. D.

---

3. **to which the figure to be constructed must be similar,** literally " to which it is required to construct (one) similar," ᾧ δεῖ ὅμοιον συστήσασθαι.

This is the highly important problem which Pythagoras is credited with having solved.   Compare the passage from Plutarch (*Symp.* VIII. 2, 4) quoted in the note on I. 44 above, Vol. I. pp. 343—4.

We are bidden to construct a rectilineal figure which shall have the *form* of one and the *size* of another rectilineal figure.   The corresponding proposition of the *Data*, Prop. 55, asserts that, "if an area (χωρίον) be given in form (εἴδει) and in magnitude, its sides will also be given in magnitude."

Simson sees signs of corruption in the text of this proposition also.   In the first place, the proof speaks of the *triangle ABC*, though, according to the enunciation, the figure for which ABC is taken may be *any* rectilineal figure, εὐθύγραμμον "rectilineal figure" would be more correct, or εἶδος, "figure"; the mistake, however, of using τρίγωνον is not one of great importance, being no doubt due to the accident by which the figure was drawn as a triangle in the diagram.

The other observation is more important.   After Euclid has proved that

(fig. ABC) : (fig. KGH) = (BE) : (EF),

he might have inferred *directly* from V. 14 that, since ABC is equal to BE, KGH is equal to EF.   For V. 14 includes the proof of the fact that, if A is to B as C is to D, and A is equal to C, then B is equal to D, or that of four proportional magnitudes, if the first is equal to the third, the second is equal to the fourth.   Instead of proceeding in this way, Euclid first permutes the proportion by V. 16 into

(fig. ABC) : (BE) = (fig. KGH) : (EF),

and then infers, as if the inference were easier in this form, that, since the *first* is equal to the *second*, the *third* is equal to the *fourth*.   Yet there is no proposition to this effect in Euclid.   The same unnecessary step of permutation is also found in the Greek text of XI. 23 and XII. 2, 5, 11, 12 and 18.   In reproducing the proofs we may simply leave out the steps and refer to V. 14.

PROPOSITION 26.

If from a parallelogram there be taken away a parallelogram similar and similarly situated to the whole and having a common angle with it, it is about the same diameter with the whole

For from the parallelogram $ABCD$ let there be taken away the parallelogram AF similar and similarly situated to $ABCD$, and having the angle DAB common with it;

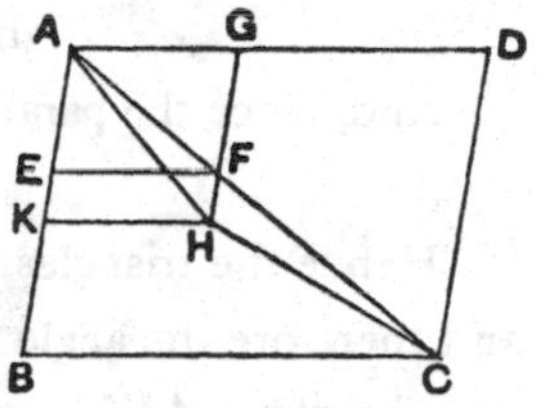

I say that $ABCD$ is about the same diameter with AF.

For suppose it is not, but, if possible, let AHC be the diameter < of $ABCD$ >, let GF be produced and carried through to H, and let HK be drawn through H parallel to either of the straight lines AD, BC.   [I. 31]

Since, then, $ABCD$ is about the same diameter with KG, therefore, as DA is to AB, so is GA to AK.   [VI. 24]

But also, because of the similarity of $ABCD$, EG,

as DA is to AB, so is GA to AE;

therefore also, as GA is to AK, so is GA to AE.   [V. 11]

Therefore GA has the same ratio to each of the straight lines AK, AE.

Therefore AE is equal to AK [v. 9], the less to the greater: which is impossible.

Therefore $ABCD$ cannot but be about the same diameter with AF;

therefore the parallelogram $ABCD$ is about the same diameter with the parallelogram AF.

Therefore etc.

Q. E. D.

"For suppose it is not, but, if possible, let AHC be the diameter." What is meant is "For, if AFC is not the diameter of the parallelogram AC, let AHC be its diameter." The Greek text has ἔστω αὐτῶν διάμετρος ἡ ΑΘΓ; but clearly αὐτῶν is wrong, as we cannot assume that one straight line is the diameter of both parallelograms, which is just what we have to prove. F and V omit the αὐτῶν, and Heiberg prefers this correction to substituting αὐτοῦ after Peyrard. I have inserted "$<$of $ABCD>$" to make the meaning clear.

If the straight line AHC does not pass through F, it must meet either GF or GF produced in some point H. The reading in the text "and let GF be *produced and carried through to* H" (καὶ ἐκβληθεῖσα ἡ ΗΖ διήχθω ἐπὶ τὸ Θ) corresponds to the supposition that H is on GF produced. The words were left out by Theon, evidently because in the figure of the MSS. the letters E, Z and K, Θ were interchanged. Heiberg therefore, following August, has preferred to retain the words and to correct the figure, as well as the passage in the text where AE, AK were interchanged to be in accord with the MS. figure.

It is of course possible to prove the proposition directly, as is done by Dr Lachlan. Let AF, AC be the diagonals, and let us make no assumption as to how they fall.

Then, since EF is parallel to AG and therefore to BC,

the angles AEF, ABC are equal.

And, since the parallelograms are similar,

$$AE : EF = AB : BC.$$ [VI. Def. 1]

Hence the triangles AEF, ABC are similar,   [VI. 6]

and therefore the angle FAE is equal to the angle CAB.

Therefore AF falls on AC.

The proposition is equally true if the parallelogram which is similar and similarly situated to the given parallelogram is not "taken away" from it, but is so placed that it is entirely outside the other, while two sides form an angle vertically opposite to an angle of the other. In this case the diameters are not "the same," in the words of the enunciation, but are in a straight line with one another. This extension of the proposition is, as will be seen, necessary for obtaining, according to the method adopted by Euclid in his solution of the problem in VI. 28, the second solution of that problem.

PROPOSITION 27.

Of all the parallelograms applied to the same straight line and deficient by parallelogrammic figures similar and similarly situated to that described on the half of the straight line, that parallelogram is greatest which is applied to the half of the straight line and is similar to the defect.

Let AB be a straight line and let it be bisected at C; let there be applied to the straight
line AB the parallelogram AD deficient by the parallelogrammic figure DB described on the half of AB, that is, CB;
I say that, of all the parallelograms applied to AB and deficient by parallelogrammic figures similar and similarly situated to DB, AD is greatest.

For let there be applied to the straight line AB the parallelogram AF deficient by the parallelogrammic figure FB similar and similarly situated to DB;
I say that AD is greater than AF.

For, since the parallelogram DB is similar to the parallelogram FB,

they are about the same diameter.   [VI. 26]

Let their diameter DB be drawn, and let the figure be described.

Then, since CF is equal to FE,   [I. 43]
and FB is common,

therefore the whole CH is equal to the whole KE.

But CH is equal to CG, since AC is also equal to CB.

[I. 36]

Therefore GC is also equal to EK.
Let CF be added to each;

therefore the whole AF is equal to the gnomon LMN;
so that the parallelogram DB, that is, AD, is greater than the parallelogram AF.

Therefore etc.

Q. E. D.

We have already (note on I. 44) seen the significance, in Greek geometry, of the theory of "the application of areas, their exceeding and their falling-short." In I. 44 it was a question of "applying to a given straight line (exactly, without 'excess' or 'defect') a parallelogram equal to a given rectilineal figure, in a given angle." Here, in VI. 27—29, it is a question of parallelograms applied to a straight line but "*deficient (or exceeding) by parallelograms similar and similarly situated to a given parallelogram.*" Apart from size, it is easy to construct any number of parallelograms "deficient" or "exceeding" in the manner described. Given the straight line AB to which the parallelogram has to be applied, we describe on the base CB, where C is on AB, or on BA

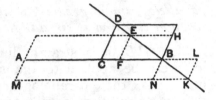

produced beyond A, any parallelogram "similarly situated" and either equal or similar to the given parallelogram (Euclid takes the similar and similarly situated parallelogram on half the line), draw the diagonal BD, take on it (produced if necessary) any points as E, K, draw EF, or KL, parallel to CD to meet AB or AB produced and complete the parallelograms, as AH, ML.

If the point E is taken on BD or BD produced beyond D, it must be so taken that EF meets AB between A and B. Otherwise the parallelogram AE would not be applied to AB itself, as it is required to be.

The parallelograms BD, BE, being about the same diameter, are similar [VI. 24], and BE is the defect of the parallelogram AE relatively to AB. AE is then a parallelogram applied to AB but deficient by a parallelogram similar and similarly situated to BD.

If K is on DB produced, the parallelogram BK is similar to BD, but it is the *excess* of the parallelogram AK relatively to the base AB. AK is a parallelogram applied to AB but exceeding by a parallelogram similar and similarly situated to BD.

Thus it is seen that BD produced both ways is the *locus* of points, such as E or K, which determine, with the direction of CD, the position of A, and the direction of AB, parallelograms applied to AB and *deficient* or *exceeding* by parallelograms similar and similarly situated to the given parallelogram.

The importance of VI. 27—29 from a historical point of view cannot be overrated. They give the geometrical equivalent of the algebraical solution of the most general form of quadratic equation when that equation has a real and positive root. It will also enable us to find a real *negative* root of a quadratic equation; for such an equation can, by altering the sign of x, be turned into another with a real *positive* root, when the geometrical method again becomes applicable. It will also, as we shall see, enable us to represent *both* roots when both are real and positive, and therefore to represent both roots when both are real but either positive or negative.

The method of these propositions was constantly used by the Greek geometers in the solution of problems, and they constitute the foundation of Book X. of the *Elements* and of Apollonius' treatment of the conic sections. Simson's observation on the subject is entirely justified. He says namely on VI. 28, 29: "These two problems, to the first of which the 27th Prop. is necessary, are the most general and useful of all in the Elements, and are most frequently made use of by the ancient geometers in the solution of other problems; and therefore are very ignorantly left out by Tacquet and

Dechales in their editions of the Elements, who pretend that they are scarce of any use."

It is strange that, with this observation before him, even Todhunter should have written as follows. "We have omitted in the sixth Book Propositions 27, 28, 29 and the first solution which Euclid gives of Proposition 30, as they appear now to be never required, and have been condemned as useless by various modern commentators; see Austin, Walker and Lardner."

VI. 27 contains the διορισμός, the condition for a real solution, of the problem contained in the proposition following it. The maximum of all the parallelograms having the given property which can be applied to a given straight line is that which is described upon half the line (τὸ ἀπὸ τῆς ἡμισείας ἀναγραφόμενον). This corresponds to the condition that an equation of the form

$$ax - px^2 = A$$

may have a real root. The correctness of the result may be seen by taking the case in which the parallelograms are rectangles, which enables us to leave out of account the *sine of the angle* of the parallelograms without any real loss of generality. Suppose the sides of the rectangle to which the *defect* is to be similar to be as b to c, b corresponding to the side of the defect which lies along AB. Suppose that $AKFG$ is any parallelogram applied to AB having the given property, that $AB = a$, and that $FK = x$. Then

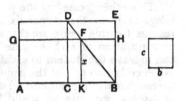

$$KB = \frac{b}{c}x, \text{ and therefore } AK = a - \frac{b}{c}x.$$

Hence $\left(a - \frac{b}{c}x\right)x = S$, where S is the area of the rectangle $AKFG$.

Thus, given the equation

$$ax - \frac{b}{c}x^2 = S,$$

where S is undetermined, VI. 27 tells us that, if x is to have a real value, S cannot be greater than the rectangle CE.

Now $CB = \frac{a}{2}$, and therefore $CD = \frac{c}{b} \cdot \frac{a}{2}$;

whence $S \not> \frac{c}{b} \cdot \frac{a^2}{4}$,

which is just the same result as we obtain by the algebraical method.

In the particular case where the *defect* of the parallelogram is to be a *square*, the condition becomes the statement of the fact that, *if a straight line be divided into two parts, the rectangle contained by the parts cannot exceed the square on half the line.*

Now suppose that, instead of taking F on BD as in the figure of the proposition, we take F on BD produced beyond D but so that DF is less than BD.

Complete the figure, as shown, after the manner of the construction in the proposition.

Then the parallelogram $FKBH$ is similar to the given parallelogram to which the defect is to be similar. Hence the parallelogram $GAKF$ is also a parallelogram applied to AB and satisfying the given condition.

We can now prove that $GAKF$ is less than CE or AD.

Let ED produced meet AG in O.

Now, since BF is the diagonal of the parallelogram KH, the complements KD, DH are equal.

But

$DH = DG$, and DG is greater than OF.

Therefore $KD > OF$.

Add OK to each;

and AD, or CE, $> AF$.

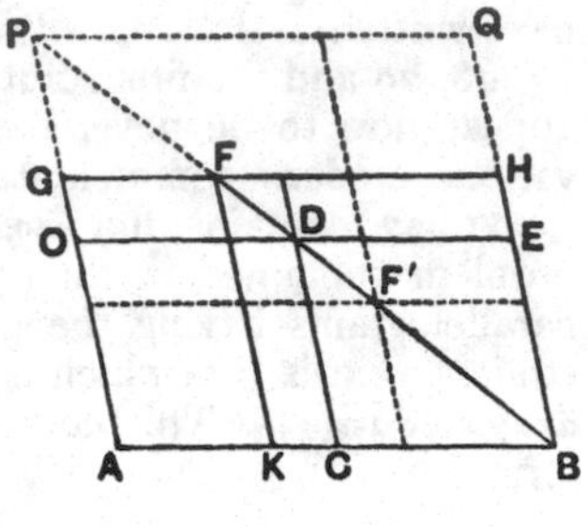

This other " case " of the proposition is found in all the MSS., but Heiberg relegates it to the Appendix as being very obviously interpolated. The reasons for this course are that it is not in Euclid's manner to give a separate demonstration of such a " case "; it is rather his habit to give one case only and to leave the student to satisfy himself about any others (cf. I. 7). Internal evidence is also against the genuineness of the separate proof. It is put after the *conclusion* of the proposition instead of before it, and, if Euclid had intended to discuss two cases, he would have distinguished them at the beginning of the proposition, as it was his invariable practice to do. Moreover the second " case " is the less worth giving because it can be so easily reduced to the first. For suppose F' to be taken on BD so that $FD = F'D$. Produce BF to meet AG produced in P. Complete the parallelogram $BAPQ$, and draw through F' straight lines parallel to and meeting its opposite sides.

Then the complement $F'Q$ is equal to the complement AF'.

And it is at once seen that AF, $F'Q$ are equal and similar. Hence the solution of the problem represented by AF or $F'Q$ gives a parallelogram of the same size as AF' arrived at as in the first " case."

It is worth noting that the actual difference between the parallelogram AF and the maximum area AD that it can possibly have is represented in the figure. The difference is the small parallelogram DF.

PROPOSITION 28.

To a given straight line to apply a parallelogram equal to a given rectilineal figure and deficient by a parallelogrammic figure similar to a given one : thus the given rectilineal figure must not be greater than the parallelogram described on the half of the straight line and similar to the defect.

Let AB be the given straight line, C the given rectilineal figure to which the figure to be applied to AB is required to be equal, not being greater than the parallelogram described on the half of AB and similar to the defect, and D the parallelogram to which the defect is required to be similar;

thus it is required to apply to the given straight line AB a parallelogram equal to the given rectilineal figure C and deficient by a parallelogrammic figure which is similar to D.

Let AB be bisected at the point E, and on EB let $EBFG$ be described similar and similarly situated to D;   [VI. 18] let the parallelogram AG be completed.

If then AG is equal to C, that which was enjoined will have been done;

for there has been applied to the given straight line AB the parallelogram AG equal to the given rectilineal figure C and deficient by a parallelogrammic figure GB which is similar to D.

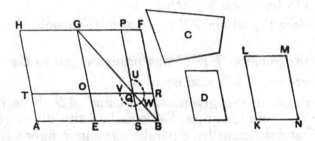

But, if not, let HE be greater than C.
Now HE is equal to GB;
     therefore GB is also greater than C.

Let $KLMN$ be constructed at once equal to the excess by which GB is greater than C and similar and similarly situated to D.   [VI. 25]
But D is similar to GB;
     therefore KM is also similar to GB.   [VI. 21]
Let, then, KL correspond to GE, and LM to GF.
Now, since GB is equal to C, KM,
     therefore GB is greater than KM;
therefore also GE is greater than KL, and GF than LM.

Let GO be made equal to KL, and GP equal to LM; and let the parallelogram $OGPQ$ be completed;
     therefore it is equal and similar to KM.

Therefore GQ is also similar to GB;   [VI. 21]
therefore GQ is about the same diameter with GB.   [VI. 26]

Let GQB be their diameter, and let the figure be described.

Then, since BG is equal to C, KM,

and in them GQ is equal to KM,

therefore the remainder, the gnomon UWV, is equal to the remainder C.

And, since PR is equal to OS,

let QB be added to each;

therefore the whole PB is equal to the whole OB.

But OB is equal to TE, since the side AE is also equal to the side EB;   [I. 36]

therefore TE is also equal to PB.

Let OS be added to each;

therefore the whole TS is equal to the whole, the gnomon VWU.

But the gnomon VWU was proved equal to C;

therefore TS is also equal to C.

Therefore to the given straight line AB there has been applied the parallelogram ST equal to the given rectilineal figure C and deficient by a parallelogrammic figure QB which is similar to D.

Q. E. F.

The second part of the enunciation of this proposition which states the διορισμός appears to have been considerably amplified, but not improved in the process, by Theon. His version would read as follows. "But the given rectilineal figure, that namely to which the applied parallelogram must be equal (ᾧ δεῖ ἴσον παραβαλεῖν), must not be greater than that applied to the half (παραβαλλομένου instead of ἀναγραφομένου), the defects being similar, (namely) that (of the parallelogram applied) to the half and that (of the required parallelogram) which must have a similar defect" (ὁμοίων ὄντων τῶν ἐλλειμμάτων τοῦ τε ἀπὸ τῆς ἡμισείας καὶ ᾧ δεῖ ὅμοιον ἐλλείπειν). The first amplification "that to which the applied parallelogram must be equal" is quite unnecessary, since "the given rectilineal figure" could mean nothing else. The above attempt at a translation will show how difficult it is to make sense of the words at the end; they speak of *two* defects apparently and, while one may well be the "defect on the half," the other can hardly be *the given parallelogram* "to which the defect (of the required parallelogram) must be similar." Clearly the reading given above (from P) is by far the better.

In this proposition and the next there occurs the tacit assumption (already alluded to in the note on VI. 22) that *if, of two similar parallelograms, one is greater than the other, either side of the greater is greater than the corresponding side of the less.*

As already remarked, VI. 28 is the geometrical equivalent of the solution of the quadratic equation

$$ax - \frac{b}{c}x^2 = S,$$

subject to the condition necessary to admit of a real solution, namely that

$$S \not> \frac{c}{b} \cdot \frac{a^2}{4}.$$

The corresponding proposition in the *Data* is (Prop. 58), *If a given (area) be applied* (i.e. in the form of a parallelogram) *to a given straight line and be deficient by a figure* (i.e. a parallelogram) *given in species, the breadths of the defect are given.*

To exhibit the exact correspondence between Euclid's geometrical and the ordinary algebraical method of solving the equation we will, as before (in order to avoid bringing in a constant dependent on the sine of the angle of the parallelograms), suppose the parallelograms to be rectangles. To solve the equation algebraically we change the signs and write it

$$\frac{b}{c}x^2 - ax = -S.$$

We may now complete the square by adding $\frac{c}{b} \cdot \frac{a^2}{4}$.

Thus
$$\frac{b}{c}x^2 - ax + \frac{c}{b} \cdot \frac{a^2}{4} = \frac{c}{b} \cdot \frac{a^2}{4} - S;$$

and, extracting the square root, we have

$$\sqrt{\frac{b}{c}} \cdot x - \sqrt{\frac{c}{b}} \cdot \frac{a}{2} = \pm \sqrt{\frac{c}{b} \cdot \frac{a^2}{4} - S},$$

and
$$x = \frac{c}{b} \cdot \frac{a}{2} \pm \sqrt{\frac{c}{b}\left(\frac{c}{b} \cdot \frac{a^2}{4} - S\right)}.$$

Now let us observe Euclid's method.

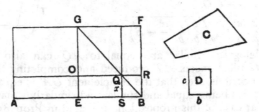

He first describes $GEBF$ on EB (half of AB) similar to the given parallelogram D.

He then places in one angle FGE of $GEBF$ a similar and similarly situated parallelogram GQ, equal to the difference between the parallelogram GB and the area C.

With our notation, $GO : OQ = c : b,$

whence $OQ = GO \cdot \frac{b}{c}.$

Similarly
$$\frac{a}{2} = EB = GE \cdot \frac{b}{c},$$

so that
$$GE = \frac{c}{b} \cdot \frac{a}{2}.$$

Therefore the parallelogram $GQ = GO^2 \cdot \frac{b}{c}$,

and   the parallelogram $GB = \frac{c}{b} \cdot \frac{a^2}{4}.$

Thus, in taking the parallelogram GQ equal to $(GB - S)$, Euclid really finds GO from the equation

$$GO^2 \cdot \frac{b}{c} = \frac{c}{b} \cdot \frac{a^2}{4} - S.$$

The value which he finds is

$$GO = \sqrt{\frac{c}{b}\left(\frac{c}{b} \cdot \frac{a^2}{4} - S\right)},$$

and he finds QS (or x) by *subtracting* GO from GE; whence

$$x = \frac{c}{b} \cdot \frac{a}{2} - \sqrt{\frac{c}{b}\left(\frac{c}{b} \cdot \frac{a^2}{4} - S\right)}.$$

It will be observed that Euclid only gives one solution, that corresponding to the *negative* sign before the radical. But the reason must be the same as that for which he only gives one "case" in VI. 27. He cannot have failed to see how to *add* GO to GE would give another solution. As shown under the last proposition, the other solution can be arrived at

(1) by placing the parallelogram $GOQP$ in the angle vertically opposite to FGE so that GQ' lies along BG *produced*. The parallelogram AQ' then gives the second solution. The side of this parallelogram lying along AB is equal to SB. The other side is what we have called x, and in this case

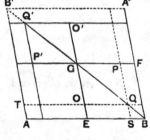

$$x = EG + GO$$
$$= \frac{c}{b} \cdot \frac{a}{2} + \sqrt{\frac{c}{b}\left(\frac{c}{b} \cdot \frac{a^2}{4} - S\right)}.$$

(2) A parallelogram similar and equal to AQ' can also be obtained by producing BG till it meets AT produced and completing the parallelogram $B'ABA'$, whence it is seen that the complement QA' is equal to the complement AQ, besides being equal and similar and similarly situated to AQ'.

A particular case of this proposition, indicated in Prop. 85 of the *Data*, is that in which the sides of the defect are equal, so that the defect is a *rhombus* with a given angle. Prop. 85 proves that, *If two straight lines contain a given area in a given angle, and the sum of the straight lines be given, each of them will be given also.* AB, BC being the given straight lines "containing a given area AC in a given angle ABC," one side CB is produced to D so that BD is equal to AB, and the parallelograms are completed. Then, by hypothesis, CD is of given length, and AC is a parallelo-

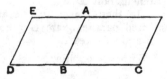

gram applied to CD falling short by a rhombus (AD) with a given angle EDB. The case is thus a particular case of Prop. 58 of the *Data* quoted above (p. 263) as corresponding to VI. 28.

A particular case of the last, that namely in which the defect is a *square*, corresponding to the equation

$$ax - x^2 = b^2,$$

is important. This is the problem of *applying to a given straight line a rectangle equal to a given area and falling short by a square*; and it can be solved, without the aid of Book VI., as shown above under II. 5 (Vol. I. pp. 383—4).

Proposition 29.

To a given straight line to apply a parallelogram equal to a given rectilineal figure and exceeding by a parallelogrammic figure similar to a given one.

Let AB be the given straight line, C the given rectilineal figure to which the figure to be applied to AB is required to be equal, and D that to which the excess is required to be similar;

thus it is required to apply to the straight line AB a parallelogram equal to the rectilineal figure C and exceeding by a parallelogrammic figure similar to D.

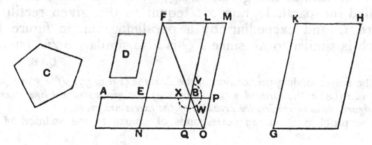

Let AB be bisected at E;

let there be described on EB the parallelogram BF similar and similarly situated to D;

and let GH be constructed at once equal to the sum of BF, C and similar and similarly situated to D.   [VI. 25]

Let KH correspond to FL and KG to FE.

Now, since GH is greater than FB,

therefore KH is also greater than FL, and KG than FE.

Let FL, FE be produced,
let FLM be equal to KH, and FEN to KG,
and let MN be completed;

therefore MN is both equal and similar to GH.

But GH is similar to EL;

therefore MN is also similar to EL;   [VI. 21]
therefore EL is about the same diameter with MN.   [VI. 26]

Let their diameter FO be drawn, and let the figure be described.

Since GH is equal to EL, C,
while GH is equal to MN,
therefore MN is also equal to EL, C.

Let EL be subtracted from each;

therefore the remainder, the gnomon XWV, is equal to C.

Now, since AE is equal to EB,

AN is also equal to NB [I. 36], that is, to LP [I. 43].

Let EO be added to each;

therefore the whole AO is equal to the gnomon VWX.

But the gnomon VWX is equal to C;

therefore AO is also equal to C.

Therefore to the given straight line AB there has been applied the parallelogram AO equal to the given rectilineal figure C and exceeding by a parallelogrammic figure QP which is similar to D, since PQ is also similar to EL [VI. 24].

Q. E. F.

The corresponding proposition in the *Data* is (Prop. 59), *If a given (area) be applied* (i.e. in the form of a parallelogram) *to a given straight line exceeding by a figure given in species, the breadths of the excess are given.*

The problem of VI. 29 corresponds of course to the solution of the quadratic equation

$$ax + \frac{b}{c} x^2 = S.$$

The algebraical solution of this equation gives

$$x = -\frac{c}{b} \cdot \frac{a}{2} \pm \sqrt{\frac{c}{b} \left(\frac{c}{b} \cdot \frac{a^2}{4} + S \right)}.$$

The exact correspondence of Euclid's method to the algebraical solution may be seen, as in the case of VI. 28, by supposing the parallelograms to be *rectangles*. In this case Euclid's construction on EB of the parallelogram EL similar to D is equivalent to finding that

$$FE = \frac{c}{b} \cdot \frac{a}{2}, \text{ and } EL = \frac{c}{b} \cdot \frac{a^2}{4}.$$

His determination of the similar parallelogram MN equal to the sum of EL and S corresponds to proving that

$$FN^2 \cdot \frac{b}{c} = \frac{c}{b} \cdot \frac{a^2}{4} + S,$$

or

$$FN = \sqrt{\frac{c}{b}\left(\frac{c}{b} \cdot \frac{a^2}{4} + S\right)},$$

whence x is found as

$$x = FN - FE = \sqrt{\frac{c}{b}\left(\frac{c}{b} \cdot \frac{a^2}{4} + S\right)} - \frac{c}{b} \cdot \frac{a}{2}.$$

Euclid takes, in this case, the solution corresponding to the *positive* sign before the radical because, from his point of view, that would be the *only* solution.

No διορισμός is necessary because a real geometrical solution is always possible whatever be the size of S.

Again the *Data* has a proposition indicating the particular case in which the *excess* is a *rhombus* with a given angle. Prop. 84 proves that, *If two straight lines contain a given area in a given angle, and one of the straight lines is greater than the other by a given straight line, each of the two straight lines is given also.* The proof reduces the proposition to a particular case of *Data*, Prop. 59, quoted above as corresponding to VI. 29.

Again there is an important particular case which can be solved by means of Book II. only, as shown under II. 6 above (Vol. I. pp. 386—8), the case namely in which the excess is a *square*, corresponding to the solution of the equation

$$ax + x^2 = b^2.$$

This is the problem of *applying to a given straight line a rectangle equal to a given area and exceeding by a square.*

PROPOSITION 30.

To cut a given finite straight line in extreme and mean ratio.

Let AB be the given finite straight line;

thus it is required to cut AB in extreme and mean ratio.

On AB let the square BC be described;

and let there be applied to AC the parallelogram CD equal to BC and exceeding by the figure AD similar to BC. [VI. 29]

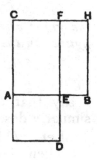

Now BC is a square;

therefore AD is also a square.

And, since BC is equal to CD,

let CE be subtracted from each;

therefore the remainder BF is equal to the remainder AD.

But it is also equiangular with it ;

therefore in *BF*, *AD* the sides about the equal angles are reciprocally proportional ;   [VI. 14]

therefore, as *FE* is to *ED*, so is *AE* to *EB*.

But *FE* is equal to *AB*, and *ED* to *AE*.

Therefore, as *BA* is to *AE*, so is *AE* to *EB*.

And *AB* is greater than *AE* ;

therefore *AE* is also greater than *EB*.

Therefore the straight line *AB* has been cut in extreme and mean ratio at *E*, and the greater segment of it is *AE*.

Q. E. F.

It will be observed that the construction in the text is a direct application of the preceding Prop. 29 in the particular case where the *excess* of the parallelogram which is applied is a *square*. This fact coupled with the position of VI. 30 is a sufficient indication that the construction is Euclid's.

In one place Theon appears to have amplified the argument. The text above says "But *FE* is equal to *AB*," while the MSS. B, F, V and p have "But *FE* is equal to *AC*, that is, to *AB*."

The MSS. give after ὅπερ ἔδει ποιῆσαι an alternative construction which Heiberg relegates to the Appendix. The text-books give this construction alone and leave out the other. It will be remembered that the alternative proof does no more than refer to the equivalent construction in II. 11.

"Let *AB* be cut at *C* so that the rectangle *AB*, *BC* is equal to the square on *CA*.   [II. 11]

Since then the rectangle *AB*, *BC* is equal to the square on *CA*, therefore, as *BA* is to *AC*, so is *AC* to *CB*.   [VI. 17]

Therefore *AB* has been cut in extreme and mean ratio at *C*."

It is intrinsically improbable that this alternative construction was added to the other by Euclid himself. It is however just the kind of interpolation that might be expected from an editor. If Euclid had preferred the alternative construction, he would have been more likely to give it alone.

Proposition 31.

In right-angled triangles the figure on the side subtending the right angle is equal to the similar and similarly described figures on the sides containing the right angle.

Let *ABC* be a right-angled triangle having the angle *BAC* right ;

I say that the figure on *BC* is equal to the similar and similarly described figures on *BA*, *AC*.

Let *AD* be drawn perpendicular.

Then since, in the right-angled triangle *ABC*, *AD* has

been drawn from the right angle at *A* perpendicular to the base *BC*,

the triangles *ABD*, *ADC* adjoining the perpendicular are similar both to the whole *ABC* and to one another.   [VI. 8]

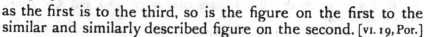

And, since *ABC* is similar to *ABD*,

therefore, as *CB* is to *BA*, so is *AB* to *BD*.   [VI. Def. 1]

And, since three straight lines are proportional,

as the first is to the third, so is the figure on the first to the similar and similarly described figure on the second. [VI. 19, Por.]

Therefore, as *CB* is to *BD*, so is the figure on *CB* to the similar and similarly described figure on *BA*.

For the same reason also,

as *BC* is to *CD*, so is the figure on *BC* to that on *CA* ;

so that, in addition,

as *BC* is to *BD*, *DC*, so is the figure on *BC* to the similar and similarly described figures on *BA*, *AC*.

But *BC* is equal to *BD*, *DC* ;

therefore the figure on *BC* is also equal to the similar and similarly described figures on *BA*, *AC*.

Therefore etc.

Q. E. D.

As we have seen (note on I. 47), this extension of I. 47 is credited by Proclus to Euclid personally.

There is one inference in the proof which requires examination. Euclid proves that

$$CB : BD = (\text{figure on } CB) : (\text{figure on } BA),$$

and that $$BC : CD = (\text{figure on } BC) : (\text{figure on } CA),$$

and then infers directly that

$$BC : (BD + CD) = (\text{fig. on } BC) : (\text{sum of figs. on } BA \text{ and } AC).$$

Apparently V. 24 must be relied on as justifying this inference. But it is not directly applicable ; for what it proves is that, if

$$a : b = c : d,$$

and $$e : b = f : d,$$

then $$(a + e) : b = (c + f) : d.$$

Thus we should *invert* the first two proportions given above (by Simson's

Prop. B which, as we have seen, is a direct consequence of the definition of proportion), and thence infer by v. 24 that

$$(BD + CD) : BC = (\text{sum of figs. on } BA, AC) : (\text{fig. on } BC).$$

But $\qquad\qquad\qquad BD + CD$ is equal to BC;

therefore (by Simson's Prop. A, which again is an immediate consequence of the definition of proportion) the sum of the figures on BA, AC is equal to the figure on BC.

The MSS. again give an alternative proof which Heiberg places in the Appendix. It first shows that the similar figures on the three sides have the same ratios to one another as the *squares* on the sides respectively. Whence, by using I. 47 and the same argument based on v. 24 as that explained above, the result is obtained.

If it is considered essential to have a proof which does not use Simson's Props. B and A or any proposition but those actually given by Euclid, no method occurs to me except the following.

Eucl. v. 22 proves that, if a, b, c are three magnitudes, and d, e, f three others, such that

$$a : b = d : e,$$
$$b : c = e : f,$$

then, *ex aequali*, $\qquad\qquad a : c = d : f.$

If now in addition $\qquad\qquad a : b = b : c,$

so that, also, $\qquad\qquad\quad d : e = e : f,$

the ratio $a : c$ is duplicate of the ratio $a : b$, and the ratio $d : f$ duplicate of the ratio $d : e$, whence *the ratios which are duplicate of equal ratios are equal*.

Now (fig. on AC) : (fig. on AB) = the ratio duplicate of $AC : AB$
$$= \text{the ratio duplicate of } CD : DA$$
$$= CD : BD.$$

Hence $\quad$ (sum of figs. on AC, AB) : (fig. on AB) = $BC : BD$. $\qquad$ [v. 18]

But $\qquad\qquad$ (fig. on BC) : (fig. on AB) = $BC : BD$
$$\qquad\qquad\qquad\qquad\qquad\qquad (\text{as in Euclid's proof}).$$

Therefore the sum of the figures on AC, AB has to the figure on AB the same ratio as the figure on BC has to the figure on AB, whence

the figures on AC, AB are together equal to the figure on BC. $\quad$ [v. 9]

PROPOSITION 32.

If two triangles having two sides proportional to two sides be placed together at one angle so that their corresponding sides are also parallel, the remaining sides of the triangles will be in a straight line.

Let ABC, DCE be two triangles having the two sides BA, AC proportional to the two sides DC, DE, so that, as AB is to AC, so is DC to DE, and AB parallel to DC, and AC to DE;

I say that BC is in a straight line with CE.

For, since AB is parallel to DC,
and the straight line AC has fallen upon them,
the alternate angles BAC, ACD
are equal to one another. [I. 29]

For the same reason
the angle CDE is also
equal to the angle ACD;
so that the angle BAC is equal
to the angle CDE.

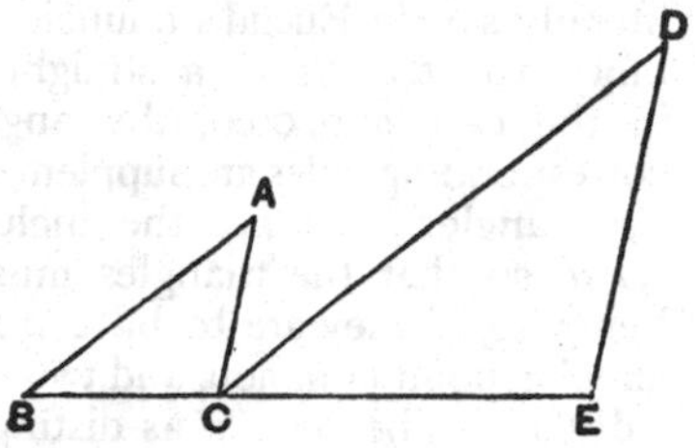

And, since ABC, DCE are
two triangles having one angle, the angle at A, equal to one
angle, the angle at D,

the sides about the equal angles proportional,
so that, as BA is to AC, so is CD to DE,

therefore the triangle ABC is equiangular with the
triangle DCE; [VI. 6]

therefore the angle ABC is equal to the angle DCE.

But the angle ACD was also proved equal to the angle
BAC;

therefore the whole angle ACE is equal to the two angles
ABC, BAC.

Let the angle ACB be added to each;

therefore the angles ACE, ACB are equal to the angles BAC,
ACB, CBA.

But the angles BAC, ABC, ACB are equal to two right
angles; [I. 32]

therefore the angles ACE, ACB are also equal to two
right angles.

Therefore with a straight line AC, and at the point C on
it, the two straight lines BC, CE not lying on the same side
make the adjacent angles ACE, ACB equal to two right
angles;

therefore BC is in a straight line with CE. [I. 14]

Therefore etc.

Q. E. D.

It has often been pointed out (e.g. by Clavius, Lardner and Todhunter)
that the enunciation of this proposition is not precise enough. Suppose that

ABC is a triangle. From *C* draw *CD* parallel to *BA* and of any length. From *D* draw *DE* parallel to *CA* and of such length that

$$CD : DE = BA : AC.$$

Then the triangles *ABC*, *ECD*, which have the angular point *C* common literally satisfy Euclid's enunciation; but by no possibility can *CE* be in a straight line with *CB* if, as in the case supposed, the angles included by the corresponding sides are supplementary (unless both are right angles). Hence the included angles must be *equal*, so that the triangles must be *similar*. That being so, if they are to have nothing more than one angular point common, and two pairs of corresponding

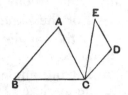

sides are to be *parallel* as distinguished from one or both being in the *same straight line*, the triangles can only be placed so that the corresponding sides in both are on the same side of the third side of either, and the sides (other than the third sides) which meet at the common angular point are not corresponding sides.

Todhunter remarks that the proposition seems of no use. Presumably he did not know that it *is* used by Euclid himself in XIII. 17. This is so however, and therefore it was not necessary, as several writers have thought, to do away with the proposition and find a substitute which should be more useful.

1. De Morgan proposes this theorem : "If two similar triangles be placed with their bases parallel, and the equal angles at the bases towards the same parts, the other sides are parallel, each to each; or one pair of sides are in the same straight line and the other pair are parallel."

2. Dr Lachlan substitutes the somewhat similar theorem, "If two similar triangles be placed so that two sides of the one are parallel to the corresponding sides of the other, the third sides are parallel."

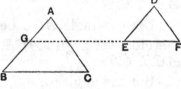

But it is to be observed that these propositions can be proved without using Book VI. at all; they can be proved from Book I., and the triangles may as well be called "equiangular" simply. It is true that Book VI. is no more than formally necessary to Euclid's proposition. He merely uses VI. 6 because his enunciation does not *say* that the triangles are similar; and he only proves them to be similar in order to conclude that they are equiangular. From this point of view Mr Taylor's substitute seems the best, viz.

3. "If two triangles have sides parallel in pairs, the straight lines joining the corresponding vertices meet in a point, or are parallel."

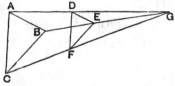

Simson has a theory (unnecessary in the circumstances) as to the possible object of VI. 32 as it stands. He points out that the enunciation of VI. 26 might be more general so as to cover the case of similar and similarly situated parallelograms with equal angles not coincident but vertically opposite. It can then be proved that the diagonals drawn

through the common angular point are in one straight line. If *ABCF, CDEG*
be similar and similarly situated parallelograms,
so that *BCG, DCF* are straight lines, and if
the diagonals *AC, CE* be drawn, the triangles
ABC, CDE are similar and *are placed exactly
as described in* VI. 32, so that *AC, CE* are in a
straight line. Hence Simson suggests that
there may have been, in addition to the indirect demonstration in VI. 26, a *direct* proof
covering the case just given which may have
used the result of VI. 32. I think however

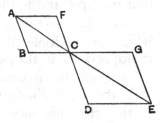

that the place given to the latter proposition in Book VI. is against this view.

PROPOSITION 33.

*In equal circles angles have the same ratio as the circum-
ferences on which they stand, whether they stand at the centres
or at the circumferences.*

Let *ABC, DEF* be equal circles, and let the angles *BGC,
EHF* be angles at their centres *G, H*, and the angles *BAC,
EDF* angles at the circumferences;

I say that, as the circumference *BC* is to the circumference
EF, so is the angle *BGC* to the angle *EHF*, and the angle
BAC to the angle *EDF*.

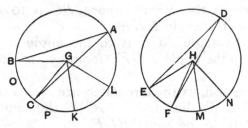

For let any number of consecutive circumferences *CK,
KL* be made equal to the circumference *BC*,

and any number of consecutive circumferences *FM, MN* equal
to the circumference *EF*;

and let *GK, GL, HM, HN* be joined.

Then, since the circumferences *BC, CK, KL* are equal
to one another,

the angles *BGC, CGK, KGL* are also equal to one another;

[III. 27]

therefore, whatever multiple the circumference BL is of BC, that multiple also is the angle BGL of the angle BGC.

For the same reason also,
whatever multiple the circumference NE is of EF, that multiple also is the angle NHE of the angle EHF.

If then the circumference BL is equal to the circumference EN, the angle BGL is also equal to the angle EHN ;   [III. 27]
if the circumference BL is greater than the circumference EN, the angle BGL is also greater than the angle EHN ;
and, if less, less.

There being then four magnitudes, two circumferences BC, EF, and two angles BGC, EHF,
there have been taken, of the circumference BC and the angle BGC equimultiples, namely the circumference BL and the angle BGL,
and of the circumference EF and the angle EHF equi-multiples, namely the circumference EN and the angle EHN.

And it has been proved that,
if the circumference BL is in excess of the circumference EN, the angle BGL is also in excess of the angle EHN ;
if equal, equal ;
and if less, less.

Therefore, as the circumference BC is to EF, so is the angle BGC to the angle EHF.   [v. Def. 5]

But, as the angle BGC is to the angle EHF, so is the angle BAC to the angle EDF ; for they are doubles respectively.

Therefore also, as the circumference BC is to the circumference EF, so is the angle BGC to the angle EHF, and the angle BAC to the angle EDF.

Therefore etc.

                                                        Q. E. D.

This proposition as generally given includes a second part relating to *sectors* of circles, corresponding to the following words added to the enunciation : "and further the sectors, as constructed at the centres" (ἔτι δὲ καὶ οἱ τομεῖς ἅτε [or οἷτε] πρὸς τοῖς κέντροις συνιστάμενοι). There is of course a corresponding addition to the "definition" or "particular statement," "and further the sector $GBOC$ to the sector $HEQF$." These additions are clearly due to Theon, as may be gathered from his own statement in his commentary on the μαθηματικὴ σύνταξις of Ptolemy, "But that sectors in equal circles are to one another as the angles on which they stand, has been proved by me in my edition of the

Elements at the end of the sixth book." Campanus omits them, and P has them only in a later hand in the margin or between the lines. Theon's proof scarcely needs to be given here in full, as it can easily be supplied. From the equality of the arcs BC, CK he infers [III. 29] the equality of the chords BC, CK. Hence, the radii being equal, the triangles GBC, GCK are equal in all respects [I. 8, 4]. Next, since the arcs BC, CK are equal, so are the arcs BAC, CAK. Therefore the angles at the circumference subtended by the latter, i.e. the angles in the segments BOC, CPK, are equal [III. 27], and the segments are therefore similar [III. Def. 11] and equal [III. 24].

Adding to the equal segments the equal triangles GBC, GCK respectively, we see that

the sectors GBC, GCK are equal.

Thus, in equal circles, sectors standing on equal arcs are equal; and the rest of the proof proceeds as in Euclid's proposition.

As regards Euclid's proposition itself, it will be noted that (1), besides quoting the theorem in III. 27 that in equal circles angles which stand on equal arcs are equal, the proof assumes that the angle standing on a greater arc is greater and that standing on a less arc is less. This is indeed a sufficiently obvious deduction from III. 27.

(2) Any equimultiples *whatever* are taken of the angle BGC and the arc BC, and any equimultiples *whatever* of the angle EHF and the arc EF. (Accordingly the words "*any equimultiples whatever*" should have been used in the step immediately preceding the inference that the angles are proportional to the arcs, where the text merely states that there have been taken of the circumference BC and the angle BGC equimultiples BL and BGL.) But, if *any multiple* of an angle is regarded as being itself an angle, it follows that the restriction in I. Deff. 8, 10, 11, 12 of the term *angle* to an angle *less than two right angles* is implicitly given up; as De Morgan says, "the angle breaks prison." Mr Dodgson (*Euclid and his Modern Rivals*, p. 193) argues that Euclid conceived of the multiple of an angle as so many separate angles not added together into one, and that, when it is inferred that, where two such multiples of an angle are equal, the arcs subtended are also equal, the argument is that the sum total of the first set of angles is equal to the sum total of the second set, and hence the second set can be broken up and put together again in such amounts as to make a set equal, each to each, to the first set, and then the sum total of the arcs will evidently be equal also. If on the other hand the multiples of the angles are regarded as single angular magnitudes, the equality of the subtending arcs is not inferrible directly from Euclid, because *his* proof of III. 26 only applies to cases where the angle is less than the sum of two right angles. (As a matter of fact, it is a question of inferring equality of angles or multiples of angles from equality of arcs, and not the converse, so that the reference should have been to III. 27, but this does not affect the question at issue.) Of course it is against this view of Mr Dodgson that Euclid speaks throughout of "*the angle* BGL" and "*the angle* EHN" (ἡ ὑπὸ BHΛ γωνία, ἡ ὑπὸ EΘN γωνία). I think the probable explanation is that here, as in III. 20, 21, 26 and 27, Euclid deliberately took no cognisance of the case in which the multiples of the angles in question would be greater than two right angles. If his attention had been called to the fact that III. 20 takes no account of the case where the segment is less than a semicircle, so that the angle in the segment is obtuse, and therefore the "angle at the centre" in that case (if the term were still applicable) would be

greater than two right angles, Euclid would no doubt have refused to regard the latter as an angle, and would have represented it otherwise, e.g. as the sum of two angles or as what is left when an *angle* in the true sense is subtracted from four right angles.   Here then, if Euclid had been asked what course he would take if the multiples of the angles in question should be greater than two right angles, he would probably have represented them, I think, as being *equal to so many right angles plus an angle less than a right angle*, or *so many times two right angles plus an angle, acute or obtuse*.   Then the equality of the arcs would be the equality of the sums of so many circumferences, semi-circumferences or quadrants plus arcs less than a semicircle or a quadrant.   Hence I agree with Mr Dodgson that VI. 33 affords no evidence of a recognition by Euclid of "angles" greater than two right angles

Theon adds to his theorem about sectors the Porism that, *As the sector is to the sector, so also is the angle to the angle*.   This corollary was used by Zenodorus in his tract περὶ ἰσομέτρων σχημάτων preserved by Theon in his commentary on Ptolemy's σύνταξις, unless indeed Theon himself interpolated the words (ὡς δ᾽ ὁ τομεὺς πρὸς τὸν τομέα, ἡ ὑπὸ ΕΘΛ γωνία πρὸς τὴν ὑπὸ ΜΘΛ).

BOOK VII.

DEFINITIONS.

1.   An **unit** is that by virtue of which each of the things that exist is called one.

2.   A **number** is a multitude composed of units.

3.   A number is **a part** of a number, the less of the greater, when it measures the greater;

4.   but **parts** when it does not measure it.

5.   The greater number is a **multiple** of the less when it is measured by the less.

6.   An **even number** is that which is divisible into two equal parts.

7.   An **odd number** is that which is not divisible into two equal parts, or that which differs by an unit from an even number.

8.   An **even-times even number** is that which is measured by an even number according to an even number.

9.   An **even-times odd number** is that which is measured by an even number according to an odd number.

10.   An **odd-times odd number** is that which is measured by an odd number according to an odd number.

11.   A **prime number** is that which is measured by an unit alone.

12.   Numbers **prime to one another** are those which are measured by an unit alone as a common measure.

13.   A **composite number** is that which is measured by some number.

14.   Numbers **composite to one another** are those which are measured by some number as a common measure.

15.   A number is said to **multiply** a number when that which is multiplied is added to itself as many times as there are units in the other, and thus some number is produced.

16.   And, when two numbers having multiplied one another make some number, the number so produced is called **plane**, and its **sides** are the numbers which have multiplied one another.

17.   And, when three numbers having multiplied one another make some number, the number so produced is **solid**, and its **sides** are the numbers which have multiplied one another.

18.   A **square number** is equal multiplied by equal, or a number which is contained by two equal numbers.

19.   And a **cube** is equal multiplied by equal and again by equal, or a number which is contained by three equal numbers.

20.   Numbers are **proportional** when the first is the same multiple, or the same part, or the same parts, of the second that the third is of the fourth.

21.   **Similar plane** and **solid** numbers are those which have their sides proportional.

22.   A **perfect number** is that which is equal to its own parts.

DEFINITION I.

Μονάς ἐστιν, καθ᾽ ἣν ἕκαστον τῶν ὄντων ἓν λέγεται.

Iamblichus (fl. *circa* 300 A.D.) tells us (*Comm. on Nicomachus*, ed. Pistelli, p. 11, 5) that the Euclidean definition of an *unit* or a *monad* was the definition given by "more recent" writers (οἱ νεώτεροι), and that it lacked the words "even though it be collective" (κᾶν συστηματικὸν ᾖ). He also gives (*ibid.* p. 11) a number of other definitions. (1) According to "some of the Pythagoreans," "an unit is the boundary between number and parts" (μονάς ἐστιν ἀριθμοῦ καὶ μορίων μεθόριον), "because from it, as from a seed and eternal root, ratios increase reciprocally on either side," i.e. on one side we have multiple ratios continually increasing and on the other (if the unit be subdivided) submultiple ratios with denominators continually increasing. (2) A somewhat similar definition is that of Thymaridas, an ancient Pythagorean, who defined a monad as "limiting quantity" (περαίνουσα ποσότης), the beginning and the end of a thing being equally an extremity (πέρας). Perhaps the words together with their explanation may best be expressed by "limit of fewness." Theon of Smyrna (p. 18, 6, ed. Hiller) adds the explanation that the monad is "that which, when the multitude is diminished by way of continued subtraction, is deprived of all number and takes an abiding position (μονήν) and rest." If, after arriving at an unit in this way, we proceed to divide the unit itself into parts, we straightway have multitude again. (3) Some, according to Iamblichus (p. 11, 16), defined it as the "form of forms" (εἰδῶν εἶδος) because it potentially comprehends all forms of number, e.g. it is a polygonal number of any number of sides from three upwards, a solid number in all forms, and so on. (We are forcibly reminded of the latest theories of number as a "Gattung" of "Mengen" or as a "class of classes.") (4) Again an unit, says Iamblichus, is the first, or smallest, in the category of *how many* (ποσόν), the common part or beginning of *how many*. Aristotle defines it as "the indivisible in the (category of) quantity," τὸ κατὰ τὸ ποσὸν ἀδιαίρετον (*Metaph.* 1089 b 35), ποσόν including in Aristotle continuous as well as discrete quantity; hence it is distinguished from a point by the fact that it has not position: "Of the indivisible in the category of, and *quâ*, quantity, that which is every way (indivisible) and destitute of position is called an *unit*, and that which is every way indivisible and has position is a *point*" (*Metaph.* 1016 b 25). (5) In accordance with the last distinction, Aristotle calls the unit "a point without position," στιγμὴ ἄθετος (*Metaph.* 1084 b 26). (6) Lastly, Iamblichus says that the school of Chrysippus defined it in a confused manner (συγκεχυμένως) as "multitude *one* (πλῆθος ἕν)," whereas it is alone contrasted with multitude. On a comparison of these definitions, it would seem that Euclid intended his to be a more *popular* one than those of his predecessors, δημώδης, as Nicomachus called Euclid's definition of an *even number*.

The etymological signification of the word μονάς is supposed by Theon of Smyrna (p. 19, 7—13) to be either (1) that it remains unaltered if it be multiplied by itself any number of times, or (2) that it is separated and *isolated* (μεμονῶσθαι) from the rest of the multitude of numbers. Nicomachus also observes (1. 8, 2) that, while any number is half the sum (1) of the adjacent numbers on each side, (2) of numbers equidistant on each side, the unit is *most solitary* (μονωτάτη) in that it has not a number on each side but only on one side, and it is half of the latter alone, i.e. of 2.

DEFINITION 2.

'Αριθμὸς δὲ τὸ ἐκ μονάδων συγκείμενον πλῆθος.

The definition of a *number* is again only one out of many that are on record. Nicomachus (I. 7, 1) combines several into one, saying that it is "a defined multitude (πλῆθος ὡρισμένον), or a collection of units (μονάδων σύστημα), or a flow of quantity made up of units" (ποσότητος χύμα ἐκ μονάδων συγκείμενον). Theon, in words almost identical with those attributed by Stobaeus (*Eclogae*, I. 1, 8) to Moderatus, a Pythagorean, says (p. 18, 3—5): "A number is a collection of units, or a progression (προποδισμός) of multitude beginning from an unit and a retrogression (ἀναποδισμός) ceasing at an unit." According to Iamblichus (p. 10) the description "collection of units" (μονάδων σύστημα) was applied to the *how many*, i.e. to number, by Thales, following the Egyptian view (κατὰ τὸ Αἰγυπτιακὸν ἀρέσκον), while it was Eudoxus the Pythagorean who said that a number was "a defined multitude" (πλῆθος ὡρισμένον). Aristotle has a number of definitions which come to the same thing: "limited multitude" (πλῆθος τὸ πεπερασμένον, *Metaph.* 1020 a 13), "multitude" (or "combination") "of units" or "multitude of indivisibles" (*ibid.* 1053 a 30, 1039 a 12, 1085 b 22), "several *ones*" (ἕνα πλείω, *Phys.* III. 7, 207 b 7), "multitude measurable by one" (*Metaph.* 1057 a 3) and "multitude measured and multitude of measures," the "measure" being unity, τὸ ἕν (*ibid.* 1088 a 5).

DEFINITION 3.

Μέρος ἐστὶν ἀριθμὸς ἀριθμοῦ ὁ ἐλάσσων τοῦ μείζονος, ὅταν καταμετρῇ τὸν μείζονα.

By *a part* Euclid means a submultiple, as he does in V. Def. 1, with which definition this one is identical except for the substitution of *number* (ἀριθμός) for *magnitude* (μέγεθος); cf. note on V. Def. 1. Nicomachus uses the word "submultiple" (ὑποπολλαπλάσιος) also. He defines it in a way corresponding to his definition of multiple (see note on Def. 5 below) as follows (I. 18, 2): "The submultiple, which is by nature first in the division of inequality (called) less, is the number which, when compared with a greater, can measure it more times than once so as to fill it exactly (πληρούντως)." Similarly *sub-double* (ὑποδιπλάσιος) is found in Nicomachus meaning *half*, and so on.

DEFINITION 4.

Μέρη δέ, ὅταν μὴ καταμετρῇ.

By the expression *parts* (μέρη, the plural of μέρος) Euclid denotes what we should call a *proper fraction*. That is, *a part* being a submultiple, the rather inconvenient term *parts* means any number of such submultiples making up a fraction less than unity. I have not found the word used in this special sense elsewhere, e.g. in Nicomachus, Theon of Smyrna or Iamblichus, except in one place of Theon (p. 79, 26) where it is used of a proper fraction, of which $\frac{2}{3}$ is an illustration.

DEFINITION 5.

Πολλαπλάσιος δὲ ὁ μείζων τοῦ ἐλάσσονος, ὅταν καταμετρῆται ὑπὸ τοῦ ἐλάσσονος.

The definition of a *multiple* is identical with that in v. Def. 2, except that the masculine of the adjectives is used agreeing with ἀριθμός understood instead of the neuter agreeing with μέγεθος understood. Nicomachus (I. 18, 1) defines a multiple as being "a species of the greater which is naturally first in order and origin, being the number which, when considered in comparison with another, contains it in itself completely more than once."

DEFINITIONS 6, 7.

6. Ἄρτιος ἀριθμός ἐστιν ὁ δίχα διαιρούμενος.

7. Περισσὸς δὲ ὁ μὴ διαιρούμενος δίχα ἢ [ὁ] μονάδι διαφέρων ἀρτίου ἀριθμοῦ.

Nicomachus (I. 7, 2) somewhat amplifies these definitions of *even* and *odd* numbers thus. "That is *even* which is capable of being divided into two equal parts without an unit falling in the middle, and that is *odd* which cannot be divided into two equal parts because of the aforesaid intervention (μεσι-τείαν) of the unit." He adds that this definition is derived "from the popular conception" (ἐκ τῆς δημώδους ὑπολήψεως). In contrast to this, he gives (I. 7, 3) the Pythagorean definition, which is, as usual, interesting. "An *even* number is that which admits of being divided, by one and the same operation, into the greatest and the least (parts), greatest in size (πηλικότητι) but least in quantity (ποσότητι)...while an *odd* number is that which cannot be so treated, but is divided into two unequal parts." That is, as Iamblichus says (p. 12, 2—9), an even number is divided into parts which are the *greatest* possible "parts," namely halves, and into the *fewest* possible, namely two, two being the first "number" or "collection of units." According to another ancient definition quoted by Nicomachus (I. 7, 4), an even number is that which can be divided both into two equal parts and into two unequal parts (except the first one, the number 2, which is only susceptible of division into equals), but, however it is divided, must have its two parts *of the same kind*, i.e. both even or both odd; while an odd number is that which can only be divided into two unequal parts, and those parts always of *different* kinds, i.e. one odd and one even. Lastly, the definition of odd and even "by means of each other" says that an odd number is that which differs by an unit from an even number on both sides of it, and an even number that which differs by an unit from an odd number on each side. This alternative definition of an odd number is the same thing as the second half of Euclid's definition, "the number which differs by an unit from an even number." This evidently pre-Euclidean definition is condemned by Aristotle as unscientific, because odd and even are coordinate, both being *differentiae* of number, so that one should not be defined by means of the other (*Topics* VI. 4, 142 b 7—10).

DEFINITION 8.

Ἀρτιάκις ἄρτιος ἀριθμός ἐστιν ὁ ὑπὸ ἀρτίου ἀριθμοῦ μετρούμενος κατὰ ἄρτιον ἀριθμόν.

Euclid's definition of an *even-times even* number differs from that given by the later writers, Nicomachus, Theon of Smyrna and Iamblichus; and the inconvenience of it is shown when we come to IX. 34, where it is proved

that a certain sort of number is *both* "even-times even" and "even-times odd."
According to the more precise classification of the three other authorities, the
"even-times even" and the "even-times odd" are mutually exclusive and are
two of three subdivisions into which even numbers fall. Of these three subdivisions the "even-times even" and the "even-times odd" form the extremes,
and the "odd-times even" is as it were intermediate, showing the character
of both extremes (cf. note on the following definition). The *even-times even* is
then the number which has its halves even, the halves of the halves even, and
so on, until unity is reached. In short the *even-times even* number is always
of the form 2^n. Hence Iamblichus (pp. 20, 21) says Euclid's definition of it
as that which is measured by an even number an even number of times is
erroneous. In support of this he quotes the number 24 which is four times 6,
or six times 4, but yet is not "even-times even" according to Euclid himself
(οὐδὲ κατ' αὐτόν), by which he must apparently mean that 24 is also 8 times 3,
which does not satisfy Euclid's definition. There can however be no doubt that
Euclid meant what he said in his definition as we have it; otherwise IX. 32,
which proves that a number of the form 2^n is *even-times even only*, would be quite
superfluous and a mere repetition of the definition, while, as already stated,
IX. 34 clearly indicates Euclid's view that a number might at the same time
be both even-times even and even-times odd. Hence the μόνως which some
editor of the commentary of Philoponus on Nicomachus found in some
copies, making the definition say that the even-times even number is *only*
measured by even numbers an even number of times, is evidently an interpolation by some one who wished to reconcile Euclid's definition with the
Pythagorean (cf. Heiberg, *Euklid-studien*, p. 200).

A consequential characteristic of the series of even-times even numbers
noted by Nicomachus brings in a curious use of the word δύναμις (generally
power in the sense of square, or square root). He says (I. 8, 6—7) that any
part, i.e. any submultiple, of an even-times even number is called by an even-
times even designation, while it also has an even-times even *value* (it is
ἀρτιάκις ἀρτιοδύναμον) when expressed as so many actual units. That is, the
$\frac{1}{2^m}$th part of 2^n (where m is less than n) is called after the even-times even
number 2^m, while its actual *value* (δύναμις) in units is 2^{n-m}, which is also an
even-times even number. Thus all the parts, or submultiples, of even-times
even numbers, as well as the even-times even numbers themselves, are connected with one kind of number only, the *even*.

DEFINITION 9.

Ἀρτιάκις δὲ περισσός ἐστιν ὁ ὑπὸ ἀρτίου ἀριθμοῦ μετρούμενος κατὰ περισσὸν
ἀριθμόν.

Euclid uses the term *even-times odd* (ἀρτιάκις περισσός), whereas Nicomachus
and the others make it one word, *even-odd* (ἀρτιοπέριττος). According to the
stricter definition given by the latter (I. 9, 1), the *even-odd* number is related to
the *even-times even* as the other extreme. It is such a number as, when once
halved, leaves as quotient an odd number; that is, it is of the form $2(2m + 1)$.
Nicomachus sets the even-odd numbers out as follows,

6, 10, 14, 18, 22, 26, 30, etc.

In this case, as Nicomachus observes, any part, or submultiple, is called by a
name *not* corresponding in kind to its actual value (δύναμις) in units. Thus,

in the case of 18, the $\frac{1}{2}$ part is called after the even number 2, but its *value* is the odd number 9, and the $\frac{1}{3}$rd part is called after the odd number 3, while its *value* is the even number 6, and so on.

The third class of even numbers according to the strict subdivision is the *odd-even* (περισσάρτιος). Numbers are of this class when they can be halved twice or more times successively, but the quotient left when they can no longer be halved is an odd number and not unity. They are therefore of the form $2^{n+1}(2m + 1)$, where n, m are integers. They are, so to say, intermediate between, or a mixture of, the extreme classes *even-times even* and *even-odd*, for the following reasons. (1) Their subdivision by 2 proceeds for some way like that of the even-times even, but ends in the way that the division of the even-odd by 2 ends. (2) The numbers after which submultiples are called and their *value* (δύναμις) in units may be both of one kind, i.e. both odd or both even (as in the case of the even-times even), or again may be one odd and one even as in the case of the even-odd. For example 24 is an odd-even number; the $\frac{1}{4}$th, $\frac{1}{12}$th, $\frac{1}{6}$th or $\frac{1}{2}$ parts of it are even, but the $\frac{1}{3}$rd part of it, or 8, is even, and the $\frac{1}{8}$th part of it, or 3, is odd. (3) Nicomachus shows (I. 10, 6—9) how to form all the numbers of the odd-even class. Set out two lines (a) of odd numbers beginning with 3, (b) of even-times even numbers beginning with 4, thus:

(a)   3, 5, 7, 9, 11, 13, 15 etc.

(b)   4, 8, 16, 32, 64, 128, 256 etc.

Now multiply each of the first numbers into each of the second row. Let the products of one of the first into all the second set make horizontal rows; we then get the rows

12, 24, 48, 96, 192, 384, 768 etc.

20, 40, 80, 160, 320, 640, 1280 etc.

28, 56, 112, 224, 448, 896, 1792 etc.

36, 72, 144, 288, 576, 1152, 2304 etc.

and so on.

Now, says Nicomachus, you will be surprised to see (φανήσεταί σοι θαυμαστῶς) that (a) the *vertical* rows have the property of the *even-odd* series, 6, 10, 14, 18, 22 etc., viz. that, if an odd number of successive numbers be taken, the middle number is half the sum of the extremes, and if an even number, the two middle numbers together are equal to the sum of the extremes, (b) the *horizontal* rows have the property of the *even-times even* series 4, 8, 16 etc., viz. that the product of the extremes of any number of successive terms is equal, if their number be odd, to the square of the middle term, or, if their number be even, to the product of the two middle terms.

Let us now return to Euclid. His 9th definition states that an *even-times odd* number is a number which, when divided by an even number, gives an odd number as quotient. Following this definition in our text comes a 10th definition which defines an *odd-times even* number; this is stated to be a number which, when divided by an odd number, gives an even number as quotient. According to these definitions any *even-times odd* number would also be *odd-times even*, and, from the fact that Iamblichus notes this, we may fairly conclude that he found Def. 10 as well as Def. 9 in the text of Euclid which he used. But, if both definitions are genuine, the enunciations of IX. 33 and IX. 34 as we have them present difficulties. IX. 33 says that "If a number have its half odd, it is even-times odd *only*"; but, on the assumption that both definitions are genuine, this would not be true, for the number would be *odd-times even* as well. IX. 34 says that "If a number neither be one of those which are continually doubled from 2, nor have its half odd, it is both even-times even and even-times odd." The term *odd-times even* (περισσάκις ἄρτιος) not occurring in these propositions, nor anywhere else after the definition, that definition becomes superfluous. Iamblichus however (p. 24, 7—14) quotes these enunciations differently. In the first he has instead of "even-times odd only" the words "*both even-times odd and odd-times even*"; and, in the second, for "both even-times even and even-times odd" he has "is both even-times even and at the same time even-times odd and odd-times even." In both cases therefore "odd-times even" is added to the enunciation as Iamblichus had it; the words cannot have been added by Iamblichus himself because he himself does not use the term *odd-times even*, but the one word *odd-even* (περισσάρτιος). In order to get over the difficulties involved by Def. 10 and these differences of reading we have practically to choose between (1) accepting Iamblichus' reading in all three places and (2) adhering to the reading of our MSS. in IX. 33, 34 and rejecting Def. 10 altogether as an interpolation. Now the readings of our text of IX. 33, 34 are those of the Vatican MS. and the Theonine MSS. as well; hence they must go back to a time before Theon, and must therefore be almost as old as those of Iamblichus. Heiberg considers it improbable that Euclid would wish to maintain a pointless distinction between *even-times odd* and *odd-times even*, and on the whole concludes that Def. 10 was first interpolated by some ignorant person who did not notice the difference between the Euclidean and Pythagorean classification, but merely noticed the absence of a definition of *odd-times even* and fabricated one as a companion to the other. When this was done, it would be easy to see that the statement in IX. 33 that the number referred to is "even-times odd *only*" was not strictly true, and that the addition of the words "and odd-times even" was necessary in IX. 33 and IX. 34 as well.

DEFINITION 10.

Περισσάκις δὲ περισσὸς ἀριθμός ἐστιν ὁ ὑπὸ περισσοῦ ἀριθμοῦ μετρούμενος κατὰ περισσὸν ἀριθμόν.

The *odd-times odd* number is not defined as such by Nicomachus and Iamblichus; for them these numbers would apparently belong to the *composite* subdivision of *odd* numbers. Theon of Smyrna on the other hand says (p. 23, 21) that *odd-times odd* was one of the names applied to *prime* numbers (excluding 2), for these have two odd factors, namely 1 and the number itself. This is certainly a curious use of the term.

DEFINITION 11.

Πρῶτος ἀριθμός ἐστιν ὁ μονάδι μόνῃ μετρούμενος.

A *prime* number (πρῶτος ἀριθμός) is called by Nicomachus, Theon, and Iamblichus a "prime *and incomposite* (ἀσύνθετος) number." Theon (p. 23, 9) defines it practically as Euclid does, viz. as a number "measured by no number, but by an unit only." Aristotle too says that a prime number is not measured by any number (*Anal. post.* II. 13, 96 a 36), an unit not being a number (*Metaph.* 1088 a 6), but only the beginning of number (Theon of Smyrna says the same thing, p. 24, 23). According to Nicomachus (I. 11, 2) the prime number is a subdivision, not of numbers, but of *odd* numbers; it is "an odd number which admits of no other part except that which is called after its own name (παρώνυμον ἑαυτῷ)." The prime numbers are 3, 5, 7 etc., and there is no submultiple of 3 except ⅓rd, no submultiple of 11 except $\frac{1}{11}$th, and so on. In all these cases the only submultiple is an unit. According to Nicomachus 3 is the first prime number, whereas Aristotle (*Topics* VIII. 2, 157 a 39) regards 2 as a prime number: "as the dyad is the only even number which is prime," showing that this divergence from the Pythagorean doctrine was earlier than Euclid. The number 2 also satisfies Euclid's definition of a prime number. Iamblichus (p. 30, 27 sqq.) makes this the ground of another attack upon Euclid. His argument (the text of which, however, leaves much to be desired) appears to be that 2 is the *only* even number which has no other part except an unit, while the subdivisions of the even, as previously explained by him (the *even-times even*, the *even-odd*, and *odd-even*), all exclude primeness, and he has previously explained that 2 is *potentially* even-odd, being obtained by multiplying by 2 the *potentially* odd, i.e. the unit; hence 2 is regarded by him as bound up with the subdivisions of even, which exclude primeness. Theon seems to hold the same view as regards 2, but supports it by an apparent circle. A prime number, he says (p. 23, 14—23), is also called *odd-times odd*; therefore only odd numbers are prime and incomposite. Even numbers are not measured by the unit alone, except 2, which therefore (p. 24, 7) is odd-*like* (περισσοειδής) without being prime.

A variety of other names were applied to prime numbers. We have already noted the curious designation of them as *odd-times odd*. According to Iamblichus (p. 27, 3—5) some called them *euthymetric* (εὐθυμετρικός), and Thymaridas *rectilinear* (εὐθυγραμμικός), the ground being that they can only be set out in one dimension with no breadth (ἁπλατὴς γὰρ ἐν τῇ ἐκθέσει ἐφ᾽ ἓν μόνον διιστάμενος). The same aspect of a prime number is also expressed by Aristotle, who (*Metaph.* 1020 b 3) contrasts the composite number with that which is only in one dimension (μόνον ἐφ᾽ ἓν ὤν). Theon of Smyrna (p. 23, 12) gives γραμμικός (*linear*) as the alternative name instead of εὐθυγραμμικός. In either case, to make the word a proper description of a prime number we have to understand the word *only*; a prime number is that which is *linear*, or *rectilinear, only*. For Nicomachus, who uses the form *linear*, expressly says (II. 13, 6) that *all* numbers are so, i.e. all can be represented as linear by dots to the required amount placed in a line.

A prime number was called *prime* or *first*, according to Nicomachus (I. 11, 3), because it can only be arrived at by putting together a certain number of units, and the unit is the beginning of number (cf. Aristotle's second sense of πρῶτος "as not being composed of *numbers*," ὡς μὴ συγκεῖσθαι ἐξ ἀριθμῶν, *Anal. Post.* II. 13, 96 a 37), and also, according to Iamblichus, because there is no number before it, being a collection of units (μονάδων σύστημα), of which it is a multiple, and it appears *first* as a basis for other numbers to be multiples of.

DEFINITION 12.

Πρῶτοι πρὸς ἀλλήλους ἀριθμοί εἰσιν οἱ μονάδι μόνῃ μετρούμενοι κοινῷ μέτρῳ.

By way of further emphasising the distinction between "prime" and "prime to one another," Theon of Smyrna (p. 23, 6—8) calls the former "prime *absolutely*" (ἁπλῶς), and the latter "prime to one another and *not absolutely*" or "*not in themselves*" (οὐ καθ᾽ αὑτούς). The latter (p. 24, 8—10) are "measured by the unit [sc. only] as common measure, even though, taken by themselves (ὡς πρὸς ἑαυτούς), they be measured by some other numbers." From Theon's illustrations it is clear that with him as with Euclid a number prime to another may be even as well as odd. In Nicomachus (I. 11, 1) and Iamblichus (p. 26, 19), on the other hand, the number which is "in itself secondary (δεύτερος) and composite (σύνθετος), but in relation to another prime and incomposite," is a subdivision of *odd*. I shall call more particular attention to this difference of classification when we have reached the definitions of "composite" and "composite to one another"; for the present it is to be noted that Nicomachus (I. 13, 1) defines a number prime *to another* after the same manner as the absolutely prime; it is a number which "is measured not only by the unit as the common measure but also by some other measure, and for this reason can also admit of a part or parts called by a different name besides that called by the same name (as itself), but, when examined in comparison with another number of similar character, is found not to be capable of being measured by a common measure in relation to the other, nor to have the same part, called by the same name as (any of) those simply (ἁπλῶς) contained in the other; e.g. 9 in relation to 25, for each of these is in itself secondary and composite, but, in comparison with one another, they have an unit alone as a common measure and no part is called by the same name in both, but the *third* in one is not in the other, nor is the *fifth* in the other found in the first."

DEFINITION 13.

Σύνθετος ἀριθμός ἐστιν ὁ ἀριθμῷ τινι μετρούμενος.

Euclid's definition of *composite* is again the same as Theon's definition of numbers "composite in relation to themselves," which (p. 24, 16) are "numbers measured by any less number," the unit being, as usual, not regarded as a number. Theon proceeds to say that "of composite numbers they call those which are contained by two numbers *plane*, as being investigated in two dimensions and, as it were, contained by a length and a breadth, while (they call) those (which are contained) by three (numbers) *solid*, as having the third dimension added to them." To a similar effect is the remark of Aristotle (*Metaph.* 1020 b 3) that certain numbers are "composite and are not only in one dimension but such as the plane and the solid (figure) are representations of (μίμημα), these numbers being so many times so many (ποσάκις ποσοί), or so many times so many times so many (ποσάκις ποσάκις ποσοί) respectively." These subdivisions of composite numbers are, of course, the subject of Euclid's definitions 17, 18 respectively. Euclid's composite numbers may be either even or odd, like those of Theon, who gives 6 as an instance, 6 being measured by both 2 and 3.

DEFINITION 14.

Σύνθετοι δὲ πρὸς ἀλλήλους ἀριθμοί εἰσιν οἱ ἀριθμῷ τινι μετρούμενοι κοινῷ μέτρῳ.

Theon (p. 24, 18), like Euclid, defines numbers *composite to one another* as "those which are measured by any common measure whatever" (excluding unity, as usual). Theon instances 8 and 6, with 2 as common measure, and 6 and 9, with 3 as common measure.

As hinted above, there is a great difference between Euclid's classification of prime and composite numbers, and of numbers prime and composite to one another, and the classification found in Nicomachus (I. 11—13) and Iamblichus. According to the latter, all these kinds of numbers are sub-divisions of the class of *odd* numbers only. As the class of *even* numbers is divided into three kinds, (1) the even-times even, (2) the even-odd, which form the extremes, and (3) the odd-even, which is, as it were, intermediate to the other two, so the class of *odd* numbers is divided into three, of which the third is again a mean between two extremes. The three are:

(1) the *prime and incomposite*, which is like Euclid's prime number except that it excludes 2;

(2) the *secondary and composite*, which is "odd because it is a distinct part of one and the same genus (διὰ τὸ ἐξ ἑνὸς καὶ τοῦ αὐτοῦ γένους διακεκρίσθαι) but has in it nothing of the nature of a first principle (ἀρχοειδές); for it arises from adding some other number (to itself), so that, besides having a part called by the same name as itself, it possesses a part or parts called by another name." Nicomachus cites 9, 15, 21, 25, 27, 33, 35, 39. It is made clear that not only must the factors be both odd, but they must all be prime numbers. This is obviously a very inconvenient restriction of the use of the word *composite*, a word of general signification.

(3) is that which is "*secondary and composite in itself but prime and incomposite to another.*" The actual words in which this is defined have been given above in the note on Def. 12. Here again all the factors must be odd and prime.

Besides the inconvenience of restricting the term *composite* to *odd* numbers which are composite, there is in this classification the further serious defect, pointed out by Nesselmann (*Die Algebra der Griechen*, 1842, p. 194), that subdivisions (2) and (3) overlap, subdivision (2) including the whole of subdivision (3). The origin of this confusion is no doubt to be found in Nicomachus' perverse anxiety to be symmetrical; by hook or by crook he must divide *odd* numbers into three kinds as he had divided the *even*. Iamblichus (p. 28, 13) carries his desire to be logical so far as to point out why there cannot be a fourth kind of number contrary in character to (3), namely a number which should be "prime and incomposite in itself, but secondary and composite to another"!

DEFINITION 15.

Ἀριθμὸς ἀριθμὸν πολλαπλασιάζειν λέγεται, ὅταν, ὅσαι εἰσὶν ἐν αὐτῷ μονάδες, τοσαυτάκις συντεθῇ ὁ πολλαπλασιαζόμενος, καὶ γένηταί τις.

This is the well known primary definition of multiplication as an abbreviation of addition.

DEFINITION 16.

Ὅταν δὲ δύο ἀριθμοὶ πολλαπλασιάσαντες ἀλλήλους ποιῶσί τινα, ὁ γενόμενος ἐπίπεδος καλεῖται, πλευραὶ δὲ αὐτοῦ οἱ πολλαπλασιάσαντες ἀλλήλους ἀριθμοί.

The words *plane* and *solid* applied to numbers are of course adapted from their use with reference to geometrical figures. A number is therefore called *linear* (γραμμικός) when it is regarded as in one dimension, as being a *length* (μῆκος). When it takes another dimension in addition, namely *breadth* (πλάτος), it is in two dimensions and becomes *plane* (ἐπίπεδος). The distinction between a *plane* and a *plane number* is marked by the use of the neuter in the former case, and the masculine, agreeing with ἀριθμός, in the latter case. So with a *square* and a *square number*, and so on. The most obvious form of a plane number is clearly that corresponding to a rectangle in geometry; the number is the product of two linear numbers regarded as *sides* (πλευραί) forming the length and breadth respectively. Such a number is, as Aristotle says, "so many times so many," and a plane is its counterpart (μίμημα). So Plato, in the *Theaetetus* (147 E—148 B), says: "We divided all numbers into two kinds, (1) that which can be expressed as equal multiplied by equal (τὸν δυνάμενον ἴσον ἰσάκις γίγνεσθαι), and which, likening its form to the square, we called *square* and equilateral; (2) that which is intermediate, and includes 3 and 5 and every number which cannot be expressed as equal multiplied by equal, but is either less times more or more times less, being always contained by a greater and a less side, which number we likened to the oblong figure (προμήκει σχήματι) and called an *oblong* number.... Such *lines* therefore as *square the equilateral and plane number* [i.e. which can form a plane number with equal sides, or a square] we defined as *length* (μῆκος); but such as square the oblong (here ἑτερομήκης) [i.e. the square of which is equal to the oblong] we called *roots* (δυνάμεις) as not being commensurable with the others in length, but only in the plane areas (ἐπιπέδοις), to which the squares on them are equal (ἃ δύνανται)." This passage seems to make it clear that Plato would have represented numbers as Euclid does, by straight lines proportional in length to the numbers they represent (so far as practicable); for, since 3 and 5 are with Plato oblong numbers, and *lines* with him represent the sides of oblong numbers (since a line represents the "root," the square on which is equal to the oblong), it follows that the *unit* representing the smaller side must have been represented as a line, and 3, the larger side, as a line of three times the length. But there is another possible way of representing numbers, not by lines of a certain length, but by *points* disposed in various ways, in straight lines or otherwise. Iamblichus tells us (p. 56, 27) that "in old days they represented the quantuplicities of number in a more natural way (φυσικώτερον) by splitting them up into units, and not, as in our day, by symbols" (συμβολικῶς). Aristotle too (*Metaph.* 1092 b 10) mentions one Eurytus as having settled what number belonged to what, such a number to a man, such a number to a horse, and so on, "copying their shapes" (reading τούτων, with Zeller) "*with pebbles* (ταῖς ψήφοις), *just as those do who arrange numbers in the forms of triangles or squares.*" We accordingly find numbers represented in Nicomachus and Theon of Smyrna by a number of α's ranged like points according to geometrical figures. According to this system, any number could be represented by points in a straight line, in which case, says Iamblichus (p. 56, 26), we shall call it rectilinear because it is *without breadth* and only advances in length (ἀπλατῶς ἐπὶ μόνον τὸ μῆκος πρόεισιν). The prime number was called by Thymaridas rectilinear *par excellence*, because it was without breadth and in one dimension *only* (ἐφ' ἓν μόνον διιστάμενος). By this must be meant the impossibility of representing, say, 3 as a plane number, in Plato's sense, i.e. as a product of two numbers corresponding to a rectangle in geometry; and this view would appear to rest simply upon the representation of a number by *points*, as distinct from lines. Three dots in a straight line would have *no* breadth; and if breadth were introduced in the sense of producing a rectangle, i.e. by placing the same number of dots in a second line below the first line, the first *plane* number would be 4, and 3 would not be a plane number at all, as Plato says it is. It seems therefore to have been the alternative representation of a number by points, and not lines, which gave rise to the different view of a plane number which we find in Nicomachus and the rest. By means of separate points we can represent numbers in geometrical forms other than rectangles and squares. One dot with two others symmetrically arranged below it shows a *triangle*, which is a figure *in two dimensions* as much as a rectangle or parallelogram is. Similarly we can arrange certain numbers in the form of regular *pentagons* or other polygons. According therefore to this mode of representation, 3 is the first *plane* number, being a *triangular* number. The method of formation of triangular, square, pentagonal and other polygonal numbers is minutely described in Nicomachus (II. 8—11), who distinguishes the separate series of *gnomons* belonging to each, i.e. gives the law determining the number which has to be added to a polygonal number with *n* in a side, in order to make it into a number of the same form but with *n* + 1 in a side (the addend being of course the gnomon). Thus the gnomonic series for triangular numbers is 1, 2, 3, 4, 5...; that for squares 1, 3, 5, 7...; that for pentagonal numbers 1, 4, 7, 10..., and so on. The subject need not detain us longer here, as we are at present only concerned with the different views of what constitutes a *plane* number.

Of *plane* numbers in the Platonic and Euclidean sense we have seen that Plato recognises *two* kinds, the *square* and the *oblong* (προμήκης or ἑτερομήκης). Here again Euclid's successors, at all events, subdivided the class more elaborately. Nicomachus, Theon of Smyrna, and Iamblichus divide *plane* numbers with unequal *sides* into (1) ἑτερομήκεις, the nearest thing to squares, viz. numbers in which the greater side exceeds the less side by 1 only, or numbers of the form *n* (*n* + 1), e.g. 1 . 2, 2 . 3, 3 . 4, etc. (according to Nicomachus), and (2) προμήκεις, or those whose sides differ by 2 or more, i.e. are of the form *n* (*n* + *m*), where *m* is not less than 2 (Nicomachus illustrates by 2 . 4, 3 . 6, etc.). Theon of Smyrna (p. 30, 8—14) makes προμήκεις include ἑτερομήκεις, saying that their sides may differ by 1 or more; he also speaks of *parallelogram-*numbers as those which have one side different from the other by 2 or more; I do not find this latter term in Nicomachus or Iamblichus, and indeed it seems superfluous, as *parallelogram* is here only another name for oblong. Iamblichus (p. 74, 23 sqq.), always critical of Euclid, attacks him again here for confusing the subject by supposing that the ἑτερομήκης number is the product of *any* two different numbers multiplied together, and by not distinguishing the oblong (προμήκης) from it : "for his definition declares the same number to be square and also ἑτερομήκης, as for example 36, 16 and many others : which would be equivalent to the odd number being the same thing as the even." No importance need be attached to this exaggerated statement ; it is in any case merely a matter of words, and it is curious that Euclid does not in fact use the word ἑτερομήκης of *numbers* at all, but only of geometrical oblong figures as opposed to squares, so that Iamblichus can apparently only have inferred that he used it in an unorthodox manner from the *geometrical* use of the term in the definitions of Book I. and from the fact that he does not give the two subdivisions of *plane numbers* which are not square, but seems only to divide *plane numbers* into square and not-square. The argument that ἑτερομήκεις numbers are a *natural*, and therefore essential, subdivision Iamblichus appears to found on the method of successive addition by which they can be evolved ; as square numbers are obtained by successively adding *odd* numbers as gnomons, so ἑτερομήκεις are obtained by adding *even* numbers as gnomons. Thus $1 \cdot 2 = 2$, $2 \cdot 3 = 2 + 4$, $3 \cdot 4 = 2 + 4 + 6$, and so on.

DEFINITION 17.

Ὅταν δὲ τρεῖς ἀριθμοὶ πολλαπλασιάσαντες ἀλλήλους ποιῶσί τινα, ὁ γενόμενος στερεός ἐστιν, πλευραὶ δὲ αὐτοῦ οἱ πολλαπλασιάσαντες ἀλλήλους ἀριθμοί.

What has been said of the two apparently different ways of regarding a *plane* number seems to apply equally, *mutatis mutandis*, to the definitions of a *solid* number. Aristotle regards it as a number which is so many times so many times so many (ποσάκις ποσάκις ποσόν). Plato finishes the passage about lines which represent the sides of *square* numbers and lines which are *roots* (δυνάμεις), i.e. the squares on which are equal to the rectangle representing a number which is oblong and not square, by adding the words, "And another similar property belongs to solids" (καὶ περὶ τὰ στερεὰ ἄλλο τοιοῦτον). That is, apparently, there would be a corresponding term to *root* (δύναμις)—practically representing a surd—to denote the side of a cube equal to a parallelepiped representing a solid number which is the product of three factors but not a cube. Such is a solid number when numbers are represented by *straight lines*: it corresponds in general to a parallelepiped and, when all the factors are equal, to a cube.

But again, if numbers be represented by *points*, we may have solid numbers (i.e. numbers in three dimensions) in the form of *pyramids* as well. The first number of this kind is 4, since we may have three points forming an equilateral triangle in one plane and a fourth point placed in another plane. The length of the sides can be increased by 1 successively; and we can have a series of pyramidal numbers, with triangles, squares or polygons as bases, made up of layers of triangles, squares or similar polygons respectively, each of which layers has one less in the side than the layer below it, until the top of the pyramid is reached, which of course is one point representing unity. Nicomachus (II. 13—16), Theon of Smyrna (p. 41—2), and Iamblichus (p. 95, 15 sqq.), all give the different kinds of *pyramidal* solid numbers in addition to the other kinds.

These three writers make the following further distinctions between solid numbers which are the product of three factors.

1. First there is the equal by equal by equal (ἰσάκις ἰσάκις ἴσος), which is, of course, the cube.

2. The other extreme is the unequal by unequal by unequal (ἀνισάκις ἀνισάκις ἄνισος), or that in which all the dimensions are different, e.g. the product of 2, 3, 4 or 2, 4, 8 or 3, 5, 12. These were, according to Nicomachus (II. 16), called *scalene*, while some called them σφηνίσκοι (*wedge-shaped*), others σφηκίσκοι (from σφήξ, a *wasp*), and others βωμίσκοι (*altar-shaped*). Theon appears to use the last term only, while Iamblichus of course gives all three names.

3. Intermediate to these, as it were, come the numbers "whose *planes* form ἑτερομήκεις numbers" (i.e. numbers of the form $n(n + 1)$). These, says Nicomachus, are called *parallelepipedal*.

Lastly come two classes of such numbers each of which has two equal dimensions but not more.

4. If the third dimension is less than the others, the number is *equal by equal by less* (ἰσάκις ἴσος ἐλαττονάκις) and is called a *plinth* (πλινθίς), e.g. 8 . 8 . 3.

5. If the third dimension is greater than the others, the number is *equal by equal by greater* (ἰσάκις ἴσος μειζονάκις) and is called a *beam* (δοκίς), e.g. 3 . 3 . 7. Another name for this latter kind of number (according to Iamblichus) was στηλίς (diminutive of στήλη).

Lastly, in connexion with pyramidal numbers, Nicomachus (II. 14, 5) distinguishes numbers corresponding to *frusta* of pyramids. These are *truncated* (κόλουροι), *twice-truncated* (δικόλουροι), *thrice-truncated* (τρικόλουροι) pyramids, and so on, the term being used mostly in theoretic treatises (ἐν συγγράμμασι μάλιστα τοῖς θεωρηματικοῖς). The *truncated* pyramid was formed by cutting off the point forming the vertex. The *twice-truncated* was that which lacked the vertex and the next plane, and so on. Theon of Smyrna (p. 42, 4) only mentions the *truncated* pyramid as "that with its vertex cut off" (ἡ τὴν κορυφὴν ἀποτετμημένη), saying that some also called it a trapezium, after the similitude of a plane trapezium formed by cutting the top off a triangle by a straight line parallel to the base.

DEFINITION 18.

Τετράγωνος ἀριθμός ἐστιν ὁ ἰσάκις ἴσος ἢ [ὁ] ὑπὸ δύο ἴσων ἀριθμῶν περιεχόμενος.

A particular kind of square distinguished by Nicomachus and the rest was the square number which ended (in the decimal notation) with the same number as its side, e.g. 1, 25, 36, which are the squares of 1, 5 and 6. These square numbers were called *cyclic* (κυκλικοί) on the analogy of circles in geometry which return again to the point from which they started.

DEFINITION 19.

Κύβος δὲ ὁ ἰσάκις ἴσος ἰσάκις ἢ [ὁ] ὑπὸ τριῶν ἴσων ἀριθμῶν περιεχόμενος.

Similarly cube numbers which ended with the same number as their sides, *and the squares of those sides also*, were called *spherical* (σφαιρικοί) or *recurrent* (ἀποκαταστατικοί). One might have expected that the term *spherical* would be applicable also to the cubes of numbers which ended with the same digit as the *side* but not necessarily with the same digit as the *square* of the side also. E.g. the cube of 4, i.e. 64, ends with the same digit as 4, but not with the same digit as 16. But apparently 64 was not called a spherical number, the only instances given by Nicomachus and the rest being those cubed from numbers ending with 5 or 6, which end with the same digit if *squared*. A *spherical* number is in fact derived from a *circular* number only, and that by adding another equal dimension. Obviously, as Nesselmann says, the names *cyclic* and *spherical* applied to numbers appeal to an entirely different principle from that on which the figured numbers so far dealt with were formed.

DEFINITION 20.

Ἀριθμοὶ ἀνάλογόν εἰσιν, ὅταν ὁ πρῶτος τοῦ δευτέρου καὶ ὁ τρίτος τοῦ τετάρτου
ἰσάκις ᾖ πολλαπλάσιος ἢ τὸ αὐτὸ μέρος ἢ τὰ αὐτὰ μέρη ὦσιν.

Euclid does not give in this Book any definition of ratio, doubtless because
it could only be the same as that given at the beginning of Book v., with
numbers substituted for "homogeneous magnitudes" and "in respect of *size*"
(πηλικότητα) omitted or altered. We do not find that Nicomachus and the
rest give any substantially different definition of a *ratio* between numbers.
Theon of Smyrna says, in fact (p. 73, 16), that "ratio in the sense of
proportion (λόγος ὁ κατ' ἀνάλογον) is a sort of relation of two homogeneous
terms to one another, as for example, double, triple." Similarly Nicomachus
says (II. 21, 3) that "a ratio is a relation of two terms to one another," the word
for "relation" being in both cases the same as Euclid's (σχέσις). Theon of
Smyrna goes on to classify ratios as greater, less, or equal, i.e. as ratios of greater
inequality, less inequality, or equality, and then to specify certain arithmetical
ratios which had special names, for which he quotes the authority of Adrastus.
The names were πολλαπλάσιος, ἐπιμόριος, ἐπιμερής, πολλαπλασιεπιμόριος,
πολλαπλασιεπιμερής (the first of which is, of course, a multiple, while the rest
are the equivalent of certain types of improper fractions as we should call
them), and the reciprocals of each of these described by prefixing ὑπό or *sub*.
After describing these particular classes of arithmetical ratios, Theon goes on
to say that numbers still have ratios to one another even if they are different
from all those previously described. We need not therefore concern ourselves
with the various types; it is sufficient to observe that any ratio between
numbers can be expressed in the manner indicated in Euclid's definition of
arithmetical proportion, for the greater is, in relation to the less, either one or
a combination of more than one of the three things, (1) a multiple, (2) a
submultiple, (3) a proper fraction.

It is when we come to the definition of *proportion* that we begin to find
differences between Euclid, Nicomachus, Theon and Iamblichus. "Proportion,"
says Theon (p. 82, 6), "is similarity or sameness of more ratios than one,"
which is of course unobjectionable if it is previously understood what a *ratio*
is; but confusion was brought in by those (like Thrasyllus) who said that
there were *three proportions* (ἀναλογίαι), the arithmetic, geometric, and
harmonic, where of course the reference is to arithmetic, geometric and
harmonic *means* (μεσότητες). Hence it was necessary to explain, as Adrastus
did (Theon, p. 106, 15), that of the several *means* "the geometric was called
both proportion *par excellence* and primary...though the other means were
also commonly called *proportions* by some writers." Accordingly we have
Nicomachus trying to extend the term "proportion" to cover the various
means as well as a proportion in three or four terms in the ordinary sense. He
says (II. 21, 2): "Proportion, *par excellence* (κυρίως), is the bringing together
(σύλληψις) to the same (point) of two or more *ratios*; or, more generally, (the
bringing together) of two or more *relations* (σχέσεων), even though they be
subjected not to the same *ratio* but to a difference or some other (law)."
Iamblichus keeps the senses of the word more distinct. He says, like Theon,
that "proportion is similarity or sameness of several ratios" (p. 98, 14), and
that "it is to be premised that it was the geometrical (proportion) which the
ancients called proportion *par excellence*, though it is now common to apply
the name generally to all the remaining means as well" (p. 100, 15). Pappus

remarks (III. p. 70, 17), "A mean differs from a proportion in this respect that, if anything is a proportion, it is also a mean, but not conversely. For there are three means, of which one is arithmetic, one geometric and one harmonic." The last remark implies plainly enough that there is only one *proportion* (ἀναλογία) in the proper sense. So, too, says Iamblichus in another place (p. 104, 19): "the second, the geometric, mean has been called *proportion par excellence* because the terms contain the same ratio, being separated according to the same proportion (ἀνὰ τὸν αὐτὸν λόγον διεστῶτες)." The natural conclusion is that of Nesselmann, that originally the geometric proportion was called ἀναλογία, the others, the arithmetic, the harmonic, etc., *means*; but later usage had obliterated the distinction.

Of proportions in the ancient and Euclidean sense Theon (p. 82, 10) distinguished the *continuous* (συνεχής) and the *separated* (διῃρημένη), using the same terms as Aristotle (*Eth. Nic.* 1131 a 32). The meaning is of course clear: in the *continuous* proportion the consequent of one ratio is the antecedent of the next; in the *separated* proportion this is not so. Nicomachus (II. 21, 5—6) uses the words *connected* (συνημμένη) and *disjoined* (διεζευγμένη) respectively. Euclid regularly speaks of numbers in continuous proportion as "proportional in order, or successively" (ἐξῆς ἀνάλογον).

DEFINITION 21.

Ὅμοιοι ἐπίπεδοι καὶ στερεοὶ ἀριθμοί εἰσιν οἱ ἀνάλογον ἔχοντες τὰς πλευράς.

Theon of Smyrna remarks (p. 36, 12) that, among plane numbers, *all* squares are similar, while of ἑτερομήκεις those are similar "whose sides, that is, the numbers containing them, are proportional." Here ἑτερομήκης must evidently be used, not in the sense of a number of the form $n(n+1)$, but as synonymous with προμήκης, *any* oblong number; so that on this occasion Theon follows the terminology of Plato and (according to Iamblichus) of Euclid. Obviously, if the strict sense of ἑτερομήκης is adhered to, no two numbers of that form can be similar unless they are also *equal*. We may compare Iamblichus' elaborate contrast of the square and the ἑτερομήκης. Since the two sides of the square are equal, a square number might, as he says (p. 82, 9), be fitly called ἰδιομήκης (Nicomachus uses ταὐτομήκης) in contrast to ἑτερομήκης; and the ancients, according to him, called square numbers "the same" and "similar" (ταὐτούς τε καὶ ὁμοίους), but ἑτερομήκεις numbers "dissimilar and other" (ἀνομοίους καὶ θατέρους).

With regard to solid numbers, Theon remarks in like manner (p. 37, 2) that *all* cube numbers are similar, while of the others those are similar whose sides are proportional, i.e. in which, as length is to length, so is breadth to breadth and height to height.

DEFINITION 22.

Τέλειος ἀριθμός ἐστιν ὁ τοῖς ἑαυτοῦ μέρεσιν ἴσος ὤν.

Theon of Smyrna (p. 45, 9 sqq.) and Nicomachus (I. 16) both give the same definition of a *perfect* number, as well as the law of formation of such numbers which Euclid proves in the later proposition, IX. 36. They add however definitions of two other kinds of numbers in contrast with it, (1) the *over-perfect* (ὑπερτελής in Nicomachus, ὑπερτέλειος in Theon), the sum of whose parts, i.e. submultiples, is greater than the number itself, e.g. 12, 24 etc., the sum of the parts of 12 being $6+4+3+2+1=16$, and the sum of the parts of 24 being $12+8+6+4+3+2+1=36$, (2) the *defective* (ἐλλιπής), the sum of whose parts is less than the whole, e.g. 8 or 14, the parts in the first case adding up to $4+2+1$, or 7, and in the second case to $7+2+1$, or 10. All three classes are however made by Theon subdivisions of numbers in general, but by Nicomachus subdivisions of *even* numbers.

The term *perfect* was used by the Pythagoreans, but in another sense, of 10; while Theon tells us (p. 46, 14) that 3 was also called perfect "because it is the first number that has beginning, middle and extremity; it is also both a *line* and a *plane* (for it is an equilateral triangle having each side made up of two units), and it is the first link and potentiality of the solid (for a solid must be conceived of in three dimensions)."

There are certain unexpressed axioms used in Book VII. as there are in earlier Books.

The following may be noted.

1. If A measures B, and B measures C, A will measure C.

2. If A measures B, and also measures C, A will measure the difference between B and C when they are unequal.

3. If A measures B, and also measures C, A will measure the sum of B and C.

It is clear, from what we know of the Pythagorean theory of numbers, of musical intervals expressed by numbers, of different kinds of *means* etc., that the substance of Euclid Books VII.—IX. was no new thing but goes back, at least, to the Pythagoreans. It is well known that the mathematics of Plato's *Timaeus* is essentially Pythagorean. It is therefore *a priori* probable (if not perhaps quite certain) that Plato πυθαγορίζει even in the passage (32 A, B) where he speaks of numbers "whether solid or square" in continued proportion, and proceeds to say that between *planes* one mean suffices, but to connect two *solids* two means are necessary. This passage has been much discussed, but I think that by "planes" and "solids" Plato certainly meant *square* and *solid numbers* respectively, so that the allusion must be to the theorems established in Eucl. VIII. 11, 12, that between two square numbers there is one mean proportional number, and between two cube numbers there are two mean proportional numbers[1].

---

[1] It is true that *similar* plane and solid numbers have the same property (Eucl. VIII. 18, 19); but, if Plato had meant similar plane and solid numbers generally, I think it would have been necessary to specify that they were "similar," whereas, seeing that the *Timaeus* is as a whole concerned with regular figures, there is nothing unnatural in allowing *regular* or *equilateral* to be understood. Further Plato speaks first of δυνάμεις and ὄγκοι and then of "planes" (ἐπίπεδα) and "solids" (στερεά) in such a way as to suggest that δυνάμεις correspond to ἐπίπεδα and ὄγκοι to στερεά. Now the regular meaning of δύναμις is *square* (or sometimes *square root*), and I think it is here used in the sense of *square*, notwithstanding that Plato seems to speak of *three* squares in continued proportion, whereas, in general, the mean between two squares as extremes would not be square but oblong. And, if δυνάμεις are squares, it is reasonable to suppose that the ὄγκοι are also *equilateral*, i.e. the "solids" are cubes. I am aware that Th. Häbler (*Bibliotheca Mathematica*, VIII₃, 1908, pp. 173—4) thinks that the passage is to be explained by reference to the problem of the duplication of the cube, and does not refer to numbers at all. Against this we have to put the evidence of Nicomachus (II. 24, 6) who, in speaking of "a certain Platonic theorem," quotes the very same results of Eucl. VIII. 11, 12. Secondly, it is worth noting that Häbler's explanation is distinctly ruled out by Democritus the Platonist (3rd cent. A.D.) who, according to Proclus

It is no less clear that, in his method and line of argument, Euclid was following earlier models, though no doubt making improvements in the exposition. The tract on the *Sectio Canonis*, κατατομὴ κανόνος (as to the genuineness of which see above, Vol. I., p. 17) is in style and in the form of the propositions generally akin to the *Elements*. In one proposition (2) the author says "*we learned* (ἐμάθομεν) that, if as many numbers as we please be in (continued) proportion, and the first measures the last, the first will also measure the intermediate numbers"; here he practically quotes *Elem.* VIII. 7. In the 3rd proposition he proves that no number can be a mean between two numbers in the ratio known as ἐπιμόριος, the ratio, that is, of $n+1$ to n, where n is any integer greater than unity. Now, fortunately, Boethius, *De institutione musica*, III. 11 (pp. 285—6, ed. Friedlein), has preserved a proof by Archytas of this same proposition; and the proof is substantially identical with that of Euclid. The two proofs are placed side by side in an article by Tannery (*Bibliotheca Mathematica*, VI₃, 1905/6, p. 227). Archytas writes the smaller term of the proportion first (instead of the greater, as Euclid does). Let, he says, A, B be the "superparticularis proportio" (ἐπιμόριον διάστημα in Euclid). Take C, DE the smallest numbers which are in the ratio of A to B. [Here DE means $D+E$: and in this respect the notation is different from that of Euclid who, as usual, takes a line DF divided into two parts at G, GF corresponding to E, and DG to D, in Archytas' notation. The step of taking C, DE, the smallest numbers in the ratio of A to B, presupposes Eucl. VII. 33.] Then DE exceeds C by an aliquot part of itself and of C [cf. the definition of ἐπιμόριος ἀριθμός in Nicomachus, I. 19, 1]. Let D be the excess [i.e. E is supposed equal to C]. "I say that D is not a number but an unit."

For, if D is a number and a part of DE, it measures DE; hence it measures E, that is, C. Thus D measures both C and DE, which is impossible; for the smallest numbers which are in the same ratio as any numbers are prime to one another. [This presupposes Eucl. VII. 22.] Therefore D is an unit; that is, DE exceeds C by an unit. Hence no number can be found which is a mean between two numbers C, DE. Therefore neither can any number be a mean between the original numbers A, B which are in the same ratio [this implies Eucl. VII. 20].

We have then here a clear indication of the existence at least as early as the date of Archytas (about 430—365 B.C.) of an *Elements of Arithmetic* in the form which we call Euclidean; and no doubt text-books of the sort existed even before Archytas, which probably Archytas himself and Eudoxus improved and developed in their turn.

(*In Platonis Timaeum commentaria*, 149 c), said that the difficulties of the passage of the *Timaeus* had *misled* some people into connecting it with the duplication of the cube, whereas it really referred to *similar* planes and solids with sides in *rational numbers*. Thirdly, I do not think that, under the supposition that the Delian problem is referred to, we get the required sense. The problem in that case is not that of finding two mean proportionals *between two cubes* but that of finding a second cube the content of which shall be equal to twice, or k times (where k is any number not a complete cube), the content of a given cube (a^3). Two mean proportionals are found, not between cubes, but between two *straight lines* in the ratio of 1 to k, or between a and ka. Unless k is a cube, there would be no point in saying that two means are necessary to connect 1 and k, and not one mean; for $\sqrt[3]{k}$ is no more natural than $\sqrt{k}$, and would be less natural in the case where k happened to be square. On the other hand, if k is a cube, so that it is a question of finding means between *cube numbers*, the dictum of Plato is perfectly intelligible; nor is any real difficulty caused by the generality of the statement that two means are *always* necessary to connect them, because any property enunciated generally of two cube numbers should obviously be true of cubes *as such*, that is, it must hold in the extreme case of two cubes which are *prime to one another*.

BOOK VII. PROPOSITIONS.

PROPOSITION 1.

Two unequal numbers being set out, and the less being continually subtracted in turn from the greater, if the number which is left never measures the one before it until an unit is left, the original numbers will be prime to one another.

For, the less of two unequal numbers AB, CD being continually subtracted from the greater, let the number which is left never measure the one before it until an unit is left;

I say that AB, CD are prime to one another, that is, that an unit alone measures AB, CD.

For, if AB, CD are not prime to one another, some number will measure them.

Let a number measure them, and let it be E; let CD, measuring BF, leave FA less than itself,

let AF, measuring DG, leave GC less than itself,

and let GC, measuring FH, leave an unit HA.

Since, then, E measures CD, and CD measures BF,

therefore E also measures BF.

But it also measures the whole BA;

therefore it will also measure the remainder AF.

But AF measures DG;

therefore E also measures DG.

But it also measures the whole $DC \cdot$

therefore it will also measure the remainder CG.

But CG measures FH;

therefore E also measures FH.

But it also measures the whole FA;

therefore it will also measure the remainder, the unit AH, though it is a number : which is impossible.

Therefore no number will measure the numbers AB, CD; therefore AB, CD are prime to one another.   [VII. Def. 12]

Q. E. D.

It is proper to remark here that the representation in Books VII. to IX. of numbers by straight lines is adopted by Heiberg from the MSS. The method of those editors who substitute *points* for lines is open to objection because it practically necessitates, in many cases, the use of specific numbers, which is contrary to Euclid's manner.

"Let CD, measuring BF, leave FA less than itself." This is a neat abbreviation for saying, measure along BA successive lengths equal to CD until a point F is reached such that the length FA remaining is less than CD; in other words, let BF be the largest exact multiple of CD contained in BA.

Euclid's method in this proposition is an application to the particular case of prime numbers of the method of finding the greatest common measure of two numbers not prime to one another, which we shall find in the next proposition. With our notation, the method may be shown thus. Supposing the two numbers to be a, b, we have, say,

$$
\begin{array}{l}
b \,) \, a \, (\, p \\
\underline{pb} \\
\quad c \,) \, b \, (\, q \\
\quad \underline{qc} \\
\qquad d \,) \, c \, (\, r \\
\qquad \underline{rd} \\
\qquad \quad \mathbf{1}
\end{array}
$$

If now a, b are not prime to one another, they must have a common measure e, where e is some integer, not unity.

And since e measures a, b, it measures $a - pb$, i.e. c.

Again, since e measures b, c, it measures $b - qc$, i.e. d,

and lastly, since e measures c, d, it measures $c - rd$, i.e. 1:

which is impossible.

Therefore there is no integer, except unity, that measures a, b, which are accordingly prime to one another.

Observe that Euclid assumes as an axiom that, if a, b are both divisible by c, so is $a - pb$. In the next proposition he assumes as an axiom that c will in the case supposed divide $a + pb$.

PROPOSITION 2.

Given two numbers not prime to one another, to find their greatest common measure.

Let *AB*, *CD* be the two given numbers not prime to one another.

Thus it is required to find the greatest common measure of *AB*, *CD*.

If now *CD* measures *AB*—and it also measures itself—*CD* is a common measure of *CD*, *AB*.

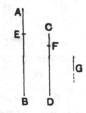

And it is manifest that it is also the greatest; for no greater number than *CD* will measure *CD*.

But, if *CD* does not measure *AB*, then, the less of the numbers *AB*, *CD* being continually subtracted from the greater, some number will be left which will measure the one before it.

For an unit will not be left; otherwise *AB*, *CD* will be prime to one another [VII. 1], which is contrary to the hypothesis.

Therefore some number will be left which will measure the one before it.

Now let *CD*, measuring *BE*, leave *EA* less than itself, let *EA*, measuring *DF*, leave *FC* less than itself, and let *CF* measure *AE*.

Since then, *CF* measures *AE*, and *AE* measures *DF*, therefore *CF* will also measure *DF*.

But it also measures itself; therefore it will also measure the whole *CD*.

But *CD* measures *BE*; therefore *CF* also measures *BE*.

But it also measures *EA*; therefore it will also measure the whole *BA*.

But it also measures *CD*; therefore *CF* measures *AB*, *CD*.

Therefore *CF* is a common measure of *AB*, *CD*.

I say next that it is also the greatest.

For, if CF is not the greatest common measure of AB, CD, some number which is greater than CF will measure the numbers AB, CD.

Let such a number measure them, and let it be G.

Now, since G measures CD, while CD measures BE, G also measures BE.

But it also measures the whole BA;

therefore it will also measure the remainder AE.

But AE measures DF;

therefore G will also measure DF.

But it also measures the whole DC;

therefore it will also measure the remainder CF, that is, the greater will measure the less: which is impossible.

Therefore no number which is greater than CF will measure the numbers AB, CD;

therefore CF is the greatest common measure of AB, CD.

PORISM. From this it is manifest that, if a number measure two numbers, it will also measure their greatest common measure.   Q. E. D..

Here we have the exact method of finding the greatest common measure given in the text-books of algebra, including the *reductio ad absurdum* proof that the number arrived at is not only a common measure but the *greatest* common measure. The process of finding the greatest common measure is simply shown thus :

$$b \,) \, a \, (\, p$$
$$\underline{pb}$$
$$c \,) \, b \, (\, q$$
$$\underline{qc}$$
$$d \,) \, c \, (\, r$$
$$\underline{rd}$$

We shall arrive, says Euclid, at some number, say d, which measures the one before it, i.e. such that $c = rd$. Otherwise the process would go on until we arrived at unity. This is impossible because in that case a, b would be prime to one another, which is contrary to the hypothesis.

Next, like the text-books of algebra, he goes on to show that d will be *some* common measure of a, b. For d measures c;

therefore it measures $qc + d$, that is, b,

and hence it measures $pb + c$, that is, a.

Lastly, he proves that d is the *greatest* common measure of a, b as follows.

Suppose that e is a common measure greater than d.

Then e, measuring a, b, must measure $a - pb$, or c.

Similarly e must measure $b-qc$, that is, d: which is impossible, since e is by hypothesis greater than d.

Therefore etc.

Euclid's proposition is thus *identical* with the algebraical proposition as generally given, e.g. in Todhunter's algebra, except that of course Euclid's numbers are integers.

Nicomachus gives the same rule (though without proving it) when he shows how to determine whether two given *odd* numbers are prime or not prime to one another, and, if they are not prime to one another, what is their common measure. We are, he says, to compare the numbers in turn by continually taking the less from the greater as many times as possible, then taking the remainder as many times as possible from the less of the original numbers, and so on; this process "will finish either at an unit or at some one and the same number," by which it is implied that the division of a greater number by a less is done by *separate subtractions* of the less. Thus, with regard to 21 and 49, Nicomachus says, "I subtract the less from the greater; 28 is left; then again I subtract from this the same 21 (for this is possible); 7 is left; I subtract this from 21, 14 is left; from which I again subtract 7 (for this is possible); 7 will be left, but 7 cannot be subtracted from 7." The last phrase is curious, but the meaning of it is obvious enough, as also the meaning of the phrase about ending "at one and the same number."

The proof of the Porism is of course contained in that part of the proposition which proves that G, a common measure different from CF, must measure CF. The supposition, thereby proved to be false, that G is greater than CF does not affect the validity of the proof that G measures CF in any case.

PROPOSITION 3.

Given three numbers not prime to one another, to find their greatest common measure.

Let A, B, C be the three given numbers not prime to one another;

thus it is required to find the greatest common measure of A, B, C.

For let the greatest common measure, D, of the two numbers A, B be taken;

[VII. 2]

then D either measures, or does not measure, C.

First, let it measure it.

But it measures A, B also;

therefore D measures A, B, C;

therefore D is a common measure of A, B, C.

I say that it is also the greatest.

For, if D is not the greatest common measure of A, B, C, some number which is greater than D will measure the numbers A, B, C.

Let such a number measure them, and let it be E.

Since then E measures A, B, C,

it will also measure A, B;

therefore it will also measure the greatest common measure of A, B.   [VII. 2, Por.]

But the greatest common measure of A, B is D;

therefore E measures D, the greater the less: which is impossible.

Therefore no number which is greater than D will measure the numbers A, B, C;

therefore D is the greatest common measure of A, B, C.

Next, let D not measure C;

I say first that C, D are not prime to one another.

For, since A, B, C are not prime to one another, some number will measure them.

Now that which measures A, B, C will also measure A, B, and will measure D, the greatest common measure of A, B.

[VII. 2, Por.]

But it measures C also;

therefore some number will measure the numbers D, C;

therefore D, C are not prime to one another.

Let then their greatest common measure E be taken.

[VII. 2]

Then, since E measures D,

and D measures A, B,

therefore E also measures A, B.

But it measures C also;

therefore E measures A, B, C;

therefore E is a common measure of A, B, C.

I say next that it is also the greatest.

For, if E is not the greatest common measure of A, B, C, some number which is greater than E will measure the numbers A, B, C.

Let such a number measure them, and let it be F.

Now, since F measures A, B, C,
it also measures A, B;
therefore it will also measure the greatest common measure
of A, B.   [VII. 2, Por.]

But the greatest common measure of A, B is D;
therefore F measures D.

And it measures C also;
therefore F measures D, C;
therefore it will also measure the greatest common measure
of D, C.   [VII. 2, Por.]

But the greatest common measure of D, C is E;
therefore F measures E, the greater the less: which is
impossible.

Therefore no number which is greater than E will measure
the numbers A, B, C;
therefore E is the greatest common measure of A, B, C.

Q. E. D.

Euclid's proof is here longer than we should make it because he
distinguishes two cases, the simpler of which is really included in the other.

Having taken the greatest common measure, say d, of a, b, two of the
three given numbers a, b, c, he distinguishes the cases

(1) in which d measures c,

(2) in which d does not measure c.

In the first case the greatest common measure of d, c is d itself; in the
second case it has to be found by a repetition of the process of VII. 2. In
either case the greatest common measure of a, b, c is the greatest common
measure of d, c.

But, after disposing of the simpler case, Euclid thinks it necessary to
prove that, if d does not measure c, d and c must necessarily *have* a greatest
common measure. This he does by means of the original hypothesis that
a, b, c are not prime to one another. Since they are not prime to one another,
they must have a common measure; any common measure of a, b is a measure
of d, and therefore any common measure of a, b, c is a common measure of
d, c; hence d, c must have a common measure, and are therefore not prime to
one another.

The proofs of cases (1) and (2) repeat exactly the same argument as we
saw in VII. 2, and it is proved separately for d in case (1) and e in case (2),
where e is the greatest common measure of d, c,

(α)   that it is a common measure of a, b, c,

(β)   that it is the *greatest* common measure.

Heron remarks (an-Nairīzī, ed. Curtze, p. 191) that the method does
not only enable us to find the greatest common measure of *three* numbers;
it can be used to find the greatest common measure of as many numbers

as we please. This is because any number measuring two numbers also measures their greatest common measure; and hence we can find the G.C.M. of pairs, then the G.C.M. of pairs of these, and so on, until only two numbers are left and we find the G.C.M. of these. Euclid tacitly assumes this extension in VII. 33, where he takes the greatest common measure of *as many numbers as we please.*

PROPOSITION 4.

Any number is either a part or parts of any number, the less of the greater.

Let A, BC be two numbers, and let BC be the less; I say that BC is either a part, or parts, of A.

For A, BC are either prime to one another or not.

First, let A, BC be prime to one another.

Then, if BC be divided into the units in it, each unit of those in BC will be some part of A; so that BC is parts of A.

Next let A, BC not be prime to one another; then BC either measures, or does not measure, A.

If now BC measures A, BC is a part of A.

But, if not, let the greatest common measure D of A, BC be taken;   [VII. 2]

and let BC be divided into the numbers equal to D, namely BE, EF, FC.

Now, since D measures A, D is a part of A.

But D is equal to each of the numbers BE, EF, FC; therefore each of the numbers BE, EF, FC is also a part of A; so that BC is parts of A.

Therefore etc.

Q. E. D.

The meaning of the enunciation is of course that, if a, b be two numbers of which b is the less, then b is either a *submultiple* or *some proper fraction* of a.

(1) If a, b are prime to one another, divide each into its units; then b contains b of the same parts of which a contains a. Therefore b is "parts" or a *proper fraction* of a.

(2) If a, b be not prime to one another, either b measures a, in which case b is a submultiple or "part" of a, or, if g be the greatest common measure of a, b, we may put $a = mg$ and $b = ng$, and b will contain n of the same parts (g) of which a contains m, so that b is again "parts," or a *proper fraction*, of a.

PROPOSITION 5.

If a number be a part of a number, and another be the same part of another, the sum will also be the same part of the sum that the one is of the one.

For let the number A be a part of BC,

and another, D, the same part of another EF that A is of BC;

I say that the sum of A, D is also the same part of the sum of BC, EF that A is of BC.

For since, whatever part A is of BC, D is also the same part of EF,

therefore, as many numbers as there are in BC equal to A, so many numbers are there also in EF equal to D.

Let BC be divided into the numbers equal to A, namely BG, GC,

and EF into the numbers equal to D, namely EH, HF;

then the multitude of BG, GC will be equal to the multitude of EH, HF.

And, since BG is equal to A, and EH to D,

therefore BG, EH are also equal to A, D.

For the same reason

GC, HF are also equal to A, D.

Therefore, as many numbers as there are in BC equal to A, so many are there also in BC, EF equal to A, D.

Therefore, whatever multiple BC is of A, the same multiple also is the sum of BC, EF of the sum of A, D.

Therefore, whatever part A is of BC, the same part also is the sum of A, D of the sum of BC, EF.

Q. E. D.

If $$a = \frac{1}{n} b, \text{ and } c = \frac{1}{n} d, \text{ then}$$

$$a + c = \frac{1}{n}(b + d).$$

The proposition is of course true for any quantity of pairs of numbers similarly related, as is the next proposition also; and both propositions are used in the extended form in VII. 9, 10.

PROPOSITION 6.

If a number be parts of a number, and another be the same parts of another, the sum will also be the same parts of the sum that the one is of the one.

For let the number AB be parts of the number C, and another, DE, the same parts of another, F, that AB is of C;
I say that the sum of AB, DE is also the same parts of the sum of C, F that AB is of C.

For since, whatever parts AB is of C, DE is also the same parts of F,
therefore, as many parts of C as there are in AB, so many parts of F are there also in DE.

Let AB be divided into the parts of C, namely AG, GB, and DE into the parts of F, namely DH, HE;
thus the multitude of AG, GB will be equal to the multitude of DH, HE.

And since, whatever part AG is of C, the same part is DH of F also,
therefore, whatever part AG is of C, the same part also is the sum of AG, DH of the sum of C, F.   [VII. 5]

For the same reason,
whatever part GB is of C, the same part also is the sum of GB, HE of the sum of C, F.

Therefore, whatever parts AB is of C, the same parts also is the sum of AB, DE of the sum of C, F.

Q. E. D.

If
$$a = \frac{m}{n} b, \text{ and } c = \frac{m}{n} d,$$

then
$$a + c = \frac{m}{n} (b + d).$$

More generally, if
$$a = \frac{m}{n} b, \; c = \frac{m}{n} d, \; e = \frac{m}{n} f, \; \dots$$

then
$$(a + c + e + g + \dots) = \frac{m}{n} (b + d + f + h + \dots).$$

In Euclid's proposition $m < n$, but the generality of the result is of course not affected. This proposition and the last are complementary to v. 1, which proves the corresponding result with *multiple* substituted for "*part*" or "*parts.*"

PROPOSITION 7.

If a number be that part of a number, which a number subtracted is of a number subtracted, the remainder will also be the same part of the remainder that the whole is of the whole.

For let the number AB be that part of the number CD which AE subtracted is of CF subtracted;

I say that the remainder EB is also the same part of the remainder FD that the whole AB is of the whole CD.

For, whatever part AE is of CF, the same part also let EB be of CG.

Now since, whatever part AE is of CF, the same part also is EB of CG,

therefore, whatever part AE is of CF, the same part also is AB of GF.   [VII. 5]

But, whatever part AE is of CF, the same part also, by hypothesis, is AB of CD;

therefore, whatever part AB is of GF, the same part is it of CD also;

therefore GF is equal to CD.

Let CF be subtracted from each;

therefore the remainder GC is equal to the remainder FD.

Now since, whatever part AE is of CF, the same part also is EB of GC,

while GC is equal to FD,

therefore, whatever part AE is of CF, the same part also is EB of FD.

But, whatever part AE is of CF, the same part also is AB of CD;

therefore also the remainder EB is the same part of the remainder FD that the whole AB is of the whole CD.

Q. E. D.

If $a = \dfrac{1}{n}b$ and $c = \dfrac{1}{n}d$, we are to prove that

$$a - c = \frac{1}{n}(b - d),$$

a result differing from that of VII. 5 in that *minus* is substituted for *plus*. Euclid's method is as follows.

Suppose that e is taken such that

$$a - c = \frac{1}{n}e. \quad \dots\dots\dots\dots\dots\dots\dots\dots\dots(1)$$

Now

$$c = \frac{1}{n}d.$$

Therefore

$$a = \frac{1}{n}(d + e), \qquad\qquad \text{[VII. 5]}$$

whence, from the hypothesis, $d + e = b$,

so that $e = b - d$,

and, substituting this value of e in (1), we have

$$a - c = \frac{1}{n}(b - d).$$

PROPOSITION 8.

If a number be the same parts of a number that a number subtracted is of a number subtracted, the remainder will also be the same parts of the remainder that the whole is of the whole.

For let the number AB be the same parts of the number CD that AE subtracted is of CF subtracted;

I say that the remainder EB is also the same parts of the remainder FD that the whole AB is of the whole CD.

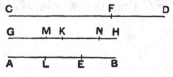

For let GH be made equal to AB.

Therefore, whatever parts GH is of CD, the same parts also is AE of CF.

Let GH be divided into the parts of CD, namely GK, KH, and AE into the parts of CF, namely AL, LE;

thus the multitude of GK, KH will be equal to the multitude of AL, LE.

Now since, whatever part GK is of CD, the same part also is AL of CF,

while. CD is greater than CF,

therefore GK is also greater than AL.

Let GM be made equal to AL.

Therefore, whatever part GK is of CD, the same part also is GM of CF;

therefore also the remainder MK is the same part of the remainder FD that the whole GK is of the whole CD. [VII. 7]

Again, since, whatever part KH is of CD, the same part also is EL of CF,

while CD is greater than CF,

therefore HK is also greater than EL.

Let KN be made equal to EL.

Therefore, whatever part KH is of CD, the same part also is KN of CF;

therefore also the remainder NH is the same part of the remainder FD that the whole KH is of the whole CD.

[VII. 7]

But the remainder MK was also proved to be the same part of the remainder FD that the whole GK is of the whole CD;

therefore also the sum of MK, NH is the same parts of DF that the whole HG is of the whole CD.

But the sum of MK, NH is equal to EB,

and HG is equal to BA;

therefore the remainder EB is the same parts of the remainder FD that the whole AB is of the whole CD.

Q. E. D.

If $\qquad\qquad a = \dfrac{m}{n} b$ and $c = \dfrac{m}{n} d,$ $\qquad (m < n)$

then $\qquad\qquad a - c = \dfrac{m}{n} (b - d).$

Euclid's proof amounts to the following.

Take e equal to $\dfrac{1}{n} b$, and f equal to $\dfrac{1}{n} d$.

Then since, by hypothesis, $b > d$,

$$e > f,$$

and, by VII. 7, $\qquad\qquad e - f = \dfrac{1}{n} (b - d).$

Repeat this for all the parts equal to e and f that there are in a, b respectively, and we have, by addition (a, b containing m of such parts respectively),

$$m(e-f) = \frac{m}{n}(b-d).$$

But　　　　　　　　$m(e-f) = a-c.$

Therefore　　　　　$a-c = \frac{m}{n}(b-d).$

The propositions VII. 7, 8 are complementary to v. 5 which gives the corresponding result with *multiple* in the place of "part" or "parts."

PROPOSITION 9.

If a number be a part of a number, and another be the same part of another, alternately also, whatever part or parts the first is of the third, the same part, or the same parts, will the second also be of the fourth.

For let the number A be a part of the number BC, and another, D, the same part of another, EF, that A is of BC;

I say that, alternately also, whatever part or parts A is of D, the same part or parts is BC of EF also.

For since, whatever part A is of BC, the same part also is D of EF,

therefore, as many numbers as there are in BC equal to A, so many also are there in EF equal to D.

Let BC be divided into the numbers equal to A, namely BG, GC,

and EF into those equal to D, namely EH, HF;

thus the multitude of BG, GC will be equal to the multitude of EH, HF.

Now, since the numbers BG, GC are equal to one another, and the numbers EH, HF are also equal to one another,

while the multitude of BG, GC is equal to the multitude of EH, HF,

therefore, whatever part or parts BG is of EH, the same part or the same parts is GC of HF also;

so that, in addition, whatever part or parts BG is of EH, the same part also, or the same parts, is the sum BC of the sum EF.

　　　　　　　　　　　　　　　　　　　　　[VII. 5, 6]

But BG is equal to A, and EH to D;

therefore, whatever part or parts A is of D, the same part or the same parts is BC of EF also.

Q. E. D.

If $a = \frac{1}{n} b$ and $c = \frac{1}{n} d$, then, whatever fraction (" part " or " parts ") a is of c, the same fraction will b be of d.

Dividing b into each of its parts equal to a, and d into each of its parts equal to c, it is clear that, whatever fraction one of the parts a is of one of the parts c, the same fraction is any other of the parts a of any other of the parts c.

And the number of the parts a is equal to the number of the parts c, viz. n.

Therefore, by VII. 5, 6, na is the same fraction of nc that a is of c, i.e. b is the same fraction of d that a is of c.

PROPOSITION 10.

If a number be parts of a number, and another be the same parts of another, alternately also, whatever parts or part the first is of the third, the same parts or the same part will the second also be of the fourth.

For let the number AB be parts of the number C, and another, DE, the same parts of another, F;

I say that, alternately also, whatever parts or part AB is of DE, the same parts or the same part is C of F also.

For since, whatever parts AB is of C, the same parts also is DE of F,

therefore, as many parts of C as there are in AB, so many parts also of F are there in DE.

Let AB be divided into the parts of C, namely AG, GB, and DE into the parts of F, namely DH, HE;

thus the multitude of AG, GB will be equal to the multitude of DH, HE.

Now since, whatever part AG is of C, the same part also is DH of F,

alternately also, whatever part or parts AG is of DH, the same part or the same parts is C of F also.   [VII. 9]

For the same reason also,

whatever part or parts GB is of HE, the same part or the same parts is C of F also;

so that, in addition, whatever parts or part AB is of DE, the same parts also, or the same part, is C of F.   [VII. 5, 6]

Q. E. D.

If $a = \dfrac{m}{n} b$ and $c = \dfrac{m}{n} d$, then, whatever fraction a is of c, the same fraction is b of d.

To prove this, a is divided into its m parts equal to b/n, and c into its m parts equal to d/n.

Then, by VII. 9, whatever fraction one of the m parts of a is of one of the m parts of c, the same fraction is b of d.

And, by VII. 5, 6, whatever fraction one of the m parts of a is of one of the m parts of c, the same fraction is the sum of the parts of a (that is, a) of the sum of the parts of c (that is, c).

Whence the result follows.

In the Greek text, after the words "so that, in addition" in the last line but one, is an additional explanation making the reference to VII. 5, 6 clearer, as follows: "whatever part or parts AG is of DH, the same part or the same parts is GB of HE also;

therefore also, whatever part or parts AG is of DH, the same part or the same parts is AB of DE also.   [VII. 5, 6]

But it was proved that, whatever part or parts AG is of DH, the same part or the same parts is C of F also;

therefore also" etc. as in the last two lines of the text.

Heiberg concludes, on the authority of P, which only has the words in the margin in a later hand, that they may be attributed to Theon.

Proposition 11.

If, as whole is to whole, so is a number subtracted to a number subtracted, the remainder will also be to the remainder as whole to whole.

As the whole AB is to the whole CD, so let AE subtracted be to CF subtracted;

I say that the remainder EB is also to the remainder FD as the whole AB to the whole CD.

Since, as AB is to CD, so is AE to CF, whatever part or parts AB is of CD, the same part or the same parts is AE of CF also;   [VII. Def. 20]

Therefore also the remainder EB is the same part or parts of FD that AB is of CD.   [VII. 7, 8]

Therefore, as EB is to FD, so is AB to CD. [VII. Def. 20]

Q. E. D.

It will be observed that, in dealing with the proportions in Props. 11—13, Euclid only contemplates the case where the first number is "a part" or "parts" of the second, while in Prop. 13 he assumes the first to be "a part" or "parts" of the third also; that is, the first number is in all three propositions assumed to be less than the second, and in Prop. 13 less than the third also. Yet the figures in Props. 11 and 13 are inconsistent with these assumptions. If the facts are taken to correspond to the figures in these propositions, it is necessary to take account of the other possibilities involved in the definition of proportion (VII. Def. 20), that the first number may also be a multiple, or a multiple *plus* "a part" or "parts" (including *once* as a multiple in this case), of each number with which it is compared. Thus a number of different cases would have to be considered. The remedy is to make the ratio which is in the lower terms the first ratio, and to invert the ratios, if necessary, in order to make "a part" or "parts" literally apply.

If $a : b = c : d$, $(a > c, b > d)$
then $(a - c) : (b - d) = a : b$.

This proposition for numbers corresponds to v. 19 for magnitudes. The enunciation is the same except that the masculine (agreeing with ἀριθμός) takes the place of the neuter (agreeing with μέγεθος).

The proof is no more than a combination of the arithmetical definition of proportion (VII. Def. 20) with the results of VII. 7, 8. The language of proportions is turned into the language of fractions by Def. 20; the results of VII. 7, 8 are then used and the language retransformed by Def. 20 into the language of proportions.

PROPOSITION 12.

If there be as many numbers as we please in proportion, then, as one of the antecedents is to one of the consequents, so are all the antecedents to all the consequents.

Let A, B, C, D be as many numbers as we please in proportion, so that,

as A is to B, so is C to D;

I say that, as A is to B, so are A, C to B, D.

For since, as A is to B, so is C to D, whatever part or parts A is of B, the same part or parts is C of D also.   [VII. Def. 20]

Therefore also the sum of A, C is the same part or the same parts of the sum of B, D that A is of B.

[VII. 5, 6]

Therefore, as A is to B, so are A, C to B, D. [VII. Def. 20]

If $a : a' = b : b' = c : c' = ...,$
then each ratio is equal to $(a + b + c + ...) : (a' + b' + c' + ...)$.

The proposition corresponds to v. 12, and the enunciation is word for word the same with that of v. 12 except that ἀριθμός takes the place of μέγεθος.

Again the proof merely connects the arithmetical definition of proportion (VII. Def. 20) with the results of VII. 5, 6, which are quoted as true for any number of numbers, and not merely for two numbers as in the enunciations of VII. 5, 6.

PROPOSITION 13.

If four numbers be proportional, they will also be proportional alternately.

Let the four numbers A, B, C, D be proportional, so that,

as A is to B, so is C to D;

I say that they will also be proportional alternately, so that,

as A is to C, so will B be to D.

For since, as A is to B, so is C to D, therefore, whatever part or parts A is of B, the same part or the same parts is C of D also.

[VII. Def. 20]

Therefore, alternately, whatever part or parts A is of C, the same part or the same parts is B of D also.

[VII. 10]

Therefore, as A is to C, so is B to D.

[VII. Def. 20]

Q. E. D.

If $a : b = c : d$,
then, alternately, $a : c = b : d$.

The proposition corresponds to v. 16 for magnitudes, and the proof consists in connecting VII. Def. 20 with the result of VII. 10.

PROPOSITION 14.

If there be as many numbers as we please, and others equal to them in multitude, which taken two and two are in the same ratio, they will also be in the same ratio ex aequali.

Let there be as many numbers as we please A, B, C, and others equal to them in multitude D, E, F, which taken two and two are in the same ratio, so that,

as A is to B, so is D to E,

and,   as B is to C, so is E to F;

I say that, *ex aequali*,

as A is to C, so also is D to F.

A	D
B	E
C	F

For, since, as A is to B, so is D to E, therefore, alternately,

as A is to D, so is B to E.   [VII. 13]

Again, since, as B is to C, so is E to F,
therefore, alternately,
$$\text{as } B \text{ is to } E, \text{ so is } C \text{ to } F. \qquad [\text{VII. 13}]$$
But, as B is to E, so is A to D;
therefore also, as A is to D, so is C to F.

Therefore, alternately,
$$\text{as } A \text{ is to } C, \text{ so is } D \text{ to } F. \qquad [id.]$$

If $\qquad\qquad\qquad\qquad a : b = d : e,$
and $\qquad\qquad\qquad\qquad b : c = e : f,$
    then, *ex aequali*, $\qquad a : c = d : f;$
and the same is true however many successive numbers are so related.

The proof is simplicity itself.

By VII. 13, alternately, $\qquad a : d = b : e,$
and $\qquad\qquad\qquad\qquad b : e = c : f.$
Therefore $\qquad\qquad\qquad a : d = c : f,$
    and, again alternately, $\qquad a : c = d : f.$

Observe that this simple method cannot be used to prove the corresponding proposition for magnitudes, V. 22, although V. 22 has been preceded by the two propositions in that Book corresponding to the propositions used here, viz. V. 16 and V. 11. The reason of this is that this method would only prove V. 22 for six magnitudes *all of the same kind*, whereas the magnitudes in V. 22 are not subject to this limitation.

Heiberg remarks in a note on VII. 19 that, while Euclid has proved several propositions of Book V. over again, by a separate proof, for numbers, he has neglected to do so in certain cases; e.g., he often uses V. 11 in these propositions of Book VII., V. 9 in VII. 19, V. 7 in the same proposition, and so on. Thus Heiberg would apparently suppose Euclid to use V. 11 in the last step of the present proof (*Ratios which are the same with the same ratio are also the same with one another*). I think it preferable to suppose that Euclid regarded the last step as axiomatic; since, by the definition of proportion, the first number is the same multiple or the same part or the same parts of the second that the third is of the fourth: the assumption is no more than an assumption that the numbers or proper fractions which are respectively equal to the same number or proper fraction are equal to one another.

Though the proposition is only proved of six numbers, the extension to as many as we please (as expressed in the enunciation) is obvious.

Proposition 15.

If an unit measure any number, and another number measure any other number the same number of times, alternately also, the unit will measure the third number the same number of times that the second measures the fourth.

For let the unit A measure any number BC,
and let another number D
measure any other number EF
the same number of times;
I say that, alternately also, the
unit A measures the number
D the same number of times that BC measures EF.

For, since the unit A measures the number BC the same
number of times that D measures EF,

therefore, as many units as there are in BC, so many numbers
equal to D are there in EF also.

Let BC be divided into the units in it, BG, GH, HC,
and EF into the numbers EK, KL, LF equal to D.

Thus the multitude of BG, GH, HC will be equal to the
multitude of EK, KL, LF.

And, since the units BG, GH, HC are equal to one another,
and the numbers EK, KL, LF are also equal to one another,
while the multitude of the units BG, GH, HC is equal to the
multitude of the numbers EK, KL, LF,

therefore, as the unit BG is to the number EK, so will the
unit GH be to the number KL, and the unit HC to the
number LF.

Therefore also, as one of the antecedents is to one of
the consequents, so will all the antecedents be to all the
consequents;　　　　　　　　　　　　　　　　　　　[VII. 12]

therefore, as the unit BG is to the number EK, so is BC to
EF.

But the unit BG is equal to the unit A,
and the number EK to the number D.

Therefore, as the unit A is to the number D, so is BC to
EF.

Therefore the unit A measures the number D the same
number of times that BC measures EF.　　　Q. E. D.

If there be four numbers 1, m, a, ma (such that 1 measures m the same
number of times that a measures ma), 1 measures a the same number of
times that m measures ma.

Except that the first number is unity and the numbers are said to *measure*
instead of being a *part* of others, this proposition and its proof do not differ
from VII. 9; in fact this proposition is a particular case of the other.

PROPOSITION 16.

If two numbers by multiplying one another make certain numbers, the numbers so produced will be equal to one another.

Let *A*, *B* be two numbers, and let *A* by multiplying *B* make *C*, and *B* by multiplying *A* make *D* ;

I say that *C* is equal to *D*.

For, since *A* by multiplying *B* has made *C*,

therefore *B* measures *C* according to the units in *A*.

But the unit *E* also measures the number *A* according to the units in it ;

therefore the unit *E* measures *A* the same number of times that *B* measures *C*.

Therefore, alternately, the unit *E* measures the number *B* the same number of times that *A* measures *C*.   [VII. 15]

Again, since *B* by multiplying *A* has made *D*,

therefore *A* measures *D* according to the units in *B*.

But the unit *E* also measures *B* according to the units in it ;

therefore the unit *E* measures the number *B* the same number of times that *A* measures *D*.

But the unit *E* measured the number *B* the same number of times that *A* measures *C* ;

therefore *A* measures each of the numbers *C*, *D* the same number of times.

Therefore *C* is equal to *D*.   Q. E. D.

2. **The numbers so produced.** The Greek has οἱ γενόμενοι ἐξ αὐτῶν, "the (numbers) produced *from them*." By "from them" Euclid means "from the original numbers," though this is not very clear even in the Greek. I think ambiguity is best avoided by leaving out the words.

This proposition proves that, *if any numbers be multiplied together, the order of multiplication is indifferent*, or *ab = ba*.

It is important to get a clear understanding of what Euclid means when he speaks of *one number multiplying another*. VII. Def. 15 states that the effect of "*a* multiplying *b*" is taking *a* times *b*. We shall always represent "*a* times *b*" by *ab* and "*b* times *a*" by *ba*. This being premised, the proof that *ab = ba* may be represented as follows in the language of proportions.

By VII. Def. 20, $1 : a = b : ab.$

Therefore, alternately, $1 : b = a : ab.$ [VII. 13]

Again, by VII. Def. 20, $1 : b = a : ba.$

Therefore $a : ab = a : ba,$

or $ab = ba.$

Euclid does not use the language of proportions but that of fractions or their equivalent measures, quoting VII. 15, a particular case of VII. 13 differently expressed, instead of VII. 13 itself.

PROPOSITION 17.

If a number by multiplying two numbers make certain numbers, the numbers so produced will have the same ratio as the numbers multiplied.

For let the number A by multiplying the two numbers B, C make D, E;

I say that, as B is to C, so is D to E.

For, since A by multiplying B has made D,

therefore B measures D according to the units in A.

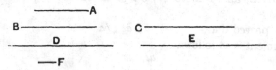

But the unit F also measures the number A according to the units in it;

therefore the unit F measures the number A the same number of times that B measures D.

Therefore, as the unit F is to the number A, so is B to D.

[VII. Def. 20]

For the same reason,

as the unit F is to the number A, so also is C to E;

therefore also, as B is to D, so is C to E.

Therefore, alternately, as B is to C, so is D to E. [VII. 13]

Q. E. D.

$$b : c = ab : ac.$$

In this case Euclid translates the language of measures into that of proportions, and the proof is exactly like that set out in the last note.

By VII. Def. 20, $1 : a = b : ab,$

and $1 : a = c : ac.$

Therefore $b : ab = c : ac,$

and, alternately, $b : c = ab : ac.$ [VII. 13]

Proposition 18.

If two numbers by multiplying any number make certain numbers, the numbers so produced will have the same ratio as the multipliers.

For let two numbers A, B by multiplying any number C make D, E;

I say that, as A is to B, so is D to E.

For, since A by multiplying C has made D,

therefore also C by multiplying A has made D.   [VII. 16]
For the same reason also
C by multiplying B has made E.

Therefore the number C by multiplying the two numbers A, B has made D, E.

Therefore, as A is to B, so is D to E.   [VII. 17]

It is here proved that $a : b = ac : bc.$
The argument is as follows.
                                $ac = ca.$ [VII. 16]
        Similarly $bc = cb.$
And $a : b = ca : cb ;$ [VII. 17]
        therefore $a : b = ac : bc.$

Proposition 19.

If four numbers be proportional, the number produced from the first and fourth will be equal to the number produced from the second and third; and, if the number produced from the first and fourth be equal to that produced from the second and third, the four numbers will be proportional.

Let A, B, C, D be four numbers in proportion, so that,
                as A is to B, so is C to D;
and let A by multiplying D make E, and let B by multiplying C make F;

I say that E is equal to F.

For let A by multiplying C make G.

Since, then, A by multiplying C has made G, and by multiplying D has made E,
the number A by multiplying the two numbers C, D has made G, E.

Therefore, as C is to D, so is G to E.

[VII. 17]

But, as C is to D, so is A to B;
therefore also, as A is to B, so is G to E.

Again, since A by multiplying C has made G,
but, further, B has also by multiplying C made F,
the two numbers A, B by multiplying a certain number C have made G, F.

Therefore, as A is to B, so is G to F.   [VII. 18]

But further, as A is to B, so is G to E also;
therefore also, as G is to E, so is G to F.

Therefore G has to each of the numbers E, F the same ratio;

therefore E is equal to F.   [cf. v. 9]

Again, let E be equal to F;
I say that, as A is to B, so is C to D.

For, with the same construction,
since E is equal to F,
therefore, as G is to E, so is G to F.   [cf. v. 7]

But, as G is to E, so is C to D,   [VII. 17]
and, as G is to F, so is A to B.   [VII. 18]

Therefore also, as A is to B, so is C to D.

Q. E. D.

If	$a : b = c : d$,	
then	$ad = bc$; and conversely.	

The proof is equivalent to the following.

(1)	$ac : ad = c : d$	[VII. 17]
	$= a : b$.	
But	$a : b = ac : bc$.	[VII. 18]
Therefore	$ac : ad = ac : bc$,	
or	$ad = bc$.	

(2) Since
$$ad = bc,$$
$$ac : ad = ac : bc.$$

But $ac : ad = c : d,$ [VII. 17]

and $ac : bc = a : b.$ [VII. 18]

Therefore $a : b = c : d.$

As indicated in the note on VII. 14 above, Heiberg regards Euclid as basing the inferences contained in the last step of part (1) of this proof and in the first step of part (2) on the propositions V. 9 and V. 7 respectively, since he has not proved those propositions separately for numbers in this Book. I prefer to suppose that he regarded the inferences as obvious and not needing proof, in view of the definition of numbers which are in proportion. E.g., if ac is the same fraction ("part" or "parts") of ad that ac is of bc, it is obvious that ad must be equal to bc.

Heiberg omits from his text here, and relegates to an Appendix, a proposition appearing in the manuscripts V, p, ϕ to the effect that, if *three* numbers be proportional, the product of the extremes is equal to the square of the mean, and conversely. It does not appear in P in the first hand, B has it in the margin only, and Campanus omits it, remarking that Euclid does not give the proposition about *three* proportionals as he does in VI. 17, since it is easily proved by the proposition just given. Moreover an-Nairīzī quotes the proposition about three proportionals *as an observation on* VII. 19 probably due to Heron (who is mentioned by name in the preceding paragraph).

PROPOSITION 20.

The least numbers of those which have the same ratio with them measure those which have the same ratio the same number of times, the greater the greater and the less the less.

For let CD, EF be the least numbers of those which have the same ratio with A, B;

I say that CD measures A the same number of times that EF measures B.

Now CD is not parts of A.

For, if possible, let it be so;

therefore EF is also the same parts of B that CD is of A.   [VII. 13 and Def. 20]

Therefore, as many parts of A as there are in CD, so many parts of B are there also in EF.

Let CD be divided into the parts of A, namely CG, GD, and EF into the parts of B, namely EH, HF;

thus the multitude of CG, GD will be equal to the multitude of EH, HF.

Now, since the numbers CG, GD are equal to one another, and the numbers EH, HF are also equal to one another,

while the multitude of CG, GD is equal to the multitude of EH, HF,

therefore, as CG is to EH, so is GD to HF.

Therefore also, as one of the antecedents is to one of the consequents, so will all the antecedents be to all the consequents.   [VII. 12]

Therefore, as CG is to EH, so is CD to EF.

Therefore CG, EH are in the same ratio with CD, EF, being less than they:

which is impossible, for by hypothesis CD, EF are the least numbers of those which have the same ratio with them.

Therefore CD is not parts of A;

therefore it is a part of it.   [VII. 4]

And EF is the same part of B that CD is of A;

[VII. 13 and Def. 20]

therefore CD measures A the same number of times that EF measures B.

Q. E. D.

If a, b are the least numbers among those which have the same ratio (i.e. if a/b is a fraction in its lowest terms), and c, d are any others in the same ratio, i.e. if

$$a : b = c : d,$$

then $a = \frac{1}{n} c$ and $b = \frac{1}{n} d$, where n is some integer.

The proof is by *reductio ad absurdum*, thus.

[Since $a < c$, a is some proper fraction ("part" or "parts") of c, by VII. 4.]

Now a cannot be equal to $\frac{m}{n} c$, where m is an integer less than n but greater than 1.

For, if $a = \frac{m}{n} c$, $b = \frac{m}{n} d$ also.   [VII. 13 and Def. 20]

Take each of the m parts of a with each of the m parts of b, two and two; the ratio of the members of all pairs is the same ratio $\frac{1}{m} a : \frac{1}{m} b$.

Therefore

$$\frac{1}{m} a : \frac{1}{m} b = a : b.$$ [VII. 12]

But $\frac{1}{m} a$ and $\frac{1}{m} b$ are respectively less than a, b and they are in the same ratio: which contradicts the hypothesis.

Hence a can only be "a part" of c, or

$$a \text{ is of the form } \frac{1}{n}c,$$

and therefore b is of the form $\frac{1}{n}d$.

Here also Heiberg omits a proposition which was no doubt interpolated by Theon (B, V, p, φ have it as VII. 22, but P only has it in the margin and in a later hand; Campanus also omits it) proving for numbers the *ex aequali* proposition when "the proportion is perturbed," i.e. (cf. enunciation of v. 22) if

$$a : b = e : f, \quad \dots\dots\dots\dots\dots\dots\dots\dots\dots\dots(1)$$

and $b : c = d : e, \quad \dots\dots\dots\dots\dots\dots\dots\dots\dots(2)$

then $a : c = d : f.$

The proof (see Heiberg's Appendix) depends on VII. 19.

From (1) we have $af = be,$

and from (2) $be = cd.$ [VII. 19]

Therefore $af = cd,$

and accordingly $a : c = d : f.$ [VII. 19]

PROPOSITION 21.

Numbers prime to one another are the least of those which have the same ratio with them.

Let A, B be numbers prime to one another;
I say that A, B are the least of those which have the same ratio with them.

For, if not, there will be some numbers less than A, B which are in the same ratio with A, B.

Let them be C, D.

Since, then, the least numbers of those which have the same ratio measure those which have the same ratio the same number of times, the greater the greater and the less the less, that is, the antecedent the antecedent and the consequent the consequent,   [VII. 20]

therefore C measures A the same number of times that D measures B.

Now, as many times as C measures A, so many units let there be in E.

Therefore D also measures B according to the units in E.

And, since C measures A according to the units in E, therefore E also measures A according to the units in C.

[VII. 16]

For the same reason

E also measures B according to the units in D.　　[VII. 16]

Therefore E measures A, B which are prime to one another: which is impossible.　　[VII. Def. 12]

Therefore there will be no numbers less than A, B which are in the same ratio with A, B.

Therefore A, B are the least of those which have the same ratio with them.

Q. E. D.

In other words, if a, b are prime to one another, the ratio $a : b$ is "in its lowest terms."

The proof is equivalent to the following.

If not, suppose that c, d are the *least* numbers for which

$$a : b = c : d.$$

[Euclid only supposes *some* numbers c, d in the ratio of a to b such that $c < a$, and (consequently) $d < b$. It is however necessary to suppose that c, d are the *least* numbers in that ratio in order to enable VII. 20 to be used in the proof.]

Then [VII. 20] $a = mc$, and $b = md$, where m is some integer.

Therefore　　　　　　　　$a = cm$,　　$b = dm$,　　　　　　　　[VII. 16]

and m is a common measure of a, b, though these are prime to one another, which is impossible.　　[VII. Def. 12]

Thus the least numbers in the ratio of a to b cannot be less than a, b themselves.

Where I have quoted VII. 16 Heiberg regards the reference as being to VII. 15. I think the phraseology of the text combined with that of Def. 15 suggests the former rather than the latter.

PROPOSITION 22.

The least numbers of those which have the same ratio with them are prime to one another.

Let A, B be the least numbers of those which have the same ratio with them;

I say that A, B are prime to one another.

For, if they are not prime to one another, some number will measure them.

Let some number measure them, and let it be C.

A ————————————

B ——————

C ————

D ———

E ——

And, as many times as C measures A, so many units let there be in D,

and, as many times as C measures B, so many units let there be in E

Since C measures A according to the units in D,

therefore C by multiplying D has made A.　　　[VII. Def. 15]

For the same reason also

C by multiplying E has made B.

Thus the number C by multiplying the two numbers D, E has made A, B;

therefore, as D is to E, so is A to B;　　　　　[VII. 17]

therefore D, E are in the same ratio with A, B, being less than they: which is impossible.

Therefore no number will measure the numbers A, B.

Therefore A, B are prime to one another.

Q. E. D.

If $a : b$ is "in its lowest terms," a, b are prime to one another.
Again the proof is indirect.
If a, b are not prime to one another, they have some common measure c, and

$$a = mc, \quad b = nc.$$

Therefore　　　　　　　$m : n = a : b.$　　　　　[VII. 17 or 18]

But m, n are less than a, b respectively, so that $a : b$ is not in its lowest terms: which is contrary to the hypothesis.
Therefore etc.

PROPOSITION 23.

If two numbers be prime to one another, the number which measures the one of them will be prime to the remaining number.

Let A, B be two numbers prime to one another, and let any number C measure A;

I say that C, B are also prime to one another.

For, if C, B are not prime to one another,

some number will measure C, B.

Let a number measure them, and let it be D.

Since D measures C, and C measures A,

therefore D also measures A.

But it also measures B;

A　B　C　D

therefore D measures A, B which are prime to one another:
which is impossible.   [VII. Def. 12]

Therefore no number will measure the numbers C, B.

Therefore C, B are prime to one another.

Q. E. D.

If a, mb are prime to one another, b is prime to a. For, if not, some
number d will measure both a and b, and therefore both a and mb: which is
contrary to the hypothesis.

Therefore etc.

PROPOSITION 24.

*If two numbers be prime to any number, their product also
will be prime to the same.*

For let the two numbers A, B be prime to any number C,
and let A by multiplying B make D;
I say that C, D are prime to one another.

For, if C, D are not prime to one another,
some number will measure C, D.

Let a number measure them, and let it
be E.

Now, since C, A are prime to one
another,
and a certain number E measures C,
therefore A, E are prime to one another.   [VII. 23]

As many times, then, as E measures D, so many units let
there be in F;
therefore F also measures D according to the units in E.
[VII. 16]

Therefore E by multiplying F has made D.   [VII. Def. 15]

But, further, A by multiplying B has also made D;
therefore the product of E, F is equal to the product of A, B.

But, if the product of the extremes be equal to that of the
means, the four numbers are proportional;   [VII. 19]
therefore, as E is to A, so is B to F.

But A, E are prime to one another,
numbers which are prime to one another are also the least of
those which have the same ratio,   [VII. 21]
and the least numbers of those which have the same ratio
with them measure those which have the same ratio the same

number of times, the greater the greater and the less the less, that is, the antecedent the antecedent and the consequent the consequent ;   [VII. 20]

therefore E measures B.

But it also measures C;

therefore E measures B, C which are prime to one another: which is impossible.   [VII. Def. 12]

Therefore no number will measure the numbers C, D.

Therefore C, D are prime to one another.

            Q. E. D.

---

1. **their product.** ὁ ἐξ αὐτῶν γενόμενος, literally "the (number) produced from them," will henceforth be translated as "their product."

If a, b are both prime to c, then ab, c are prime to one another.

The proof is again by *reductio ad absurdum*.

If ab, c are not prime to one another, let them be measured by a and be equal to md, nd, say, respectively.

Now, since a, c are prime to one another and d measures c,

            a, d are prime to one another.   [VII. 23]

But, since $ab = md$,

            $d : a = b : m$.   [VII. 19]

Therefore [VII. 20] d measures b,

         or $b = pd$, say.

         But $c = nd$.

Therefore d measures both b and c, which are therefore not prime to one another: which is impossible.

Therefore etc.

PROPOSITION 25.

If two numbers be prime to one another, the product of one of them into itself will be prime to the remaining one.

Let A, B be two numbers prime to one another, and let A by multiplying itself make C: I say that B, C are prime to one another.

For let D be made equal to A.

Since A, B are prime to one another, and A is equal to D, therefore D, B are also prime to one another.

Therefore each of the two numbers D, A is prime to B;

therefore the product of D, A will also be prime to B. [VII. 24]

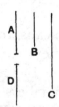

But the number which is the product of D, A is C.

Therefore C, B are prime to one another.   Q. E. D.

1. **the product of one of them into itself.** The Greek, ὁ ἐκ τοῦ ἑνὸς αὐτῶν γενόμενος, literally "the number produced from the one of them," leaves "multiplied into itself" to be understood.

If a, b are prime to one another,

$$a^2 \text{ is prime to } b.$$

Euclid takes d equal to a, so that d, a are both prime to b.

Hence, by VII. 24, da, i.e. a^2, is prime to b.

The proposition is a particular case of the preceding proposition; and the method of proof is by substitution of different numbers in the result of that proposition.

PROPOSITION 26.

If two numbers be prime to two numbers, both to each, their products also will be prime to one another.

For let the two numbers A, B be prime to the two numbers C, D; both to each, and let A by multiplying B make E, and let C by multiplying D make F; I say that E, F are prime to one another.

For, since each of the numbers A, B is prime to C, therefore the product of A, B will also be prime to C. [VII. 24]

But the product of A, B is E;

therefore E, C are prime to one another.

For the same reason

E, D are also prime to one another.

Therefore each of the numbers C, D is prime to E.

Therefore the product of C, D will also be prime to E.

[VII. 24]

But the product of C, D is F.

Therefore E, F are prime to one another.   Q. E. D.

If both a and b are prime to each of two numbers c, d, then ab, cd will be prime to one another.

Since a, b are both prime to c,

ab, c are prime to one another.   [VII. 24]

Similarly ab, d are prime to one another.

Therefore c, d are both prime to ab,

and so therefore is cd.   [VII. 24]

PROPOSITION 27.

*If two numbers be prime to one another, and each by
multiplying itself make a certain number, the products will be
prime to one another; and, if the original numbers by multiplying the products make certain numbers, the latter will also
be prime to one another* [*and this is always the case with the
extremes*].

Let *A*, *B* be two numbers prime to one another,
let *A* by multiplying itself make *C*, and by
multiplying *C* make *D*,
and let *B* by multiplying itself make *E*, and
by multiplying *E* make *F*;
I say that both *C*, *E* and *D*, *F* are prime
to one another.

For, since *A*, *B* are prime to one another,
and *A* by multiplying itself has made *C*,
therefore *C*, *B* are prime to one another.   [VII. 25]

Since then *C*, *B* are prime to one another,
and *B* by multiplying itself has made *E*,
therefore *C*, *E* are prime to one another.   [*id.*]

Again, since *A*, *B* are prime to one another,
and *B* by multiplying itself has made *E*,
therefore *A*, *E* are prime to one another.   [*id.*]

Since then the two numbers *A*, *C* are prime to the two
numbers *B*, *E*, both to each,
therefore also the product of *A*, *C* is prime to the product of
B, *E*.   [VII. 26]

And the product of *A*, *C* is *D*, and the product of *B*, *E*
is *F*.

Therefore *D*, *F* are prime to one another.

Q. E. D.

If *a*, *b* are prime to one another, so are a^2, b^2 and so are a^3, b^3; and,
generally, a^n, b^n are prime to one another.

The words in the enunciation which assert the truth of the proposition for
any powers are suspected and bracketed by Heiberg because (1) in περὶ τοὺς
ἄκρους the use of ἄκροι is peculiar, for it can only mean "the last products,"
and (2) the words have nothing corresponding to them in the proof, much
less is the generalisation proved. Campanus omits the words in the enuncia-

tion, though he adds to the proof a remark that the proposition is true of any, the same or different, powers of *a, b*.   Heiberg concludes that the words are an interpolation of date earlier than Theon.

Euclid's proof amounts to this.

Since *a, b* are prime to one another, so are a^2, *b* [VII. 25], and therefore also a^2, b^2.   [VII. 25]

Similarly [VII. 25] *a*, b^2 are prime to one another.

Therefore *a*, a^2 and *b*, b^2 satisfy the description in the enunciation of VII. 26.

Hence a^3, b^3 are prime to one another.

PROPOSITION 28.

If two numbers be prime to one another, the sum will also be prime to each of them; and, if the sum of two numbers be prime to any one of them, the original nvmbers will also be prime to one another.

For let two numbers *AB, BC* prime to one another be added;

I say that the sum *AC* is also prime to each of the numbers *AB, BC*.

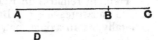

For, if *CA, AB* are not prime to onc another,

some number will measure *CA, AB*.

Let a number measure them, and let it be *D*.

Since then *D* measures *CA, AB*,

therefore it will also measure the remainder *BC*.

But it also measures *BA*;

therefore *D* measures *AB, BC* which are prime to one another: which is impossible.   [VII. Def. 12]

Therefore no number will measure the numbers *CA, AB*; therefore *CA, AB* are prime to one another.

For the same reason

AC, CB are also prime to one another.

Therefore *CA* is prime to each of the numbers *AB, BC*.

Again, let *CA, AB* be prime to one another;

I say that *AB, BC* are also prime to one another.

For, if *AB, BC* are not prime to one another,

some number will measure *AB, BC*.

Let a number measure them, and let it be D.

Now, since D measures each of the numbers AB, BC, it will also measure the whole CA.

But it also measures AB;

therefore D measures CA, AB which are prime to one another: which is impossible.   [VII. Def. 12]

Therefore no number will measure the numbers AB, BC.

Therefore AB, BC are prime to one another.

Q. E. D.

If a, b are prime to one another, $a + b$ will be prime to both a and b; and conversely.

For suppose $(a + b)$, a are not prime to one another. They must then have some common measure d.

Therefore d also divides the difference $(a + b) - a$, or b,. as well as a; and therefore a, b are not prime to one another: which is contrary to the hypothesis.

Therefore $a + b$ is prime to a.

Similarly $a + b$ is prime to b.

The converse is proved in the same way.

Heiberg remarks on Euclid's assumption that, if c measures both a and b, it also measures $a \pm b$. But it has already (VII. 1, 2) been assumed, more generally, as an axiom that, in the case supposed, c measures $a \pm pb$.

PROPOSITION 29.

Any prime number is prime to any number which it does not measure.

Let A be a prime number, and let it not measure B;
I say that B, A are prime to one another.

For, if B, A are not prime to one
another,
some number will measure them.

Let C measure them.

Since C measures B,
and A does not measure B,
therefore C is not the same with A.

Now, since C measures B, A,
therefore it also measures A which is prime, though it is not the same with it:
which is impossible.

Therefore no number will measure B, A.

Therefore A, B are prime to one another.

Q. E. D.

If a is prime and does not measure b, then a, b are prime to one another. The proof is self-evident.

PROPOSITION 30.

If two numbers by multiplying one another make some number, and any prime number measure the product, it will also measure one of the original numbers.

For let the two numbers A, B by multiplying one another make C, and let any prime number D measure C;

I say that D measures one of the numbers A, B.

For let it not measure A.

Now D is prime;

therefore A, D are prime to one another. [VII. 29]

And, as many times as D measures C, so many units let there be in E.

Since then D measures C according to the units in E, therefore D by multiplying E has made C. [VII. Def. 15]

Further, A by multiplying B has also made C; therefore the product of D, E is equal to the product of A, B.

Therefore, as D is to A, so is B to E. [VII. 19]

But D, A are prime to one another, primes are also least, [VII. 21] and the least measure the numbers which have the same ratio the same number of times, the greater the greater and the less the less, that is, the antecedent the antecedent and the consequent the consequent; [VII. 20] therefore D measures B.

Similarly we can also show that, if D do not measure B, it will measure A.

Therefore D measures one of the numbers A, B.

Q. E. D.

If c, a prime number, measure ab, c will measure either a or b.

Suppose c does not measure a.

Therefore c, a are prime to one another.   [VII. 29]

Suppose $ab = mc$.

   Therefore $c : a = b : m$.   [VII. 19]

Hence [VII. 20, 21] c measures b.

Similarly, if c does not measure b, it measures a.

Therefore it measures one or other of the two numbers a, b.

PROPOSITION 31.

Any composite number is measured by some prime number.

Let A be a composite number;

I say that A is measured by some prime number.

For, since A is composite,

5 some number will measure it.

Let a number measure it, and let it be B.

Now, if B is prime, what was enjoined will have been done.

10 But if it is composite, some number will measure it.

Let a number measure it, and let it be C.

Then, since C measures B,

and B measures A,

therefore C also measures A.

15 And, if C is prime, what was enjoined will have been done.

But if it is composite, some number will measure it.

Thus, if the investigation be continued in this way, some prime number will be found which will measure the number 20 before it, which will also measure A.

For, if it is not found, an infinite series of numbers will measure the number A, each of which is less than the other: which is impossible in numbers.

Therefore some prime number will be found which will 25 measure the one before it, which will also measure A.

Therefore any composite number is measured by some prime number.

Q. E. D.

8. **if B is prime, what was enjoined will have been done**, i.e. the implied *problem* of finding a prime number which measures *A*.

18. **some prime number will be found which will measure.** In the Greek the sentence stops here, but it is necessary to add the words "the number before it, which will also measure *A*," which are found a few lines further down. It is possible that the words may have fallen out of P here by a simple mistake due to ὁμοιοτέλευτον (Heiberg).

Heiberg relegates to the Appendix an alternative proof of this proposition, to the following effect. Since *A* is composite, some number will measure it. Let *B* be the *least* such number. I say that *B* is prime. For, if not, *B* is composite, and some number will measure it, say *C*; so that *C* is less than *B*. But, since *C* measures *B*, and *B* measures *A*, *C* must measure *A*. And *C* is less than *B*: which is contrary to the hypothesis.

PROPOSITION 32.

Any number either is prime or is measured by some prime number.

Let *A* be a number;
I say that *A* either is prime or is measured by some prime number.

If now *A* is prime, that which was   A————————
enjoined will have been done.

But if it is composite, some prime number will measure it.

[VII. 31]

Therefore any number either is prime or is measured by some prime number.

Q. E. D.

PROPOSITION 33.

Given as many numbers as we please, to find the least of those which have the same ratio with them.

Let *A*, *B*, *C* be the given numbers, as many as we please;
thus it is required to find the least of
5 those which have the same ratio with
A, *B*, *C*.
    A, *B*, *C* are either prime to one
another or not.
    Now, if *A*, *B*, *C* are prime to one
10 another, they are the least of those
which have the same ratio with them.

[VII. 21]

But, if not, let *D* the greatest common measure of *A*, *B*, *C*
be taken,   [VII. 3]

and, as many times as D measures the numbers A, B, C
15 respectively, so many units let there be in the numbers
E, F, G respectively.

Therefore the numbers E, F, G measure the numbers A,
B, C respectively according to the units in D.   [VII. 16]

Therefore E, F, G measure A, B, C the same number of
20 times ;

therefore E, F, G are in the same ratio with A, B, C.

[VII. Def. 20]

I say next that they are the least that are in that ratio.

For, if E, F, G are not the least of those which have the
same ratio with A, B, C,

25 there will be numbers less than E, F, G which are in the
same ratio with A, B, C.

Let them be H, K, L ;

therefore H measures A the same number of times that the
numbers K, L measure the numbers B, C respectively.

30   Now, as many times as H measures A, so many units let
there be in M ;

therefore the numbers K, L also measure the numbers B, C
respectively according to the units in M.

And, since H measures A according to the units in M,
35 therefore M also measures A according to the units in H.

[VII. 16]

For the same reason

M also measures the numbers B, C according to the units in
the numbers K, L respectively ;

Therefore M measures A, B, C.

40   Now, since H measures A according to the units in M,
therefore H by multiplying M has made A.   [VII. Def. 15]

For the same reason also

E by multiplying D has made A.

Therefore the product of E, D is equal to the product of
45 H, M.

Therefore, as E is to H, so is M to D.   [VII. 19]

But E is greater than H ;

therefore M is also greater than D.

And it measures A, B, C :

50 which is impossible, for by hypothesis D is the greatest common measure of A, B, C.

Therefore there cannot be any numbers less than E, F, G which are in the same ratio with A, B, C.

Therefore E, F, G are the least of those which have the 55 same ratio with A, B, C.

Q. E. D.

17. the numbers E, F, G measure the numbers A, B, C respectively, literally (as usual) "each of the numbers E, F, G measures each of the numbers A, B, C."

Given any numbers a, b, c, ..., to find the least numbers that are in the same ratio.

Euclid's method is the obvious one, and the result is verified by *reductio ad absurdum*.

We will, like Euclid, take three numbers only, a, b, c.

Let g, their greatest common measure, be found [VII. 3], and suppose that

$$a = mg, \text{ i.e. } gm, \qquad\qquad [\text{VII. } 16]$$
$$b = ng, \text{ i.e. } gn,$$
$$c = pg, \text{ i.e. } gp.$$

It follows, by VII. Def. 20, that

$$m : n : p = a : b : c.$$

m, n, p shall be the numbers required.

For, if not, let x, y, z be the least numbers in the same ratio as a, b, c, being less than m, n, p.

Therefore
$$a = kx \text{ (or } xk, \text{ VII. } 16),$$
$$b = ky \text{ (or } yk),$$
$$c = kz \text{ (or } zk),$$

where k is some integer.   [VII. 20]

Thus
$$mg = a = xk.$$

Therefore
$$m : x = k : g. \qquad\qquad [\text{VII. } 19]$$

And $m > x$; therefore $k > g$.

Since then k measures a, b, c, it follows that g is not the greatest common measure: which contradicts the hypothesis.

Therefore etc.

It is to be observed that Euclid merely supposes that x, y, z are smaller numbers than m, n, p in the ratio of a, b, c; but, in order to justify the next inference, which apparently can only depend on VII. 20, x, y, z must also be assumed to be the *least* numbers in the ratio of a, b, c.

The inference from the last proportion that, since $m > x$, $k > g$ is supposed by Heiberg to depend upon VII. 13 and V. 14 together. I prefer to regard Euclid as making the inference quite independently of Book v. E.g., the proportion could just as well be written

$$x : m = g : k,$$

when the definition of proportion in Book VII. (Def. 20) gives all that we want, since, whatever proper fraction x is of m, the same proper fraction is g of k.

PROPOSITION 34.

Given two numbers, to find the least number which they measure.

Let A, B be the two given numbers;
thus it is required to find the least number which they measure.

Now A, B are either prime to one another or not.

First, let A, B be prime to one another, and let A by multiplying B make C;
therefore also B by multiplying A has made C.   [VII. 16]

Therefore A, B measure C

I say next that it is also the least number they measure.

For, if not, A, B will measure some number which is less than C.

Let them measure D.

Then, as many times as A measures D, so many units let there be in E,
and, as many times as B measures D, so many units let there be in F;
therefore A by multiplying E has made D,
and B by multiplying F has made D;   [VII. Def. 15]
therefore the product of A, E is equal to the product of B, F.

Therefore, as A is to B, so is F to E.   [VII. 19]

But A, B are prime,   [VII. 21]
primes are also least,
and the least measure the numbers which have the same ratio the same number of times, the greater the greater and the less the less;   [VII. 20]
therefore B measures E, as consequent consequent.

And, since A by multiplying B, E has made C, D,
therefore, as B is to E, so is C to D.   [VII. 17]

But B measures E;
therefore C also measures D, the greater the less:
which is impossible.

Therefore A, B do not measure any number less than C; therefore C is the least that is measured by A, B.

Next, let A, B not be prime to one another, and let F, E, the least numbers of those which have the same ratio with A, B, be taken;   [VII. 33] therefore the product of A, E is equal to the product of B, F.
                    [VII. 19]

And let A by multiplying E make C;

therefore also B by multiplying F has made C;

therefore A, B measure C.

I say next that it is also the least number that they measure.

For, if not, A, B will measure some number which is less than C.

Let them measure D.

And, as many times as A measures D, so many units let there be in G,

and, as many times as B measures D, so many units let there be in H.

Therefore A by multiplying G has made D, and B by multiplying H has made D.

Therefore the product of A, G is equal to the product of B, H;

therefore, as A is to B, so is H to G.   [VII. 19]

But, as A is to B, so is F to E.

Therefore also, as F is to E, so is H to G.

But F, E are least, and the least measure the numbers which have the same ratio the same number of times, the greater the greater and the less the less;   [VII. 20] therefore E measures G.

And, since A by multiplying E, G has made C, D, therefore, as E is to G, so is C to D.   [VII. 17]

But E measures G; therefore C also measures D, the greater the less: which is impossible.

Therefore A, B will not measure any number which is less than C.

Therefore C is the least that is measured by A, B.

Q. E. D.

This is the problem of finding the *least common multiple* of two numbers, as a, b.

I. If a, b be prime to one another, the L.C.M. is ab.

For, if not, let it be d, some number less than ab.

Then $\qquad\qquad d = ma = nb$, where m, n are integers.

Therefore $\qquad\qquad a : b = n : m$, $\qquad\qquad$ [VII. 19]

and hence, a, b being prime to one another,

$\qquad\qquad\qquad b$ measures m. $\qquad\qquad$ [VII. 20, 21]

But $\qquad\qquad b : m = ab : am$ $\qquad\qquad$ [VII. 17]

$\qquad\qquad\qquad\quad = ab : d$.

Therefore ab measures d: which is impossible.

II. If a, b be not prime to one another, find the numbers which are the least of those having the ratio of a to b, say m, n; $\qquad$ [VII. 33]

then $\qquad\qquad\quad a : b = m : n$,

and $\qquad\qquad\quad an = bm \; (= c, \text{ say})$; $\qquad\qquad$ [VII. 19]

c is then the L.C.M.

For, if not, let it be $d \; (< c)$, so that

$\qquad\qquad ap = bq = d$, where p, q are integers.

Then $\qquad\qquad a : b = q : p$, $\qquad\qquad$ [VII. 19]

whence $\qquad\qquad m : n = q : p$,

so that $\qquad\qquad n$ measures p. $\qquad\qquad$ [VII. 20, 21]

And $\qquad\qquad n : p = an : ap = c : d$,

so that $\qquad\qquad c$ measures d:

which is impossible.

Therefore etc.

By VII. 33, $\qquad m = \dfrac{a}{g}$
$\qquad\qquad\qquad\qquad\qquad\qquad$, where g is the G.C.M. of a, b.
$\qquad\qquad\qquad n = \dfrac{b}{g}$

Hence the L.C.M. is $\dfrac{ab}{g}$.

PROPOSITION 35.

If two numbers measure any number, the least number measured by them will also measure the same.

For let the two numbers A, B measure any number CD, and let E be the least that they measure;

I say that E also measures CD.

For, if E does not measure CD, let E, measuring DF, leave CF less than itself.

Now, since A, B measure E,

and E measures DF,

therefore A, B will also measure DF.

But they also measure the whole CD;

therefore they will also measure the remainder CF which is less than E:

which is impossible.

Therefore E cannot fail to measure CD;

therefore it measures it.

Q. E. D.

The *least* common multiple of any two numbers must measure any other common multiple.

The proof is obvious, depending on the fact that, if any number divides a and b, it also divides $a - pb$.

PROPOSITION 36.

Given three numbers, to find the least number which they measure.

Let A, B, C be the three given numbers; thus it is required to find the least number which they measure.

Let D, the least number measured by the two numbers A, B, be taken.   [VII. 34]

Then C either measures, or does not measure, D.

First, let it measure it.

But A, B also measure D;

therefore A, B, C measure D.

I say next that it is also the least that they measure.

For, if not, A, B, C will measure some number which is less than D.

Let them measure E.

Since A, B, C measure E,

therefore also A, B measure E.

Therefore the least number measured by A, B will also measure E.   [VII. 35]

But D is the least number measured by A, B;

therefore D will measure E, the greater the less:

which is impossible.

Therefore A, B, C will not measure any number which is less than D;

therefore D is the least that A, B, C measure.

Again, let C not measure D,

and let E, the least number measured by C, D, be taken.   [VII. 34]

Since A, B measure D,

and D measures E,

therefore also A, B measure E.

But C also measures E;

therefore also A, B, C measure E.

I say next that it is also the least that they measure.

For, if not, A, B, C will measure some number which is less than E.

Let them measure F.

Since A, B, C measure F,

therefore also A, B measure F;

therefore the least number measured by A, B will also measure F.   [VII. 35]

But D is the least number measured by A, B;

therefore D measures F.

But C also measures F;

therefore D, C measure F,

so that the least number measured by D, C will also measure F.

But E is the least number measured by C, D;
therefore E measures F, the greater the less:
which is impossible.

Therefore A, B, C will not measure any number which is
less than E.

Therefore E is the least that is measured by A, B, C.

Q. E. D.

Euclid's rule for finding the L.C.M. of *three* numbers a, b, c is the rule with
which we are familiar.   The L.C.M. of a, b is first found, say d, and then the
L.C.M. of d and c is found.

Euclid distinguishes the cases (1) in which c measures d, (2) in which c
does not measure d.   We need only reproduce the proof of the general case
(2).   The method is that of *reductio ad absurdum*.

Let e be the L.C.M. of d, c.

Since a, b both measure d, and d measures e,

$$a, b \text{ both measure } e.$$

So does c.

Therefore e is *some* common multiple of a, b, c.

If it is not the *least*, let f be the L.C.M.

Now a, b both measure f;

therefore d, their L.C.M., also measures f.   [VII. 35]

Thus d, c both measure f;

therefore e, their L.C.M., measures f: [VII. 35]
which is impossible, since $f < e$.

Therefore etc.

The process can be continued *ad libitum*, so that we can find the L.C.M.,
not only of three, but of as many numbers as we please.

PROPOSITION 37.

*If a number be measured by any number, the number which
is measured will have a part called by the same name as the
measuring number.*

For let the number A be measured by any number B;
I say that A has a part called by the same
name as B.

For, as many times as B measures A,
so many units let there be in C.

Since B measures A according to the
units in C,
and the unit D also measures the number C according to the
units in it,

therefore the unit D measures the number C the same number of times as B measures A.

Therefore, alternately, the unit D measures the number B the same number of times as C measures A;   [VII. 15] therefore, whatever part the unit D is of the number B, the same part is C of A also.

But the unit D is a part of the number B called by the same name as it;

therefore C is also a part of A called by the same name as B, so that A has a part C which is called by the same name as B.

Q. E. D.

If b measures a, then $\frac{1}{b}$th of a is a whole number.

Let $a = m \cdot b.$

Now $m = m \cdot 1.$

Thus 1, m, b, a satisfy the enunciation of VII. 15; therefore m measures a the same number of times that 1 measures b.

But 1 is $\frac{1}{b}$th part of b;

therefore m is $\frac{1}{b}$th part of a.

PROPOSITION 38.

If a number have any part whatever, it will be measured by a number called by the same name as the part.

For let the number A have any part whatever, B, and let C be a number called by the same name as the part B; I say that C measures A.

For, since B is a part of A called by the same name as C, and the unit D is also a part of C called by the same name as it, therefore, whatever part the unit D is of the number C, the same part is B of A also;

therefore the unit D measures the number C the same number of times that B measures A.

Therefore, alternately, the unit D measures the number B the same number of times that C measures A.   [VII. 15]

Therefore C measures A.

Q. E. D.

This proposition is practically a restatement of the preceding proposition. It asserts that, if b is $\frac{1}{m}$ th part of a,

i.e., if
$$b = \frac{1}{m} a,$$

then m measures a.

We have
$$b = \frac{1}{m} a,$$

and
$$1 = \frac{1}{m} m.$$

Therefore 1, m, b, a, satisfy the enunciation of VII. 15, and therefore m measures a the same number of times as 1 measures b, or

$$m = \frac{1}{b} a.$$

PROPOSITION 39.

To find the number which is the least that will have given parts.

Let A, B, C be the given parts;

thus it is required to find the number which is the least that will have the parts A, B, C.

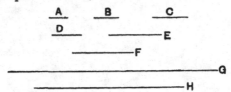

Let D, E, F be numbers called by the same name as the parts A, B, C,

and let G, the least number measured by D, E, F, be taken.

[VII. 36]

Therefore G has parts called by the same name as D, E, F.

[VII. 37]

But A, B, C are parts called by the same name as D, E, F; therefore G has the parts A, B, C.

I say next that it is also the least number that has.

For, if not, there will be some number less than G which will have the parts A, B, C.

Let it be H.

Since H has the parts A, B, C,

therefore H will be measured by numbers called by the same name as the parts A, B, C.   [VII. 38]

But D, E, F are numbers called by the same name as the parts A, B, C;

therefore H is measured by D, E, F.

And it is less than G: which is impossible.

Therefore there will be no number less than G that will have the parts A, B, C.

Q. E. D.

This again is practically a restatement in another form of the problem of finding the L.C.M.

To find a number which has $\frac{1}{a}$th, $\frac{1}{b}$th and $\frac{1}{c}$th parts.

Let d be the L.C.M. of a, b, c.

Thus d has $\frac{1}{a}$th, $\frac{1}{b}$th and $\frac{1}{c}$th parts.   [VII. 37]

If it is not the least number which has, let the least such number be e.

Then, since e has those parts,

e is measured by a, b, c; and $e < d$:

which is impossible.

BOOK VIII.

If there be as many numbers as we please in continued proportion, and the extremes of them be prime to one another, the numbers are the least of those which have the same ratio with them.

Let there be as many numbers as we please, A, B, C, D, in continued proportion,
and let the extremes of them A, D be prime to one another;
I say that A, B, C, D are the least of those which have the same ratio with them.

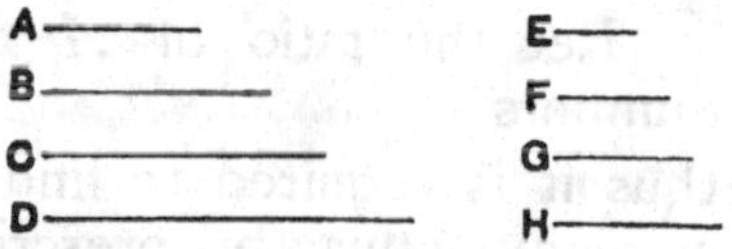

For, if not, let E, F, G, H be less than A, B, C, D, and in the same ratio with them.

Now, since A, B, C, D are in the same ratio with E, F, G, H,

and the multitude of the numbers A, B, C, D is equal to the multitude of the numbers E, F, G, H,

therefore, *ex aequali*,

　　　　　as A is to D, so is E to H. 　　　　　[VII. 14]

But A, D are prime,

primes are also least, 　　　　　[VII. 21]

and the least numbers measure those which have the same ratio the same number of times, the greater the greater and the less the less, that is, the antecedent the antecedent and the consequent the consequent. 　　　　　[VII. 20]

Therefore A measures E, the greater the less :
which is impossible.

Therefore E, F, G, H which are less than A, B, C, D are not in the same ratio with them.

Therefore A, B, C, D are the least of those which have the same ratio with them.

Q. E. D.

What we call a *geometrical progression* is with Euclid a series of terms "in continued proportion" (ἑξῆς ἀνάλογον).

This proposition proves that, if a, b, c, ... k are a series of numbers in geometrical progression, and if a, k are prime to one another, the series is in the lowest terms possible with the same common ratio.

The proof is in *form* by *reductio ad absurdum*. We should no doubt desert this *form* while retaining the substance. If a', b', c', ... k' be *any* other series of numbers in G.P. with the same common ratio as before, we have, *ex aequali*,

$$a : k = a' : k', \qquad\qquad\text{[VII. 14]}$$

whence, since a, k are prime to one another, a, k measure a', k' respectively, so that a', k' are *greater* than a, k respectively.

PROPOSITION 2.

To find numbers in continued proportion, as many as may be prescribed, and the least that are in a given ratio.

Let the ratio of A to B be the given ratio in least numbers ;

thus it is required to find numbers in continued proportion, as many as may be prescribed, and the least that are in the ratio of A to B.

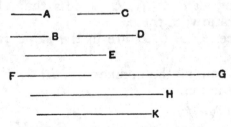

Let four be prescribed ;
let A by multiplying itself make C, and by multiplying B let it make D ;
let B by multiplying itself make E ;
further, let A by multiplying C, D, E make F, G, H,
and let B by multiplying E make K.

Now, since A by multiplying itself has made C,
and by multiplying B has made D,
therefore, as A is to B, so is C to D.　　　　　　　[VII. 17]

Again, since A by multiplying B has made D,
and B by multiplying itself has made E,
therefore the numbers A, B by multiplying B have made the numbers D, E respectively.

Therefore, as A is to B, so is D to E.　　　　　[VII. 18]
But, as A is to B, so is C to D;
therefore also, as C is to D, so is D to E.

And, since A by multiplying C, D has made F, G,
therefore, as C is to D, so is F to G.　　　　　[VII. 17]

But, as C is to D, so was A to B;
therefore also, as A is to B, so is F to G.

Again, since A by multiplying D, E has made G, H,
therefore, as D is to E, so is G to H.　　　　　[VII. 17]

But, as D is to E, so is A to B.
Therefore also, as A is to B, so is G to H.
And, since A, B by multiplying E have made H, K,
therefore, as A is to B, so is H to K.　　　　　[VII. 18]

But, as A is to B, so is F to G, and G to H.
Therefore also, as F is to G, so is G to H, and H to K;
therefore C, D, E, and F, G, H, K are proportional in the ratio of A to B.

I say next that they are the least numbers that are so.
For, since A, B are the least of those which have the same ratio with them,
and the least of those which have the same ratio are prime to one another,　　　　　　　[VII. 22]
therefore A, B are prime to one another.

And the numbers A, B by multiplying themselves respectively have made the numbers C, E, and by multiplying the numbers C, E respectively have made the numbers F, K;
therefore C, E and F, K are prime to one another respectively.
　　　　　　　　　　　　　　　　　　　　　[VII. 27]

But, if there be as many numbers as we please in continued proportion, and the extremes of them be prime to one another, they are the least of those which have the same ratio with them. [VIII. 1]

Therefore C, D, E and F, G, H, K are the least of those which have the same ratio with A, B. Q. E. D.

PORISM. From this it is manifest that, if three numbers in continued proportion be the least of those which have the same ratio with them, the extremes of them are squares, and, if four numbers, cubes.

To find a series of numbers in geometrical progression and in the least terms which have a given common ratio (understanding by that term *the ratio of one term to the next*).

Reduce the given ratio to its lowest terms, say, $a : b$. (This can be done by VII. 33.)

Then a^n, $a^{n-1}b$, $a^{n-2}b^2$, ... a^2b^{n-2}, ab^{n-1}, b^n

is the required series of numbers if $(n + 1)$ terms are required.

That this is a series of terms with the given common ratio is clear from VII. 17, 18.

That the G.P. is in the smallest terms possible is proved thus.

a, b are prime to one another, since the ratio $a : b$ is in its lowest terms. [VII. 22]

Therefore a^2, b^2 are prime to one another; so are a^3, b^3 and, generally, a^n, b^n. [VII. 27]

Whence the G.P. is in the smallest possible terms, by VIII. 1.

The Porism observes that, if there are n terms in the series, the extremes are $(n - 1)$th powers.

PROPOSITION 3.

If as many numbers as we please in continued proportion be the least of those which have the same ratio with them, the extremes of them are prime to one another.

Let as many numbers as we please, A, B, C, D, in continued proportion be the least of those which have the same ratio with them;

I say that the extremes of them A, D are prime to one another.

For let two numbers E, F, the least that are in the ratio of A, B, C, D, be taken,   [VII. 33]

then three others G, H, K with the same property ;

and others, more by one continually,   [VIII. 2]

until the multitude taken becomes equal to the multitude of the numbers A, B, C, D.

Let them be taken, and let them be L, M, N, O.

Now, since E, F are the least of those which have the same ratio with them, they are prime to one another.   [VII. 22]

And, since the numbers E, F by multiplying themselves respectively have made the numbers G, K, and by multiplying the numbers G, K respectively have made the numbers L, O,

[VIII. 2, Por.]

therefore both G, K and L, O are prime to one another.   [VII. 27]

And, since A, B, C, D are the least of those which have the same ratio with them,

while L, M, N, O are the least that are in the same ratio with A, B, C, D,

and the multitude of the numbers A, B, C, D is equal to the multitude of the numbers L, M, N, O,

therefore the numbers A, B, C, D are equal to the numbers L, M, N, O respectively ;

therefore A is equal to L, and D to O.

And L, O are prime to one another.

Therefore A, D are also prime to one another.

Q. E. D.

The proof consists in merely equating the given numbers to the terms of a series found in the manner of VIII. 2.

If a, b, c, ... k (n terms) be a geometrical progression in the lowest terms having a given common ratio, the terms must respectively be of the form

$$\alpha^{n-1}, \quad \alpha^{n-2}\beta, \; ... \; \alpha^2\beta^{n-3}, \quad \alpha\beta^{n-2}, \quad \beta^{n-1}$$

found by VIII. 2, where $\alpha : \beta$ is the ratio $a : b$ expressed in its lowest terms, so that α, β are prime to one another [VII. 22], and hence α^{n-1}, β^{n-1} are prime to one another [VII. 27].

But the two series must be the same, so that

$$a = \alpha^{n-1}, \quad b = \beta^{n-1}$$

PROPOSITION 4.

Given as many ratios as we please in least numbers, to find numbers in continued proportion which are the least in the given ratios.

Let the given ratios in least numbers be that of A to B,
5 that of C to D, and that of E to F;

thus it is required to find numbers in continued proportion which are the least that are in the ratio of A to B, in the ratio of C to D, and in the ratio of E to F.

```
A——          B———
C——          D———
E——          F———
N——                    ———————— G
O————        ————— H
M—————————         K
P—————————    ————————— L
```

Let G, the least number measured by B, C, be taken.

[VII. 34]

10   And, as many times as B measures G, so many times also let A measure H,

and, as many times as C measures G, so many times also let D measure K.

Now E either measures or does not measure K.

15   First, let it measure it.

And, as many times as E measures K, so many times let F measure L also.

Now, since A measures H the same number of times that B measures G,

20 therefore, as A is to B, so is H to G.   [VII. Def. 20, VII. 13]

For the same reason also,

as C is to D, so is G to K,

and further, as E is to F, so is K to L;

therefore H, G, K, L are continuously proportional in the
25 ratio of A to B, in the ratio of C to D, and in the ratio of E to F.

I say next that they are also the least that have this property.

For, if H, G, K, L are not the least numbers continuously
30 proportional in the ratios of A to B, of C to D, and of E
to F, let them be N, O, M, P.

Then since, as A is to B, so is N to O,

while A, B are least,

and the least numbers measure those which have the same
35 ratio the same number of times, the greater the greater and
the less the less, that is, the antecedent the antecedent and the
consequent the consequent ;

therefore B measures O.   [VII. 20]

For the same reason

40 C also measures O ;

therefore B, C measure O ;

therefore the least number measured by B, C will also
measure O.   [VII. 35]

But G is the least number measured by B, C ;

45 therefore G measures O, the greater the less :

which is impossible.

Therefore there will be no numbers less than H, G, K, L
which are continuously in the ratio of A to B, of C to D, and
of E to F.

50   Next, let E not measure K.

Let M, the least number measured by E, K, be taken.

And, as many times as K measures M, so many times let
H, G measure N, O respectively,

and, as many times as E measures M, so many times let F
55 measure P also.

Since H measures N the same number of times that G
measures O,

therefore, as H is to G, so is N to O.   [VII. 13 and Def. 20]

But, as H is to G, so is A to B;

60 therefore also, as A is to B, so is N to O.

For the same reason also,

as C is to D, so is O to M.

Again, since E measures M the same number of times that F measures P,

65 therefore, as E is to F, so is M to P;　　　　[VII. 13 and Def. 20]

therefore N, O, M, P are continuously proportional in the ratios of A to B, of C to D, and of E to F.

I say next that they are also the least that are in the ratios $A:B$, $C:D$, $E:F$.

70 For, if not, there will be some numbers less than N, O, M, P continuously proportional in the ratios $A:B$, $C:D$, $E:F$.

Let them be Q, R, S, T.

Now since, as Q is to R, so is A to B,

75 while A, B are least,

and the least numbers measure those which have the same ratio with them the same number of times, the antecedent the antecedent and the consequent the consequent,　　　　[VII. 20]

therefore B measures R.

80 For the same reason C also measures R;

therefore B, C measure R.

Therefore the least number measured by B, C will also measure R.　　　　[VII. 35]

But G is the least number measured by B, C;

85 therefore G measures R.

And, as G is to R, so is K to S:　　　　[VII. 13]

therefore K also measures S.

But E also measures S;

therefore E, K measure S.

90 Therefore the least number measured by E, K will also measure S.　　　　[VII. 35]

But M is the least number measured by E, K;

therefore M measures S, the greater the less:

which is impossible.

95 Therefore there will not be any numbers less than N, O, M, P continuously proportional in the ratios of A to B, of C to D, and of E to F;

therefore N, O, M, P are the least numbers continuously proportional in the ratios $A:B$, $C:D$, $E:F$.   Q. E. D.

69, 71, 99. **the ratios A : B, C : D, E : F.** This abbreviated expression is in the Greek οἱ ΑΒ, ΓΔ, ΕΖ λόγοι.

The term "in continued proportion" is here not used in its proper sense, since a geometrical progression is not meant, but a series of terms each of which bears to the succeeding term a *given*, but not the *same*, ratio.

The proposition furnishes a good example of the cumbrousness of the Greek method of dealing with non-determinate numbers. The proof in fact is not easy to follow without the help of modern symbolical notation. If this be used, the reasoning can be made clear enough.

Euclid takes *three* given ratios and therefore requires to find *four* numbers. We will leave out the simpler particular case which he puts first, that namely in which E accidentally measures K, the multiple of D found in the first few lines; and we will reproduce the general case with *three* ratios.

Let the ratios in their lowest terms be

$$a:b, \quad c:d, \quad e:f$$

Take l_1, the L.C.M. of b, c, and suppose that

$$l_1 = mb = nc.$$

Form the numbers ma, $mb \left. \begin{matrix} \\ = nc \end{matrix} \right\}$, nd.

These are in the ratios of a to b and of c to d respectively.

Next, let l_2 be the L.C.M. of nd, e, and let

$$l_2 = pnd = qe.$$

Now form the numbers

$$pma, \quad pmb \left. \begin{matrix} \\ = pnc \end{matrix} \right\}, \quad pnd \left. \begin{matrix} \\ = qe \end{matrix} \right\}, \quad qf,$$

and these are the four numbers required.

If they are *not* the least in the given ratios, let

$$x, \quad y, \quad z, \quad u$$

be less numbers in the given ratios.

Since $a:b$ is in its lowest terms, and

$$a:b = x:y,$$

b measures y.

Similarly, since $c:d = y:z$,

c measures y.

Therefore l_1, the L.C.M. of b, c, measures y.

But $l_1 : nd \, [= c:d] = y:z$.

Therefore nd measures z.

And, since $e:f = z:u$,

e measures z.

Therefore l_2, the L.C.M. of nd, e, measures z: which is impossible, since $z < l_2$ or pnd.

The step (line 86) inferring that $G:R = K:S$ is of course *alternando* from $G:K \, [= C:D] = R:S$.

It will be observed that VIII. 4 corresponds to the portion of VI. 23 which shows how to *compound* two ratios between straight lines.

PROPOSITION 5.

Plane numbers have to one another the ratio compounded of the ratios of their sides.

Let A, B be plane numbers, and let the numbers C, D be the sides of A, and E, F of B;

5 I say that A has to B the ratio compounded of the ratios of the sides.

For, the ratios being given which C has to E and D to F, let the least numbers G, H, K that are continuously

10 in the ratios $C:E$, $D:F$ be taken, so that,

as C is to E, so is G to H,

and,　　　　　　as D is to F, so is H to K.　　　　[VIII. 4]

And let D by multiplying E make L.

15　　Now, since D by multiplying C has made A, and by multiplying E has made L,

therefore, as C is to E, so is A to L.　　　　[VII. 17]

But, as C is to E, so is G to H;

therefore also, as G is to H, so is A to L.

20　　Again, since E by multiplying D has made L, and further by multiplying F has made B,

therefore, as D is to F, so is L to B.　　　　[VII. 17]

But, as D is to F, so is H to K;

therefore also, as H is to K, so is L to B.

25　　But it was also proved that,

as G is to H, so is A to L;

therefore, *ex aequali*,

as G is to K, so is A to B.　　　　[VII. 14]

But G has to K the ratio compounded of the ratios of the

30 sides;

therefore A also has to B the ratio compounded of the ratios of the sides.　　　　　　　　　　　　　Q. E. D.

1, 5, 29, 31. **compounded of the ratios of their sides.** As in VI. 23, the Greek has the less exact phrase, "compounded of their sides."

If　　　　　　　　　　　$a = cd$,　　$b = ef$,

then a has to b the ratio compounded of $c : e$ and $d : f$.

Take three numbers the least which are continuously in the given ratios.

If l is the L.C.M. of e, d and $l = me = nd$, the three numbers are

$$mc, \quad \begin{matrix} me \\ = nd \end{matrix} \Big\}, \quad nf. \qquad \text{[VIII. 4]}$$

Now $\qquad\qquad\qquad dc : de = c : e \qquad\qquad\qquad$ [VII. 17]

$$= mc : me = mc : nd.$$

Also $\qquad\qquad\qquad ed : ef = d : f \qquad\qquad\qquad$ [VII. 17]

$$= nd : nf.$$

Therefore, *ex aequali*, $\qquad cd : ef = mc : nf$

$$= \text{(ratio compounded of } c : e \text{ and } d : f).$$

It will be seen that this proof follows exactly the method of VI. 23 for parallelograms.

PROPOSITION 6.

If there be as many numbers as we please in continued proportion, and the first do not measure the second, neither will any other measure any other.

Let there be as many numbers as we please, A, B, C, D, E, in continued proportion, and let A not measure B;
I say that neither will any other measure any other.

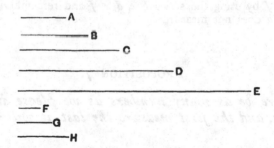

Now it is manifest that A, B, C, D, E do not measure one another in order; for A does not even measure B.

I say, then, that neither will any other measure any other.

For, if possible, let A measure C.

And, however many A, B, C are, let as many numbers F, G, H, the least of those which have the same ratio with A, B, C, be taken. [VII. 33]

Now, since F, G, H are in the same ratio with A, B, C, and the multitude of the numbers A, B, C is equal to the multitude of the numbers F, G, H,

therefore, *ex aequali*, as A is to C, so is F to H. [VII. 14]

And since, as A is to B, so is F to G,
while A does not measure B,
therefore neither does F measure G;   [VII. Def. 20]
therefore F is not an unit, for the unit measures any number.

Now F, H are prime to one another.   [VIII. 3]

And, as F is to H, so is A to C;
therefore neither does A measure C.

Similarly we can prove that neither will any other measure any other.

Q. E. D.

Let $a, b, c \ldots k$ be a geometrical progression in which a does not measure b.
Suppose, if possible, that a measures some term of the series, as f.
Take x, y, z, u, v, w the *least* numbers in the ratio a, b, c, d, e, f.
Since $\qquad\qquad x : y = a : b$,
and a does not measure b,
x does not measure y; therefore x cannot be unity.
And, *ex aequali*, $\qquad x : w = a : f$.
Now x, w are prime to one another.   [VIII. 3]
Therefore a does not measure f.

We can of course prove that an intermediate term, as b, does not measure a later term f by using the series b, c, d, e, f and remembering that, since $b : c = a : b$, b does not measure c.

PROPOSITION 7.

If there be as many numbers as we please in continued proportion, and the first measure the last, it will measure the second also.

Let there be as many numbers as we please, A, B, C, D, in continued proportion; and
let A measure D;
I say that A also measures B.

For, if A does not measure B, neither will any other of the numbers measure any other.   [VIII. 6]

But A measures D.

Therefore A also measures B.

Q. E. D.

An obvious proof by *reductio ad absurdum* from VIII. 6.

PROPOSITION 8.

If between two numbers there fall numbers in continued proportion with them, then, however many numbers fall between them in continued proportion, so many will also fall in continued proportion between the numbers which have the same ratio with the original numbers.

Let the numbers C, D fall between the two numbers A, B in continued proportion with them, and let E be made in the same ratio to F as A is to B;

I say that, as many numbers as have fallen between A, B in continued proportion, so many will also fall between E, F in continued proportion.

For, as many as A, B, C, D are in multitude, let so many numbers G, H, K, L, the least of those which have the same ratio with A, C, D, B, be taken; [VII. 33]

therefore the extremes of them G, L are prime to one another. [VIII. 3]

Now, since A, C, D, B are in the same ratio with G, H, K, L,

and the multitude of the numbers A, C, D, B is equal to the multitude of the numbers G, H, K, L,

therefore, *ex aequali*, as A is to B, so is G to L. [VII. 14]

But, as A is to B, so is E to F;

therefore also, as G is to L, so is E to F.

But G, L are prime,

primes are also least, [VII. 21]

and the least numbers measure those which have the same ratio the same number of times, the greater the greater and the less the less, that is, the antecedent the antecedent and the consequent the consequent. [VII. 20]

Therefore G measures E the same number of times as L measures F.

Next, as many times as G measures E, so many times let H, K also measure M, N respectively;

therefore G, H, K, L measure E, M, N, F the same number of times.

Therefore G, H, K, L are in the same ratio with E, M, N, F.   [VII. Def. 20]

But G, H, K, L are in the same ratio with A, C, D, B;

therefore A, C, D, B are also in the same ratio with E, M, N, F.

But A, C, D, B are in continued proportion;

therefore E, M, N, F are also in continued proportion.

Therefore, as many numbers as have fallen between A, B in continued proportion with them, so many numbers have also fallen between E, F in continued proportion.

Q. E. D.

1. **fall.** The Greek word is ἐμπίπτειν, "fall *in*" = "can be interpolated."

If $a : b = e : f$, and between a, b there are any number of geometric means c, d, there will be as many such means between e, f.

Let a, β, γ, ..., δ be the least possible terms in the same ratio as a, c, d, ... b.

Then a, δ are prime to one another,   [VIII. 3]

and, *ex aequali*, $a : \delta = a : b$

$$= e : f.$$

Therefore $e = ma$, $f = m\delta$, where m is some integer.   [VII. 20]

Take the numbers ma, $m\beta$, $m\gamma$, ... $m\delta$.

This is a series in the given ratio, and we have the same number of geometric means between ma, $m\delta$, or e, f, that there are between a, b.

PROPOSITION 9.

If two numbers be prime to one another, and numbers fall between them in continued proportion, then, however many numbers fall between them in continued proportion, so many will also fall between each of them and an unit in continued proportion.

Let A, B be two numbers prime to one another, and let C, D fall between them in continued proportion,

and let the unit E be set out;

I say that, as many numbers as fall between A, B in continued proportion, so many will also fall between either of the numbers A, B and the unit in continued proportion.

For let two numbers F, G, the least that are in the ratio of A, C, D, B, be taken,

three numbers H, K, L with the same property,

and others more by one continually, until their multitude is equal to the multitude of A, C, D, B. [VIII. 2]

A ——————————   H ——
C ————————————   K — —
D ——————————————   L ———
B ———————————————

E -   M ————————————
F —   N ——————————————
G —   O ————————————————
        P ——————————————————————

Let them be taken, and let them be M, N, O, P.

It is now manifest that F by multiplying itself has made H and by multiplying H has made M, while G by multiplying itself has made L and by multiplying L has made P. [VIII. 2, Por.]

And, since M, N, O, P are the least of those which have the same ratio with F, G,

and A, C, D, B are also the least of those which have the same ratio with F, G, [VIII. 1]

while the multitude of the numbers M, N, O, P is equal to the multitude of the numbers A, C, D, B,

therefore M, N, O, P are equal to A, C, D, B respectively;

therefore M is equal to A, and P to B.

Now, since F by multiplying itself has made H,

therefore F measures H according to the units in F.

But the unit E also measures F according to the units in it;

therefore the unit E measures the number F the same number of times as F measures H.

Therefore, as the unit E is to the number F, so is F to H. [VII. Def. 20]

Again, since F by multiplying H has made M,

therefore H measures M according to the units in F.

But the unit E also measures the number F according to the units in it;

therefore the unit E measures the number F the same number of times as H measures M.

Therefore, as the unit E is to the number F, so is H to M.

But it was also proved that, as the unit E is to the number F, so is F to H;

therefore also, as the unit E is to the number F, so is F to H, and H to M.

But M is equal to A;

therefore, as the unit E is to the number F, so is F to H, and H to A.

For the same reason also,

as the unit E is to the number G, so is G to L and L to B.

Therefore, as many numbers as have fallen between A, B in continued proportion, so many numbers also have fallen between each of the numbers A. B and the unit E in continued proportion.

Q. E. D.

Suppose there are n geometric means between a, b, two numbers prime to one another; there are the same number (n) of geometric means between 1 and a and between 1 and b.

If c, d... are the n means between a, b,

$$a,\quad c,\quad d \ldots b$$

are the least numbers in that ratio, since a, b are prime to one another. [VIII. 1]

The terms are therefore respectively identical with

$$a^{n+1},\quad a^n\beta,\quad a^{n-1}\beta^2 \ldots a\beta^n,\quad \beta^{n+1},$$

where a, β is the common ratio in its lowest terms.   [VIII. 2, Por.]

Thus $a = a^{n+1},\quad b = \beta^{n+1}.$

Now $1 : a = a : a^2 = a^2 : a^3 \ldots = a^n : a^{n+1},$

and $1 : \beta = \beta : \beta^2 = \beta^2 : \beta^3 \ldots = \beta^n : \beta^{n+1};$

whence there are n geometric means between 1, a, and between 1, b.

PROPOSITION 10.

If numbers fall between each of two numbers and an unit in continued proportion, however many numbers fall between each of them and an unit in continued proportion, so many also will fall between the numbers themselves in continued proportion.

For let the numbers D, E and F, G respectively fall between the two numbers A, B and the unit C in continued proportion;

I say that, as many numbers as have fallen between each of the numbers A, B and the unit C in continued proportion, so many numbers will also fall between A, B in continued proportion.

For let D by multiplying F make H, and let the numbers D, F by multiplying H make K, L respectively.

```
C —                    A ———
                       B ——————————

D —
E ——                   H ————
F ——                   K ————————
G ————                 L ——————————————
```

Now, since, as the unit C is to the number D, so is D to E, therefore the unit C measures the number D the same number of times as D measures E.   [VII. Def. 20]

But the unit C measures the number D according to the units in D;

therefore the number D also measures E according to the units in D;

therefore D by multiplying itself has made E.

Again, since, as C is to the number D, so is E to A,

therefore the unit C measures the number D the same number of times as E measures A.

But the unit C measures the number D according to the units in D;

therefore E also measures A according to the units in D;

therefore D by multiplying E has made A.

For the same reason also

F by multiplying itself has made G, and by multiplying G has made B.

And, since D by multiplying itself has made E and by multiplying F has made H,

therefore, as D is to F, so is E to H.   [VII. 17]

For the same reason also,

as D is to F, so is H to G.   [VII. 18]

Therefore also, as E is to H, so is H to G.

Again, since D by multiplying the numbers E, H has made A, K respectively,

therefore, as E is to H, so is A to K.   [VII. 17]

But, as E is to H, so is D to F;

therefore also, as D is to F, so is A to K.

Again, since the numbers D, F by multiplying H have made K, L respectively,

therefore, as D is to F, so is K to L.   [VII. 18]

But, as D is to F, so is A to K;

therefore also, as A is to K, so is K to L.

Further, since F by multiplying the numbers H, G has made L, B respectively,

therefore, as H is to G, so is L to B.   [VII. 17]

But, as H is to G, so is D to F;

therefore also, as D is to F, so is L to B.

But it was also proved that,

as D is to F, so is A to K and K to L;

therefore also, as A is to K, so is K to L and L to B.

Therefore A, K, L, B are in continued proportion.

Therefore, as many numbers as fall between each of the numbers A, B and the unit C in continued proportion, so many also will fall between A, B in continued proportion.

Q. E. D.

If there be n geometric means between 1 and a, and also between 1 and b, there will be n geometric means between a and b.

The proposition is the converse of the preceding.

The n means with the extremes form two geometric series of the form

$$1, \quad a, \quad a^2 \ldots a^n, \quad a^{n+1},$$
$$1, \quad \beta, \quad \beta^2 \ldots \beta^n, \quad \beta^{n+1},$$

where
$$a^{n+1} = a, \quad \beta^{n+1} = b.$$

By multiplying the last term in the first line by the first in the second, the last but one in the first line by the second in the second, and so on, we get the series

$$a^{n+1}, \quad a^n\beta, \quad a^{n-1}\beta^2 \ldots a^2\beta^{n-1}, \quad a\beta^n, \quad \beta^{n+1}$$

and we have the n means between a and b.

It will be observed that, when Euclid says "*For the same reason* also, as D is to F, so is H to G," the reference is really to VII. 18 instead of VII. 17.

He infers namely that $D \times F : F \times F = D : F$. But since, by VII. 16, the order of multiplication is indifferent, he is practically justified in saying "for the same reason." The same thing occurs in later propositions.

PROPOSITION 11.

Between two square numbers there is one mean proportional number, and the square has to the square the ratio duplicate of that which the side has to the side.

Let A, B be square numbers,
and let C be the side of A, and D of B;
I say that between A, B there is one mean proportional number, and A has to B the ratio duplicate of that which C has to D.

For let C by multiplying D make E.

Now, since A is a square and C is its side,
therefore C by multiplying itself has made A.

For the same reason also
D by multiplying itself has made B.

Since then C by multiplying the numbers C, D has made A, E respectively,
therefore, as C is to D, so is A to E.   [VII. 17]

For the same reason also,
        as C is to D, so is E to B.   [VII. 18]

Therefore also, as A is to E, so is E to B.

Therefore between A, B there is one mean proportional number.

I say next that A also has to B the ratio duplicate of that which C has to D.

For, since A, E, B are three numbers in proportion,
therefore A has to B the ratio duplicate of that which A has to E.   [v. Def. 9]

But, as A is to E, so is C to D.

Therefore A has to B the ratio duplicate of that which the side C has to D.   Q. E. D.

According to Nicomachus the theorems in this proposition and the next, that two squares have *one* geometric mean, and two cubes *two* geometric means, between them are Platonic. Cf. *Timaeus*, 32 A sqq. and the note thereon, p. 294 above.

a^2, b^2 being two squares, it is only necessary to form the product ab and to prove that

$$a^2, \quad ab, \quad b^2$$

are in geometrical progression. Euclid proves that

$$a^2 : ab = ab : b^2$$

by means of VII. 17, 18, as usual.

In assuming that, since a^2 is to b^2 in the duplicate ratio of a^2 to ab, a^2 is to b^2 in the duplicate ratio of a to b, Euclid assumes that ratios which are the duplicates of equal ratios are equal. This, an obvious inference from V. 22, can be inferred just as easily for numbers from VII. 14.

PROPOSITION 12.

Between two cube numbers there are two mean proportional numbers, and the cube has to the cube the ratio triplicate of that which the side has to the side.

Let A, B be cube numbers,
and let C be the side of A, and D of B;
I say that between A, B there are two mean proportional numbers, and A has to B the ratio triplicate of that which C has to D.

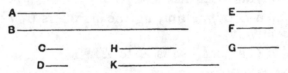

For let C by multiplying itself make E, and by multiplying D let it make F;
let D by multiplying itself make G,
and let the numbers C, D by multiplying F make H, K respectively.

Now, since A is a cube, and C its side,
and C by multiplying itself has made E,
therefore C by multiplying itself has made E and by multiplying E has made A.

For the same reason also
D by multiplying itself has made G and by multiplying G has made B.

And, since C by multiplying the numbers C, D has made E, F respectively,
therefore, as C is to D, so is E to F.   [VII. 17]

For the same reason also,

as C is to D, so is F to G.   [VII. 18]

Again, since C by multiplying the numbers E, F has made A, H respectively,

therefore, as E is to F, so is A to H.   [VII. 17]

But, as E is to F, so is C to D.

Therefore also, as C is to D, so is A to H.

Again, since the numbers C, D by multiplying F have made H, K respectively,

therefore, as C is to D, so is H to K.   [VII. 18]

Again, since D by multiplying each of the numbers F, G has made K, B respectively,

therefore, as F is to G, so is K to B.   [VII. 17]

But, as F is to G, so is C to D;

therefore also, as C is to D, so is A to H, H to K, and K to B.

Therefore H, K are two mean proportionals between A, B.

I say next that A also has to B the ratio triplicate of that which C has to D.

For, since A, H, K, B are four numbers in proportion, therefore A has to B the ratio triplicate of that which A has to H.   [v. Def. 10]

But, as A is to H, so is C to D;

therefore A also has to B the ratio triplicate of that which C has to D.

Q. E. D.

The cube numbers a^3, b^3 being given, Euclid forms the products a^2b, ab^2 and then proves, as usual, by means of VII. 17, 18 that

$$a^3, \quad a^2b, \quad ab^2, \quad b^3$$

are in continued proportion.

He assumes that, since a^3 has to b^3 the ratio triplicate of $a^3 : a^2b$, the ratio $a^3 : b^3$ is triplicate of the ratio $a : b$ which is equal to $a^3 : a^2b$. This is again an obvious inference from VII. 14.

PROPOSITION 13.

If there be as many numbers as we please in continued proportion, and each by multiplying itself make some number, the products will be proportional; and, if the original numbers by multiplying the products make certain numbers, the latter will also be proportional.

Let there be as many numbers as we please, A, B, C, in continued proportion, so that, as A is to B, so is B to C;

let A, B, C by multiplying themselves make D, E, F, and by multiplying D, E, F let them make G, H, K;

I say that D, E, F and G, H, K are in continued proportion.

```
A ———                    G ————————
B ———                    H ——————————
C ——                     K ———————————————
D ———
E ————                   M ————————
F —————                  N ——————————
L ————                   P ————————————
O ———                    Q ———————————————
```

For let A by multiplying B make L,

and let the numbers A, B by multiplying L make M. N respectively.

And again let B by multiplying C make O,

and let the numbers B, C by multiplying O make P, Q respectively.

Then, in manner similar to the foregoing, we can prove that

D, L, E and G, M, N, H are continuously proportional in the ratio of A to B,

and further E, O, F and H, P, Q, K are continuously proportional in the ratio of B to C.

Now, as A is to B, so is B to C;

therefore D, L, E are also in the same ratio with E, O, F,

and further G, M, N, H in the same ratio with H, P, Q, K.

And the multitude of D, L, E is equal to the multitude of E, O, F, and that of G, M, N, H to that of H, P, Q, K;

therefore, *ex aequali*,

         as D is to E, so is E to F,

and,   as G is to H, so is H to K.   [VII. 14]

Q. E. D.

If a, b, c ... be a series in geometrical progression, then

$$a^2, \quad b^2, \quad c^2 \ldots \Big\}$$
$$\text{and} \quad a^3, \quad b^3, \quad c^3 \ldots \Big\} \text{ are also in geometrical progression.}$$

Heiberg brackets the words added to the enunciation which extend the theorem to any powers. The words are "and this always occurs with the extremes" ($\kappa\alpha\grave{\iota}$ $\grave{\alpha}\epsilon\grave{\iota}$ $\pi\epsilon\rho\grave{\iota}$ $\tau o\grave{\upsilon}\varsigma$ $\check{\alpha}\kappa\rho o\upsilon\varsigma$ $\tau o\hat{\upsilon}\tau o$ $\sigma\upsilon\mu\beta\alpha\acute{\iota}\nu\epsilon\iota$). They seem to be rightly suspected on the same grounds as the same words added to the enunciation of VII. 27. There is no allusion to them in the proof, much less any proof of the extension.

Euclid forms, besides the squares and cubes of the given numbers, the products ab, a^2b, ab^2, bc, b^2c, bc^2. When he says that "we prove in manner similar to the foregoing," he indicates successive uses of VII. 17, 18 as in VIII. 12.

With our notation the proof is as easy to see for *any* powers as for squares and cubes.

To prove that a^n, b^n, c^n... are in geometrical progression.

Form all the means between a^n, b^n, and set out the series

$$a^n, \quad a^{n-1}b, \quad a^{n-2}b^2 \ldots ab^{n-1}, \quad b^n.$$

The common ratio of one term to the next is $a : b$.

Next take the geometrical progression

$$b^n, \quad b^{n-1}c, \quad b^{n-2}c^2 \ldots bc^{n-1}, \quad c^n,$$

the common ratio of which is $b : c$.

Proceed thus for all pairs of consecutive terms.

Now
$$a : b = b : c = \ldots$$

Therefore any pair of succeeding terms in one series are in the same ratio as any pair of succeeding terms in any other of the series.

And the number of terms in each is the same, namely $(n + 1)$.

Therefore, *ex aequali*,

$$a^n : b^n = b^n : c^n = c^n : d^n = \ldots$$

PROPOSITION 14.

If a square measure a square, the side will also measure the side; and, if the side measure the side, the square will also measure the square.

Let A, B be square numbers, let C, D be their sides, and let A measure B;

I say that C also measures D.

For let C by multiplying D make E; therefore A, E, B are continuously proportional in the ratio of C to D.   [VIII. 11]

And, since A, E, B are continuously proportional, and A measures B, therefore A also measures E.   [VIII. 7]

And, as A is to E, so is C to D;
therefore also C measures D.   [VII. Def. 20]

Again, let C measure D;
I say that A also measures B.

For, with the same construction, we can in a similar manner prove that A, E, B are continuously proportional in the ratio of C to D.

And since, as C is to D, so is A to E,
and C measures D,
therefore A also measures E.   [VII. Def. 20]

And A, E, B are continuously proportional;
therefore A also measures B.

Therefore etc.

Q. E. D.

If a^2 measures b^2, a measures b; and, if a measures b, a^2 measures b^2.

(1) a^2, ab, b^2 are in continued proportion in the ratio of a to b.

Therefore, since	a^2 measures b^2,
	a^2 measures ab. [VIII. 7]
But	$a^2 : ab = a : b$.
Therefore	a measures b.

(2) Since a measures b, a^2 measures ab.

And a^2, ab, b^2 are continuously proportional.

Thus	ab measures b^2.
And	a^2 measures ab.
Therefore	a^2 measures b^2.

It will be seen that Euclid puts the last step shortly, saying that, since a^2 measures ab, and a^2, ab, b^2 are in continued proportion, a^2 measures b^2. The same thing happens in VIII. 15, where the series of terms is one more than here.

PROPOSITION 15.

If a cube number measure a cube number, the side will also measure the side; and, if the side measure the side, the cube will also measure the cube.

For let the cube number A measure the cube B,
and let C be the side of A and D of B;
I say that C measures D.

For let C by multiplying itself make E,
and let D by multiplying itself make G;
further, let C by multiplying D make F,
and let C, D by multiplying F make H, K respectively.

A ——
B ————————————————
C –   H ————
D —   K ——————————
E —
G ————
F —

Now it is manifest that E, F, G and A, H, K, B are
continuously proportional in the ratio of C to D.   [VIII. 11, 12]
And, since A, H, K, B are continuously proportional,
and A measures B,
therefore it also measures H.   [VIII. 7]
And, as A is to H, so is C to D;
therefore C also measures D.   [VII. Def. 20]

Next, let C measure D;
I say that A will also measure B.

For, with the same construction, we can prove in a similar
manner that A, H, K, B are continuously proportional in the
ratio of C to D.
And, since C measures D,
and, as C is to D, so is A to H,
therefore A also measures H,   [VII. Def. 20]
so that A measures B also.

Q. E. D.

If a^3 measures b^3, a measures b; and *vice versa*. The proof is, *mutatis
mutandis*, the same as for squares.

(1) a^3, a^2b, ab^2, b^3 are continuously proportional in the ratio of a to b;
and a^3 measures b^3.
Therefore a^3 measures a^2b;   [VIII. 7]
and hence a measures b.

(2) Since a measures b, a^3 measures a^2b.
And, a^3, a^2b, ab^2, b^3 being continuously proportional, each term measures the
succeeding term;
therefore a^3 measures b^3.

PROPOSITION 16.

If a square number do not measure a square number, neither will the side measure the side; and, if the side do not measure the side, neither will the square measure the square.

Let A, B be square numbers, and let C, D be their sides; and let A not measure B;
I say that neither does C measure D.

For, if C measures D, A will also measure B.   [VIII. 14]

But A does not measure B;
therefore neither will C measure D.

Again, let C not measure D;
I say that neither will A measure B.

For, if A measures B, C will also measure D.   [VIII. 14]

But C does not measure D;
therefore neither will A measure B.

                                                              Q. E. D.

If a^2 does not measure b^2, a will not measure b; and, if a does not measure b, a^2 will not measure b^2.
The proof is a mere *reductio ad absurdum* using VIII. 14.

PROPOSITION 17.

If a cube number do not measure a cube number, neither will the side measure the side; and, if the side do not measure the side, neither will the cube measure the cube.

For let the cube number A not measure the cube number B,
and let C be the side of A, and D of B;
I say that C will not measure D.

For if C measures D, A will also measure B.   [VIII. 15]

But A does not measure B;
therefore neither does C measure D.

Again, let C not measure D;
I say that neither will A measure B.

For, if A measures B, C will also measure D.   [VIII. 15]

But C does not measure D;

therefore neither will A measure B.

Q. E. D.

If a^3 does not measure b^3, a will not measure b; and *vice versa.*
Proved by *reductio ad absurdum* employing VIII. 15.

PROPOSITION 18.

Between two similar plane numbers there is one mean proportional number; and the plane number has to the plane number the ratio duplicate of that which the corresponding side has to the corresponding side.

Let A, B be two similar plane numbers, and let the numbers C, D be the sides of A, and E, F of B.

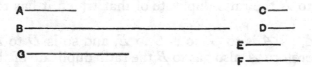

Now, since similar plane numbers are those which have their sides proportional,   [VII. Def. 21]

therefore, as C is to D, so is E to F.

I say then that between A, B there is one mean proportional number, and A has to B the ratio duplicate of that which C has to E, or D to F, that is, of that which the corresponding side has to the corresponding side.

Now since, as C is to D, so is E to F,

therefore, alternately, as C is to E, so is D to F.   [VII. 13]

And, since A is plane, and C, D are its sides,

therefore D by multiplying C has made A.

For the same reason also

E by multiplying F has made B.

Now let D by multiplying E make G.

Then, since D by multiplying C has made A, and by multiplying E has made G,

therefore, as C is to E, so is A to G.   [VII. 17]

24—2

But, as C is to E, so is D to F;

therefore also, as D is to F, so is A to G.

Again, since E by multiplying D has made G, and by multiplying F has made B,

therefore, as D is to F, so is G to B.　　　　　　　[VII. 17]

But it was also proved that,

　　　　　　as D is to F, so is A to G;

therefore also, as A is to G, so is G to B.

Therefore A, G, B are in continued proportion.

Therefore between A, B there is one mean proportional number.

I say next that A also has to B the ratio duplicate of that which the corresponding side has to the corresponding side, that is, of that which C has to E or D to F.

For, since A, G, B are in continued proportion,

A has to B the ratio duplicate of that which it has to G.

　　　　　　　　　　　　　　　　　　　　　[v. Def. 9]

And, as A is to G, so is C to E, and so is D to F.

Therefore A also has to B the ratio duplicate of that which C has to E or D to F.

　　　　　　　　　　　　　　　　　　　Q. E. D.

If ab, cd be "similar plane numbers," i.e. products of factors such that

$$a : b = c : d,$$

there is one mean proportional between ab and cd; and ab is to cd in the duplicate ratio of a to c or of b to d.

Form the product bc (or ad, which is equal to it, by VII. 19).

Then　　　　　　　　　　ab, $\left.\begin{array}{c} bc \\ = ad \end{array}\right\}$, cd

is a series of terms in geometrical progression.

For　　　　　　　　　　$a : b = c : d.$

Therefore　　　　　　　$a : c = b : d.$　　　　　　　[VII. 13]

Therefore　　　　　　　$ab : bc = bc : cd.$　　　　[VII. 17 and 16]

Thus bc (or ad) is a geometric mean between ab, cd.

And ab is to cd in the duplicate ratio of ab to bc or of bc to cd, that is, of a to c or of b to d.

PROPOSITION 19.

Between two similar solid numbers there fall two mean proportional numbers; and the solid number has to the similar solid number the ratio triplicate of that which the corresponding side has to the corresponding side.

Let A, B be two similar solid numbers, and let C, D, E be the sides of A, and F, G, H of B.

Now, since similar solid numbers are those which have their sides proportional,   [VII. Def. 21]

therefore, as C is to D, so is F to G,

and, as D is to E, so is G to H.

I say that between A, B there fall two mean proportional numbers, and A has to B the ratio triplicate of that which C has to F, D to G, and also E to H.

For let C by multiplying D make K, and let F by multiplying G make L.

Now, since C, D are in the same ratio with F, G,

and K is the product of C, D, and L the product of F, G,

K, L are similar plane numbers;   [VII. Def. 21]

therefore between K, L there is one mean proportional number.   [VIII. 18]

Let it be M.

Therefore M is the product of D, F, as was proved in the theorem preceding this.   [VIII. 18]

Now, since D by multiplying C has made K, and by multiplying F has made M,

therefore, as C is to F, so is K to M.   [VII. 17]

But, as K is to M, so is M to L.

Therefore K, M, L are continuously proportional in the ratio of C to F.

And since, as C is to D, so is F to G,
alternately therefore, as C is to F, so is D to G.   [VII. 13]

For the same reason also,

as D is to G, so is E to H.

Therefore K, M, L are continuously proportional in the ratio of C to F, in the ratio of D to G, and also in the ratio of E to H.

Next, let E, H by multiplying M make N, O respectively.

Now, since A is a solid number, and C, D, E are its sides, therefore E by multiplying the product of C, D has made A.

But the product of C, D is K;
therefore E by multiplying K has made A.

For the same reason also

H by multiplying L has made B.

Now, since E by multiplying K has made A, and further also by multiplying M has made N,
therefore, as K is to M, so is A to N.   [VII. 17]

But, as K is to M, so is C to F, D to G, and also E to H;
therefore also, as C is to F, D to G, and E to H, so is A to N.

Again, since E, H by multiplying M have made N, O respectively,
therefore, as E is to H, so is N to O.   [VII. 18]

But, as E is to H, so is C to F and D to G;
therefore also, as C is to F, D to G, and E to H, so is A to N and N to O.

Again, since H by multiplying M has made O, and further also by multiplying L has made B,
therefore, as M is to L, so is O to B.   [VII. 17]

But, as M is to L, so is C to F, D to G, and E to H.

Therefore also, as C is to F, D to G, and E to H, so not only is O to B, but also A to N and N to O.

Therefore A, N, O, B are continuously proportional in the aforesaid ratios of the sides.

I say that A also has to B the ratio triplicate of that which the corresponding side has to the corresponding side, that is, of the ratio which the number C has to F, or D to G, and also E to H.

For, since A, N, O, B are four numbers in continued proportion,

therefore A has to B the ratio triplicate of that which A has to N.   [v. Def. 10]

But, as A is to N, so it was proved that C is to F, D to G, and also E to H.

Therefore A also has to B the ratio triplicate of that which the corresponding side has to the corresponding side, that is, of the ratio which the number C has to F, D to G, and also E to H.   Q. E. D.

In other words, if $a : b : c = d : e : f$, then there are two geometric means between abc, def; and abc is to def in the triplicate ratio of a to d, or b to e, or c to f.

Euclid first takes the plane numbers ab, de (leaving out c, f) and forms the product bd. Thus, as in VIII. 18,

$$ab, \genfrac{}{}{0pt}{}{bd}{=ea}\Big\}, \; de$$

are three terms in geometrical progression in the ratio of a to d, or of b to e.

He next forms the products of c, f respectively into the mean bd.

Then abc, cbd, fbd, def

are in geometrical progression in the ratio of a to d etc.

For $abc : cbd = ab : bd = a : d$
                            $bd : fbd = c : f$ [VII. 17]
                            $fbd : def = bd : de = b : e$

And $a : d = b : e = c : f$.

The ratio of abc to def is the ratio triplicate of that of abc to cbd, i.e. of that of a to d etc.

Proposition 20.

If one mean proportional number fall between two numbers, the numbers will be similar plane numbers.

For let one mean proportional number C fall between the two numbers A, B;

5 I say that A, B are similar plane numbers.

Let D, E, the least numbers of those which have the same ratio with A, C, be taken;   [VII. 33]

therefore D measures A the same number of times that E measures C.   [VII. 20]

10   Now, as many times as D measures A, so many units let there be in F;

therefore F by multiplying D has made A,

so that A is plane, and D, F are its sides.

Again, since D, E are the least of the numbers which have
15 the same ratio with C, B,

therefore D measures C the same number of times that E
measures B.   [VII. 20]

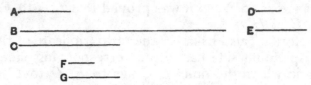

As many times, then, as E measures B, so many units let
there be in G;
20 therefore E measures B according to the units in G;

therefore G by multiplying E has made B.

Therefore B is plane, and E, G are its sides.

Therefore A, B are plane numbers.

I say next that they are also similar.

25   For, †since F by multiplying D has made A, and by
multiplying E has made C,

therefore, as D is to E, so is A to C, that is, C to B. [VII. 17]

Again,† since E by multiplying F, G has made C, B
respectively,
30 therefore, as F is to G, so is C to B.   [VII. 17]

But, as C is to B, so is D to E;

therefore also, as D is to E, so is F to G.

And alternately, as D is to F, so is E to G.   [VII. 13]

Therefore A, B are similar plane numbers; for their sides
35 are proportional.   Q. E. D.

25. **For, since F......27. C to B.** The text has clearly suffered corruption here. It
is not necessary to *infer* from other facts that, as D is to E, so is A to C; for this is part of
the hypotheses (ll. 6, 7). Again, there is no explanation of the statement (l. 25) that F by
multiplying E has made C. It is the statement and explanation of this latter fact which are
alone wanted; after which the proof proceeds as in l. 28. We might therefore substitute for
ll. 25—28 the following.

"For, since E measures C the same number of times that D measures A [l. 8], that is,
according to the units in F [l. 10], therefore F by multiplying E has made C.

And, since E by multiplying F, G," etc. etc.

This proposition is the converse of VIII. 18. If a, c, b are in geometrical
progression, a, b are "similar plane numbers."

Let $\alpha : \beta$ be the ratio $a : c$ (and therefore also the ratio $c : b$) in its lowest
terms.

Then [VII. 20]

$$a = m\alpha, \quad c = m\beta, \quad \text{where } m \text{ is some integer,}$$
$$c = n\alpha, \quad b = n\beta, \quad \text{where } n \text{ is some integer.}$$

Thus *a*, *b* are both products of two factors, i.e. plane.

Again, $a : \beta = a : c = c : b$

                                $= m : n.$ [VII. 18]

Therefore, alternately, $a : m = \beta : n,$ [VII. 13]

and hence *ma*, *nβ* are *similar* plane numbers.

[Our notation makes the second part still more obvious, for $c = m\beta = na$.]

PROPOSITION 21.

If two mean proportional numbers fall between two numbers, the numbers are similar solid numbers.

For let two mean proportional numbers *C*, *D* fall between the two numbers *A*, *B*;

I say that *A*, *B* are similar solid numbers.

For let three numbers *E*, *F*, *G*, the least of those which have the same ratio with *A*, *C*, *D*, be taken;   [VII. 33 or VIII. 2]
therefore the extremes of them *E*, *G* are prime to one another.
                                                          [VIII. 3]

Now, since one mean proportional number *F* has fallen between *E*, *G*,
therefore *E*, *G* are similar plane numbers.   [VIII. 20]

Let, then, *H*, *K* be the sides of *E*, and *L*, *M* of *G*.

Therefore it is manifest from the theorem before this that *E*, *F*, *G* are continuously proportional in the ratio of *H* to *L* and that of *K* to *M*.

Now, since *E*, *F*, *G* are the least of the numbers which have the same ratio with *A*, *C*, *D*,

and the multitude of the numbers *E*, *F*, *G* is equal to the multitude of the numbers *A*, *C*, *D*,

therefore, *ex aequali*, as *E* is to *G*, so is *A* to *D*.   [VII. 14]

But *E*, *G* are prime,

primes are also least,   [VII. 21]

and the least measure those which have the same ratio with

them the same number of times, the greater the greater and
the less the less, that is, the antecedent the antecedent and the
consequent the consequent ;   [VII. 20]

therefore E measures A the same number of times that G
measures D.

Now, as many times as E measures A, so many units let
there be in N.

Therefore N by multiplying E has made A.

But E is the product of H, K ;

therefore N by multiplying the product of H, K has made A.

Therefore A is solid, and H, K, N are its sides.

Again, since E, F, G are the least of the numbers which
have the same ratio as C, D, B,

therefore E measures C the same number of times that G
measures B.

Now, as many times as E measures C, so many units let
there be in O.

Therefore G measures B according to the units in O ;

therefore O by multiplying G has made B.

But G is the product of L, M ;

therefore O by multiplying the product of L, M has made B.

Therefore B is solid, and L, M, O are its sides ;

therefore A, B are solid.

I say that they are also similar.

For since N, O by multiplying E have made A, C,

therefore, as N is to O, so is A to C, that is, E to F. [VII. 18]

But, as E is to F, so is H to L and K to M ;

therefore also, as H is to L, so is K to M and N to O.

And H, K, N are the sides of A, and O, L, M the sides
of B.

Therefore A, B are similar solid numbers.   Q. E. D.

The converse of VIII. 19.   If a, c, d, b are in geometrical progression, a, b
are "similar solid numbers."

Let a, β, γ be the least numbers in the ratio of a, c, d (and therefore also
of c, d, b).   [VII. 33 or VIII. 2]

Therefore a, γ are prime to one another.   [VIII. 3]
They are also "similar plane numbers."   [VIII. 20]

Let $a = mn, \quad \gamma = pq,$

where $m : n = p : q.$

Then, by the proof of VIII. 20,

$$a : \beta = m : p = n : q.$$

Now, *ex aequali*, $\qquad a : d = a : \gamma,$ [VII. 14]

and, since a, γ are prime to one another,

$$a = ra, \quad d = r\gamma, \quad \text{where } r \text{ is an integer.}$$

But $\qquad\qquad\qquad a = mn :$

therefore $a = rmn$, and therefore a is "solid."

Again, *ex aequali*, $\qquad c : b = a : \gamma,$

and therefore $\qquad c = sa, \quad b = s\gamma, \quad \text{where } s \text{ is an integer.}$

Thus $b = spq$, and b is therefore "solid."

Now $\qquad\qquad a : \beta = a : c = ra : sa$

$$= r : s.$$ [VII. 18]

And, from above, $\qquad a : \beta = m : p = n : q.$

Therefore $\qquad\qquad r : s = m : p = n : q,$

and hence a, b are *similar* solid numbers.

PROPOSITION 22.

If three numbers be in continued proportion, and the first be square, the third will also be square.

Let A, B, C be three numbers in continued proportion, and let A the first be square;
I say that C the third is also square.

For, since between A, C there is one mean proportional number, B,
therefore A, C are similar plane numbers. [VIII. 20]

But A is square;

therefore C is also square. Q. E. D.

A mere application of VIII. 20 to the particular case where one of the "similar plane numbers" is square.

PROPOSITION 23.

If four numbers be in continued proportion, and the first be cube, the fourth will also be cube.

Let A, B, C, D be four numbers in continued proportion, and let A be cube;
I say that D is also cube.

For, since between A, D there are two mean proportional numbers B, C,
therefore A, D are similar solid numbers. [VIII. 21]

But A is cube;
therefore D is also cube.

Q. E. D.

A mere application of VIII. 21 to the case where one of the "similar solid numbers" is a cube.

PROPOSITION 24.

If two numbers have to one another the ratio which a square number has to a square number, and the first be square, the second will also be square.

For let the two numbers A, B have to one another the ratio which the square number C has to the square number D, and let A be square;
I say that B is also square.

For, since C, D are square,
C, D are similar plane numbers.
Therefore one mean proportional number falls between C, D. [VIII. 18]
And, as C is to D, so is A to B;
therefore one mean proportional number falls between A, B also. [VIII. 8]
And A is square;
therefore B is also square. [VIII. 22]

Q. E. D.

If $a : b = c^2 : d^2$, and a is a square, then b is also a square.
For c^2, d^2 have one mean proportional cd. [VIII. 18]
Therefore a, b, which are in the same ratio, have one mean proportional. [VIII. 8]
And, since a is square, b must also be a square. [VIII. 22]

PROPOSITION 25.

If two numbers have to one another the ratio which a cube number has to a cube number, and the first be cube, the second will also be cube.

For let the two numbers A, B have to one another the ratio which the cube number C has to the cube number D, and let A be cube;
I say that B is also cube.

For, since C, D are cube,
C, D are similar solid numbers.

Therefore two mean proportional numbers fall between
C, D.   [VIII. 19]

```
A ——
B ————            E ———
C ——              F ————
D ————
```

And, as many numbers as fall between C, D in continued
proportion, so many will also fall between those which have
the same ratio with them;   [VIII. 8]
so that two mean proportional numbers fall between A, B
also.

Let E, F so fall.

Since, then, the four numbers A, E, F, B are in continued
proportion,
and A is cube,
therefore B is also cube.   [VIII. 23]

Q. E. D.

If $a : b = c^3 : d^3$, and a is a cube, then b is also a cube.
For c^3, d^3 have two mean proportionals.   [VIII. 19]
Therefore a, b also have two mean proportionals.   [VIII. 8]
And a is a cube:
therefore b is a cube.   [VIII. 23]

PROPOSITION 26.

*Similar plane numbers have to one another the ratio which
a square number has to a square number.*

Let A, B be similar plane numbers;
I say that A has to B the ratio which a square number has
to a square number.

```
A ———        B ————————
        C ———
D ——    E ———      F ———
```

For, since A, B are similar plane numbers,
therefore one mean proportional number falls between A, B.
                                                                [VIII. 18]

Let it so fall, and let it be C;

and let D, E, F, the least numbers of those which have the same ratio with A, C, B, be taken;   [VII. 33 or VIII. 2]

therefore the extremes of them D, F are square.   [VIII. 2, Por.]

And since, as D is to F, so is A to B,

and D, F are square,

therefore A has to B the ratio which a square number has to a square number.

Q. E. D.

If a, b are similar "plane numbers," let c be the mean proportional between them.   [VIII. 18]

Take a, β, γ the smallest numbers in the ratio of a, c, b. [VII. 33 or VIII. 2]

Then a, γ are squares.   [VIII. 2, Por.]

Therefore a, b are in the ratio of a square to a square.

PROPOSITION 27.

Similar solid numbers have to one another the ratio which a cube number has to a cube number.

Let A, B be similar solid numbers;

I say that A has to B the ratio which a cube number has to a cube number.

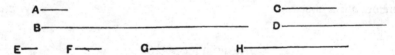

For, since A, B are similar solid numbers,

therefore two mean proportional numbers fall between A, B.

[VIII. 19]

Let C, D so fall,

and let E, F, G, H, the least numbers of those which have the same ratio with A, C, D, B, and equal with them in multitude, be taken;   [VII. 33 or VIII. 2]

therefore the extremes of them E, H are cube.   [VIII. 2, Por.]

And, as E is to H, so is A to B;

therefore A also has to B the ratio which a cube number has to a cube number.

Q. E. D.

The same thing as VIII. 26 with cubes.　It is proved in the same way except that VIII. 19 is used instead of VIII. 18.

The last note of an-Nairīzī in which the name of Heron is mentioned is on this proposition.　Heron is there stated (p. 194—5, ed. Curtze) to have added the two propositions that,

1. *If two numbers have to one another the ratio of a square to a square, the numbers are similar plane numbers;*

2. *If two numbers have to one another the ratio of a cube to a cube, the numbers are similar solid numbers.*

The propositions are of course the converses of VIII. 26, 27 respectively. They are easily proved.

(1)　If　　　　　　　　　　　$a : b = c^2 : d^2$,

then, since there is one mean proportional (cd) between c^2, d^2,

　　　　　　　　　　　　　　　　　　　　　　[VIII. 11 or 18]

there is also one mean proportional between a, b.　　　　　[VIII. 8]

Therefore a, b are similar plane numbers.　　　　　　[VIII. 20]

(2) is similarly proved by the use of VIII. 12 or 19, VIII. 8, VIII. 21.

The insertion by Heron of the first of the two propositions, the converse of VIII. 26, is perhaps an argument in favour of the correctness of the text of IX. 10, though (as remarked in the note on that proposition) it does not give the easiest proof　Cf. Heron's extension of VII. 3 tacitly assumed by Euclid in VII. 33.

BOOK IX.

PROPOSITION 1.

If two similar plane numbers by multiplying one another make some number, the product will be square.

Let A, B be two similar plane numbers, and let A by multiplying B make C;

I say that C is square.

For let A by multiplying itself make D.

Therefore D is square.

Since then A by multiplying itself has made D, and by multiplying B has made C,

therefore, as A is to B, so is D to C.   [VII. 17]

And, since A, B are similar plane numbers,

therefore one mean proportional number falls between A, B.
                    [VIII. 18]

But, if numbers fall between two numbers in continued proportion, as many as fall between them, so many also fall between those which have the same ratio;   [VIII. 8]

so that one mean proportional number falls between D, C also.

And D is square;

therefore C is also square.   [VIII. 22]

Q. E. D.

The product of two similar plane numbers is a square.

Let a, b be two similar plane numbers.

Now $a : b = a^2 : ab$.   [VII. 17]

And between a, b there is one mean proportional.   [VIII. 18]

Therefore between $a^2 : ab$ there is one mean proportional.   [VIII. 8]

And a^2 is square;

therefore ab is square.   [VIII. 22]

PROPOSITION 2.

If two numbers by multiplying one another make a square number, they are similar plane numbers.

Let A, B be two numbers, and let A by multiplying B make the square number C;

I say that A, B are similar plane numbers.

For let A by multiplying itself make D;

therefore D is square.

Now, since A by multiplying itself has made D, and by multiplying B has made C,

therefore, as A is to B, so is D to C. 　　　　[VII. 17]

And, since D is square, and C is so also

therefore D, C are similar plane numbers.

Therefore one mean proportional number falls between D, C. 　　　　[VIII. 18]

And, as D is to C, so is A to B;

therefore one mean proportional number falls between A, B also. 　　　　[VIII. 8]

But, if one mean proportional number fall between two numbers, they are similar plane numbers; 　　　　[VIII. 20]

therefore A, B are similar plane numbers.

　　　　　　　　　　　　　　　　　　　　Q. E. D.

If ab is a square number, a, b are similar plane numbers. (The converse of IX. 1.)

For 　　　　　　　　　　$a : b = a^2 : ab$. 　　　　[VII. 17]

And a^2, ab being square numbers, and therefore similar plane numbers, they have one mean proportional. 　　　　[VIII. 18]

Therefore a, b also have one mean proportional. 　　　　[VIII. 8]

whence a, b are similar plane numbers. 　　　　[VIII. 20]

PROPOSITION 3.

If a cube number by multiplying itself make some number, the product will be cube.

For let the cube number A by multiplying itself make B;
I say that B is cube.

For let C, the side of A, be taken, and let C by multiplying itself make D.

It is then manifest that C by multiplying D has made A.

Now, since C by multiplying itself has made D,

therefore C measures D according to the units in itself.

But further the unit also measures C according to the units in it;

therefore, as the unit is to C, so is C to D.   [VII. Def. 20]

Again, since C by multiplying D has made A,

therefore D measures A according to the units in C.

But the unit also measures C according to the units in it;

therefore, as the unit is to C, so is D to A.

But, as the unit is to C, so is C to D;

therefore also, as the unit is to C, so is C to D, and D to A.

Therefore between the unit and the number A two mean proportional numbers C, D have fallen in continued proportion.

Again, since A by multiplying itself has made B,

therefore A measures B according to the units in itself.

But the unit also measures A according to the units in it;

therefore, as the unit is to A, so is A to B.   [VII. Def. 20]

But between the unit and A two mean proportional numbers have fallen;

therefore two mean proportional numbers will also fall between A, B.   [VIII. 8]

But, if two mean proportional numbers fall between two numbers, and the first be cube, the second will also be cube.   [VIII. 23]

And A is cube;

therefore B is also cube.   Q. E. D.

The product of a^3 into itself, or $a^3 \cdot a^3$, is a cube.

For $1 : a = a : a^2 = a^2 : a^3$.

Therefore between 1 and a^3 there are two mean proportionals.

Also $1 : a^3 = a^3 : a^3 \cdot a^3$.

Therefore two mean proportionals fall between a^3 and $a^3 \cdot a^3$.   [VIII. 8]

(It is true that VIII. 8 is only enunciated of two pairs of numbers, but the proof is equally valid if one number of one pair is unity.)

And a^3 is a cube number:

therefore $a^3 \cdot a^3$ is also cube.   [VIII. 23]

PROPOSITION 4.

If a cube number by multiplying a cube number make some number, the product will be cube.

For let the cube number A by multiplying the cube number B make C;

I say that C is cube.

For let A by multiplying itself make D;

therefore D is cube.   [IX. 3]

And, since A by multiplying itself has made D, and by multiplying B has made C

therefore, as A is to B, so is D to C.   [VII. 17]

And, since A, B are cube numbers, A, B are similar solid numbers.

Therefore two mean proportional numbers fall between A, B;   [VIII. 19]

so that two mean proportional numbers will fall between D, C also.   [VIII. 8]

And D is cube;

therefore C is also cube   [VIII. 23]

Q. E. D.

The product of two cubes, say $a^3 \cdot b^3$, is a cube.

For $a^3 : b^3 = a^3 \cdot a^3 : a^3 \cdot b^3$.   [VII. 17]

And two mean proportionals fall between a^3, b^3 which are similar solid numbers.   [VIII. 19]

Therefore two mean proportionals fall between $a^3 \cdot a^3$, $a^3 \cdot b^3$ [VIII. 8]

But $a^3 \cdot a^3$ is a cube: [IX. 3]

therefore $a^3 \cdot b^3$ is a cube.   [VIII. 23]

PROPOSITION 5.

If a cube number by multiplying any number make a cube number, the multiplied number will also be cube.

For let the cube number A by multiplying any number B make the cube number C;

I say that B is cube.

For let A by multiplying itself make D;

therefore D is cube.   [IX. 3]

Now, since A by multiplying itself has made D, and by multiplying B has made C,

therefore, as A is to B, so is D to C.   [VII. 17]

And since D, C are cube,

they are similar solid numbers.

Therefore two mean proportional numbers fall between D, C.   [VIII. 19]

And, as D is to C, so is A to B;

therefore two mean proportional numbers fall between A, B also.   [VIII. 8]

And A is cube;

therefore B is also cube.   [VIII. 23]

If the product a^3b is a cube number, b is cube.

By IX. 3, the product $a^3 . a^3$ is a cube.

And $a^3 . a^3 : a^3b = a^3 : b.$ [VII. 17]

The first two terms are cubes, and therefore "similar solids"; therefore there are two mean proportionals between them.   [VIII. 19]

Therefore there are two mean proportionals between a^3, b.   [VIII. 8]

And a^3 is a cube:

therefore b is a cube number.   [VIII. 23]

PROPOSITION 6.

If a number by multiplying itself make a cube number, it will itself also be cube.

For let the number A by multiplying itself make the cube number B;

I say that A is also cube.

For let A by multiplying B make C.

Since, then, A by multiplying itself has made B, and by multiplying B has made C,

therefore C is cube.

And, since A by multiplying itself has made B,

therefore A measures B according to the units in itself.

But the unit also measures A according to the units in it.

Therefore, as the unit is to A, so is A to B.   [VII. Def. 20]

And, since A by multiplying B has made C,

therefore B measures C according to the units in A.

But the unit also measures A according to the units in it.

Therefore, as the unit is to A, so is B to C.   [VII. Def. 20]

But, as the unit is to A, so is A to B;

therefore also, as A is to B, so is B to C.

And, since B, C are cube,

they are similar solid numbers.

Therefore there are two mean proportional numbers between B, C.   [VIII. 19]

And, as B is to C, so is A to B.

Therefore there are two mean proportional numbers between A, B also.   [VIII. 8]

And B is cube;

therefore A is also cube.   [cf. VIII. 23]

Q. E. D.

If a^2 is a cube number, a is also a cube.

For $1 : a = a : a^2 = a^2 : a^3$.

Now a^2, a^3 are both cubes, and therefore "similar solids"; therefore there are two mean proportionals between them.   [VIII. 19]

Therefore there are two mean proportionals between a, a^2.   [VIII. 8]

And a^2 is a cube:

therefore a is also a cube number.   [VIII. 23]

It will be noticed that the last step is not an exact quotation of the result of VIII. 23, because it is there the *first* of four terms which is known to be a cube, and the *last* which is proved to be a cube; here the case is reversed. But there is no difficulty. Without inverting the proportions, we have only to refer to VIII. 21 which proves that a, a^2, having two mean proportionals between them, are two similar solid numbers; whence, since a^2 is a cube, a is also a cube.

PROPOSITION 7.

If a composite number by multiplying any number make some number, the product will be solid.

For let the composite number A by multiplying any number B make C;

I say that C is solid.

For, since A is composite, it will be measured by some number.   [VII. Def. 13]

Let it be measured by D;

and, as many times as D measures A, so many units let there be in E.

Since then D measures A according to the units in E,
therefore E by multiplying D has made A.   [VII. Def. 15]

And, since A by multiplying B has made C,
and A is the product of D, E,
therefore the product of D, E by multiplying B has made C.

Therefore C is solid, and D, E, B are its sides.

Q. E. D.

Since a composite number is the product of two factors, the result of multiplying it by another number is to produce a number which is the product of three factors, i.e. a "solid number."

PROPOSITION 8.

If as many numbers as we please beginning from an unit be in continued proportion, the third from the unit will be square, as will also those which successively leave out one; the fourth will be cube, as will also all those which leave out two; and the seventh will be at once cube and square, as will also those which leave out five.

Let there be as many numbers as we please, A, B, C, D, E, F, beginning from an unit and in continued proportion;
I say that B, the third from the unit, is square, as are also all those which leave out one; C, the fourth, is cube, as are also all those which leave out two; and F, the seventh, is at once cube and square, as are also all those which leave out five.

For since, as the unit is to A, so is A to B,
therefore the unit measures the number A the same number of times that A measures B.   [VII. Def. 20]

But the unit measures the number A according to the units in it;
therefore A also measures B according to the units in A.

Therefore A by multiplying itself has made B;
therefore B is square.

And, since B, C, D are in continued proportion, and B is square,
therefore D is also square.   [VIII. 22]

For the same reason

F is also square.

Similarly we can prove that all those which leave out one are square.

I say next that C, the fourth from the unit, is cube, as are also all those which leave out two.

For since, as the unit is to A, so is B to C,

therefore the unit measures the number A the same number of times that B measures C.

But the unit measures the number A according to the units in A;

therefore B also measures C according to the units in A.

Therefore A by multiplying B has made C.

Since then A by multiplying itself has made B, and by multiplying B has made C,

therefore C is cube.

And, since C, D, E, F are in continued proportion, and C is cube,

therefore F is also cube.   [VIII. 23]

But it was also proved square;

therefore the seventh from the unit is both cube and square.

Similarly we can prove that all the numbers which leave out five are also both cube and square.

Q. E. D.

If 1, a, a_2, a_3, ... be a geometrical progression, then a_2, a_4, a_6, ... are squares;

a_3, a_6, a_9, ... are cubes;

a_6, a_{12}, ... are both squares and cubes.

Since $1 : a = a : a_2,$
$$a_2 = a^2$$

And, since a_2, a_3, a_4 are in geometrical progression and $a_2 (= a^2)$ is a square,
a_4 is a square.   [VIII. 22]

Similarly a_6, a_8, ... are squares.

Next, $1 : a = a_2 : a_3$
$$= a^2 : a_3,$$

whence $a_3 = a^3$, a cube number.

And, since a_3, a_4, a_5, a_6 are in geometrical progression, and a_3 is a cube,
a_6 is a cube.   [VIII. 23]

Similarly a_9, a_{12}, ... are cubes.

Clearly then a_6, a_{12}, a_{18}, ... are both squares and cubes.

The whole result is of course obvious if the geometrical progression is written, with our notation, as

$$1, a, a^2, a^3, a^4, \ldots a^n.$$

PROPOSITION 9.

If as many numbers as we please beginning from an unit be in continued proportion, and the number after the unit be square, all the rest will also be square. And, if the number after the unit be cube, all the rest will also be cube.

Let there be as many numbers as we please, A, B, C, D, E, F, beginning from an unit and in continued proportion, and let A, the number after the unit, be square;

I say that all the rest will also be square.

Now it has been proved that B, the third from the unit, is square, as are also all those which leave out one;   [IX. 8]

I say that all the rest are also square.

For, since A, B, C are in continued proportion,
and A is square,
therefore C is also square.   [VIII. 22]

Again, since B, C, D are in continued proportion,
and B is square,
D is also square.   [VIII. 22]

Similarly we can prove that all the rest are also square.

Next, let A be cube;
I say that all the rest are also cube.

Now it has been proved that C, the fourth from the unit, is cube, as also are all those which leave out two;   [IX. 8]
I say that all the rest are also cube.

For, since, as the unit is to A, so is A to B,
therefore the unit measures A the same number of times as A measures B.

But the unit measures A according to the units in it;
therefore A also measures B according to the units in itself;
therefore A by multiplying itself has made B.

And A is cube.

But, if a cube number by multiplying itself make some number, the product is cube.   [IX. 3]

Therefore B is also cube.

And, since the four numbers A, B, C, D are in continued proportion,

and A is cube,

D also is cube.   [VIII. 23]

For the same reason

E is also cube, and similarly all the rest are cube.

Q. E. D.

If $1, a^2, a_2, a_3, a_4, \ldots$ are in geometrical progression, $a_2, a_3, a_4, \ldots$ are all squares;

and, if $1, a^3, a_2, a_3, a_4, \ldots$ are in geometrical progression, $a_2, a_3, \ldots$ are all cubes.

(1) By IX. 8, $a_2, a_4, a_6, \ldots$ are all squares.

And, a^2, a_2, a_3 being in geometrical progression, and a^2 being a square,

$$a_3 \text{ is a square.} \qquad [\text{VIII. 22}]$$

For the same reason $a_5, a_7, \ldots$ are all squares.

(2) By IX. 8, $a_3, a_6, a_9, \ldots$ are all cubes.

Now $1 : a^3 = a^3 : a_2.$

Therefore $a_2 = a^3 \cdot a^3$, which is a cube, by IX. 3.

And, a^3, a_2, a_3, a_4 being in geometrical progression, and a^3 being cube,

$$a_4 \text{ is cube.} \qquad [\text{VIII. 23}]$$

Similarly we prove that a_5 is cube, and so on.

The results are of course obvious in our notation, the series being

$$(1) \quad 1, a^2, a^4, a^6, \ldots a^{2n},$$
$$(2) \quad 1, a^3, a^9, a^{12}, \ldots a^{3n}.$$

Proposition 10.

If as many numbers as we please beginning from an unit be in continued proportion, and the number after the unit be not square, neither will any other be square except the third from the unit and all those which leave out one. And, if the number after the unit be not cube, neither will any other be cube except the fourth from the unit and all those which leave out two.

Let there be as many numbers as we please, A, B, C, D. E, F, beginning from an unit and in continued proportion, and let A, the number after the unit, not be square;

I say that neither will any other be square except the third from the unit < and those which leave out one >.

For, if possible, let C be square.

But B is also square;   [IX. 8]

[therefore B, C have to one another the ratio which a square number has to a square number].

And, as B is to C, so is A to B;

therefore A, B have to one another the ratio which a square number has to a square number;

[so that A, B are similar plane numbers].   [VIII. 26, converse]

And B is square;

therefore A is also square:

which is contrary to the hypothesis.

Therefore C is not square.

Similarly we can prove that neither is any other of the numbers square except the third from the unit and those which leave out one.

Next, let A not be cube.

I say that neither will any other be cube except the fourth from the unit and those which leave out two.

For, if possible, let D be cube.

Now C is also cube; for it is fourth from the unit.   [IX. 8]

And, as C is to D, so is B to C;

therefore B also has to C the ratio which a cube has to a cube.

And C is cube;

therefore B is also cube.   [VIII. 25]

And since, as the unit is to A, so is A to B,

and the unit measures A according to the units in it,

therefore A also measures B according to the units in itself;

therefore A by multiplying itself has made the cube number B.

But, if a number by multiplying itself make a cube number, it is also itself cube.   [IX. 6]

Therefore A is also cube:

which is contrary to the hypothesis.

Therefore D is not cube.

Similarly we can prove that neither is any other of the numbers cube except the fourth from the unit and those which leave out two.

Q. E. D.

If $1, a, a_2, a_3, a_4, \ldots$ be a geometrical progression, then (1), if a is not a square, none of the terms will be square except $a_2, a_4, a_6, \ldots$;

and (2), if a is not a cube, none of the terms will be cube except $a_3, a_6, a_9, \ldots$.

With reference to the first part of the proof, viz. that which proves that, if a_3 is a square, a must be a square, Heiberg remarks that the words which I have bracketed are perhaps spurious; for it is easier to use VIII. 24 than the *converse* of VIII. 26, and a use of VIII. 24 would correspond better to the use of VIII. 25 in the second part relating to cubes. I agree in this view and have bracketed the words accordingly. (See however note, p. 383, on converses of VIII. 26, 27 given by Heron.) If this change be made, the proof runs as follows.

(1)   If possible, let a_3 be square.

Now $a_2 : a_3 = a : a_2$.

But a_2 is a square.   [IX. 8]

Therefore a is to a_2 in the ratio of a square to a square.

And a_2 is square;

therefore a is square [VIII. 24]: which is impossible.

(2)   If possible, let a_4 be a cube.

Now $a_3 : a_4 = a_2 : a_3$.

And a_3 is a cube.   [IX. 8]

Therefore a_2 is to a_3 in the ratio of a cube to a cube.

And a_3 is a cube:

therefore a_2 is a cube.   [VIII. 25]

But, since $1 : a = a : a_2$,

                               $a_2 = a^2$.

And, since a^2 is a cube,

a must be a cube [IX. 6]: which is impossible.

The propositions VIII. 24, 25 are here not quoted in their exact form in that the *first* and *second* squares, or cubes, change places. But there is no difficulty, since the method by which the theorems are proved shows that either inference is equally correct.

PROPOSITION 11.

If as many numbers as we please beginning from an unit be in continued proportion, the less measures the greater according to some one of the numbers which have place among the proportional numbers.

Let there be as many numbers as we please, B, C, D, E, beginning from the unit A and in continued proportion ;

I say that B, the least of the numbers B, C, D, E, measures E according to some one of the numbers C, D.

For since, as the unit A is to B, so is D to E,

therefore the unit A measures the number B the same number of times as D measures E ;

therefore, alternately, the unit A measures D the same number of times as B measures E.   [VII. 15]

But the unit A measures D according to the units in it ;

therefore B also measures E according to the units in D ;

so that B the less measures E the greater according to some number of those which have place among the proportional numbers.—

PORISM. And it is manifest that, whatever place the measuring number has, reckoned from the unit, the same place also has the number according to which it measures, reckoned from the number measured, in the direction of the number before it.—

                                                                Q. E. D.

The proposition and the porism together assert that, if 1, a, a_2, ... a_n be a geometrical progression, a_r measures a_n and gives the quotient a_{n-r} ($r < n$).

Euclid only proves that $a_n = a \cdot a_{n-1}$, as follows.

$$1 : a = a_{n-1} : a_n.$$

Therefore 1 measures a the same number of times as a_{n-1} measures a_n.

Hence 1 measures a_{n-1} the same number of times as a measures a_n ;

                                                                [VII. 15]

that is, $a_n = a \cdot a_{n-1}.$

We can supply the proof of the porism as follows.

$$1 : a = a_r : a_{r+1},$$
$$a : a_2 = a_{r+1} : a_{r+2},$$
$$\dots\dots\dots\dots\dots\dots$$
$$a_{n-r-1} : a_{n-r} = a_{n-1} : a_n,$$

whence, *ex aequali*,

$$1 : a_{n-r} = a_r : a_n.$$ [VII. 14]

It follows, by the same argument as before, that

$$a_n = a_r \cdot a_{n-r}.$$

With our notation, we have the theorem of indices that

$$a^{m+n} = a^m \cdot a^n.$$

PROPOSITION 12.

If as many numbers as we please beginning from an unit be in continued proportion, by however many prime numbers the last is measured, the next to the unit will also be measured by the same.

Let there be as many numbers as we please, A, B, C, D, beginning from an unit, and in continued proportion;
I say that, by however many prime numbers D is measured, A will also be measured by the same.

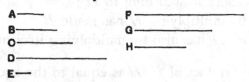

For let D be measured by any prime number E;
I say that E measures A.

For suppose it does not;
now E is prime, and any prime number is prime to any which it does not measure;   [VII. 29]
therefore E, A are prime to one another.

And, since E measures D, let it measure it according to F,
therefore E by multiplying F has made D.

Again, since A measures D according to the units in C,
[IX. 11 and Por.]
therefore A by multiplying C has made D.

But, further, E has also by multiplying F made D;
therefore the product of A, C is equal to the product of E, F.

Therefore, as A is to E, so is F to C.   [VII. 19]

But A, E are prime,
primes are also least,   [VII. 21]
and the least measure those which have the same ratio the same number of times, the antecedent the antecedent and the consequent the consequent;   [VII. 20]
therefore E measures C.

Let it measure it according to G;
therefore E by multiplying G has made C.

But, further, by the theorem before this,
A has also by multiplying B made C.   [IX. 11 and Por.]

Therefore the product of A, B is equal to the product of E, G.

Therefore, as A is to E, so is G to B.   [VII. 19]

But A, E are prime,

primes are also least,   [VII. 21]

and the least numbers measure those which have the same ratio with them the same number of times, the antecedent the antecedent and the consequent the consequent: [VII. 20]

therefore E measures B.

Let it measure it according to H;

therefore E by multiplying H has made B.

But further A has also by multiplying itself made B;

[IX. 8]

therefore the product of E, H is equal to the square on A.

Therefore, as E is to A, so is A to H.   [VII. 19]

But A, E are prime,

primes are also least,   [VII. 21]

and the least measure those which have the same ratio the same number of times, the antecedent the antecedent and the consequent the consequent;   [VII. 20]

therefore E measures A, as antecedent antecedent.

But, again, it also does not measure it:

which is impossible.

Therefore E, A are not prime to one another.

Therefore they are composite to one another.

But numbers composite to one another are measured by some number.   [VII. Def. 14]

And, since E is by hypothesis prime,

and the prime is not measured by any number other than itself,

therefore E measures A, E,

so that E measures A.

[But it also measures D;

therefore E measures A, D.]

Similarly we can prove that, by however many prime numbers D is measured, A will also be measured by the same.

Q. E. D.

If 1, a, a_2, ... a_n be a geometrical progression, and a_n be measured by any prime number p, a will also be measured by p.

For, if possible, suppose that p does not measure a; then, p being prime, p, a are prime to one another.   [VII. 29]

Suppose $a_n = m \cdot p$.

Now $a_n = a \cdot a_{n-1}$.   [IX. 11]

Therefore $a \cdot a_{n-1} = m \cdot p$,

and $a : p = m : a_{n-1}$.   [VII. 19]

Hence, a, p being prime to one another,

p measures a_{n-1}.   [VII. 20, 21]

By a repetition of the same process, we can prove that p measures a_{n-2} and therefore a_{n-3}, and so on, and finally that p measures a.

But, by hypothesis, p does not measure a: which is impossible.

Hence p, a are not prime to one another:

therefore they have some common factor.   [VII. Def. 14]

But p is the only number which measures p;

therefore p measures a.

Heiberg remarks that, as, in the ἔκθεσις, Euclid sets himself to prove that E measures A, the words bracketed above are unnecessary and therefore perhaps interpolated.

PROPOSITION 13.

If as many numbers as we please beginning from an unit be in continued proportion, and the number after the unit be prime, the greatest will not be measured by any except those which have a place among the proportional numbers.

Let there be as many numbers as we please, A, B, C, D, beginning from an unit and in continued proportion, and let A, the number after the unit, be prime;

I say that D, the greatest of them, will not be measured by any other number except A, B, C.

A ——————   E ——
B ——————   F ———
C ——————   G —
D ——————   H ——

For, if possible, let it be measured by E, and let E not be the same with any of the numbers A, B, C.

It is then manifest that E is not prime.

For, if E is prime and measures D,

it will also measure A [IX. 12], which is prime, though it is not the same with it:

which is impossible.

Therefore E is not prime.

Therefore it is composite.

But any composite number is measured by some prime number;   [VII. 31]

therefore E is measured by some prime number.

I say next that it will not be measured by any other prime except A.

For, if E is measured by another,

and E measures D,

that other will also measure D;

so that it will also measure A [IX. 12], which is prime, though it is not the same with it:

which is impossible.

Therefore A measures E.

And, since E measures D, let it measure it according to F.

I say that F is not the same with any of the numbers A, B, C.

For, if F is the same with one of the numbers A, B, C,

and measures D according to E,

therefore one of the numbers A, B, C also measures D according to E.

But one of the numbers A, B, C measures D according to some one of the numbers A, B, C;   [IX. 11]

therefore E is also the same with one of the numbers A, B, C: which is contrary to the hypothesis.

Therefore F is not the same as any one of the numbers A. B, C.

Similarly we can prove that F is measured by A, by proving again that F is not prime.

For, if it is, and measures D,

it will also measure A [IX. 12], which is prime, though it is not the same with it:

which is impossible;

therefore F is not prime.

Therefore it is composite.

But any composite number is measured by some prime number;   [VII. 31]

therefore F is measured by some prime number.

I say next that it will not be measured by any other prime except *A*.

For, if any other prime number measures *F*, and *F* measures *D*,

that other will also measure *D*;

so that it will also measure *A* [IX. 12], which is prime, though it is not the same with it:

which is impossible.

Therefore *A* measures *F*.

And, since *E* measures *D* according to *F*,

therefore *E* by multiplying *F* has made *D*.

But, further, *A* has also by multiplying *C* made *D*; [IX. 11]

therefore the product of *A*, *C* is equal to the product of *E*, *F*.

Therefore, proportionally, as *A* is to *E*, so is *F* to *C*.

[VII. 19]

But *A* measures *E*;

therefore *F* also measures *C*.

Let it measure it according to *G*.

Similarly, then, we can prove that *G* is not the same with any of the numbers *A*, *B*, and that it is measured by *A*.

And, since *F* measures *C* according to *G*

therefore *F* by multiplying *G* has made *C*.

But, further, *A* has also by multiplying *B* made *C*; [IX. 11]

therefore the product of *A*, *B* is equal to the product of *F*, *G*.

Therefore, proportionally, as *A* is to *F*, so is *G* to *B*.

[VII. 19]

But *A* measures *F*;

therefore *G* also measures *B*.

Let it measure it according to *H*.

Similarly then we can prove that *H* is not the same with *A*.

And, since *G* measures *B* according to *H*,

therefore *G* by multiplying *H* has made *B*.

But further *A* has also by multiplying itself made *B*;

[IX. 8]

therefore the product of *H*, *G* is equal to the square on *A*.

Therefore, as *H* is to *A*, so is *A* to *G*.   [VII. 19]

But A measures G;

therefore H also measures A, which is prime, though it is not the same with it:

which is absurd.

Therefore D the greatest will not be measured by any other number except A, B, C.

Q. E. D.

If $1, a, a_2, \ldots a_n$ be a geometrical progression, and if a is prime, a_n will not be measured by any numbers except the preceding terms of the series.

If possible, let a_n be measured by b, a number different from all the preceding terms.

Now b cannot be prime, for, if it were, it would measure a.   [IX. 12]

Therefore b is composite, and hence will be measured by *some* prime number [VII. 31], say p.

Thus p must measure a_n and therefore a [IX. 12]; so that p cannot be different from a, and b is not measured by any prime number except a.

Suppose that $a_n = b \cdot c.$

Now c cannot be identical with any of the terms $a, a_2, \ldots a_{n-1}$; for, if it were, b would be identical with another of them: [IX. 11]

which is contrary to the hypothesis.

We can now prove (just as for b) that c cannot be prime and cannot be measured by any prime number except a.

Since $b \cdot c = a_n = a \cdot a_{n-1},$ [IX. 11]

$a : b = c : a_{n-1},$

whence, since a measures b,

c measures $a_{n-1}.$

Let $a_{n-1} = c \cdot d.$

We now prove in the same way that d is not identical with any of the terms $a, a_2, \ldots a_{n-2}$, is not prime, and is not measured by any prime except a, and also that

d measures $a_{n-2}.$

Proceeding in this way, we get a last factor, say k, which measures a though different from it:

which is absurd, since a is prime.

Thus the original supposition that a_n can be measured by a number b different from all the terms $a, a_2, \ldots a_{n-1}$ must be incorrect.

Therefore etc.

PROPOSITION 14.

If a number be the least that is measured by prime numbers, it will not be measured by any other prime number except those originally measuring it.

For let the number A be the least that is measured by the prime numbers B, C, D;

I say that A will not be measured by any other prime number except B, C, D.

For, if possible, let it be measured by the prime number E, and let E not be the same with any one of the numbers B, C, D.

```
A ———————————        B —
E ————————           C ———
                     D ——
F ————
```

Now, since E measures A, let it measure it according to F;

therefore E by multiplying F has made A.

And A is measured by the prime numbers B, C, D.

But, if two numbers by multiplying one another make some number, and any prime number measure the product, it will also measure one of the original numbers;   [VII. 30]

therefore B, C, D will measure one of the numbers E, F.

Now they will not measure E;

for E is prime and not the same with any one of the numbers B, C, D.

Therefore they will measure F, which is less than A:

which is impossible, for A is by hypothesis the least number measured by B, C, D.

Therefore no prime number will measure A except B, C, D.

Q. E. D.

In other words, a number can be resolved into prime factors in only one way.

Let a be the least number measured by each of the prime numbers b, c, d, ... k.

If possible, suppose that a has a prime factor p different from b, c, d, ... k.

Let $$a = p \cdot m.$$

Now b, c, d, ... k, measuring a, must measure one of the two factors p, m.   [VII. 30]

They do not, by hypothesis, measure p;

therefore they must measure m, a number less than a:

which is contrary to the hypothesis.

Therefore a has no prime factors except b, c, d, ... k.

PROPOSITION 15.

If three numbers in continued proportion be the least of those which have the same ratio with them, any two whatever added together will be prime to the remaining number.

Let A, B, C, three numbers in continued proportion, be the least of those which have the same ratio with them;

I say that any two of the numbers A, B, C whatever added together are prime to the remaining number, namely A, B to C; B, C to A; and further A, C to B.

For let two numbers DE, EF, the least of those which have the same ratio with A, B, C, be taken.   [VIII. 2]

It is then manifest that DE by multiplying itself has made A, and by multiplying EF has made B, and, further, EF by multiplying itself has made C.   [VIII. 2]

Now, since DE, EF are least,
they are prime to one another.   [VII. 22]

But, if two numbers be prime to one another,
their sum is also prime to each;   [VII. 28]
therefore DF is also prime to each of the numbers DE, EF.

But further DE is also prime to EF;
therefore DF, DE are prime to EF.

But, if two numbers be prime to any number,
their product is also prime to the other;   [VII. 24]
so that the product of FD, DE is prime to EF;
hence the product of FD, DE is also prime to the square on EF.   [VII. 25]

But the product of FD, DE is the square on DE together with the product of DE, EF;   [II. 3]
therefore the square on DE together with the product of DE, EF is prime to the square on EF.

And the square on DE is A,
the product of DE, EF is B,
and the square on EF is C;
therefore A, B added together are prime to C.

Similarly we can prove that B, C added together are prime to A.

I say next that A, C added together are also prime to B.

For, since DF is prime to each of the numbers DE, EF, the square on DF is also prime to the product of DE, EF.

[VII. 24, 25]

But the squares on DE, EF together with twice the product of DE, EF are equal to the square on DF;   [II. 4] therefore the squares on DE, EF together with twice the product of DE, EF are prime to the product of DE, EF.

Separando, the squares on DE, EF together with once the product of DE, EF are prime to the product of DE, EF.

Therefore, *separando* again, the squares on DE, EF are prime to the product of DE, EF.

And the square on DE is A,

the product of DE, EF is B,

and the square on EF is C.

Therefore A, C added together are prime to B.

Q. E. D.

If a, b, c be a geometrical progression in the least terms which have a given common ratio, $(b + c)$, $(c + a)$, $(a + b)$ are respectively prime to a, b, c.

Let $a : \beta$ be the common ratio in its lowest terms, so that the geometrical progression is

$$a^2, \quad a\beta, \quad \beta^2.$$ [VIII. 2]

Now, a, β being prime to one another,

$a + \beta$ is prime to both a and β.   [VII. 28]

Therefore $(a + \beta)$, a are both prime to β.

Hence $(a + \beta) a$ is prime to β,   [VII. 24]

     and therefore to β^2;   [VII. 25]

i.e. $a^2 + a\beta$ is prime to β^2,

or $a + b$ is prime to c.

Similarly, $a\beta + \beta^2$ is prime to a^2,

or $b + c$ is prime to a.

Lastly, $a + \beta$ being prime to both a and β,

     $(a + \beta)^2$ is prime to $a\beta$,   [VII. 24, 25]

or $a^2 + \beta^2 + 2a\beta$ is prime to $a\beta$:

whence $a^2 + \beta^2$ is prime to $a\beta$.

The latter inference, made in two steps, may be proved by *reductio ad absurdum* as Commandinus proves it.

If $a^2 + \beta^2$ is not prime to $a\beta$, let x measure them ;

therefore x measures $a^2 + \beta^2 + 2a\beta$ as well as $a\beta$;

hence $a^2 + \beta^2 + 2a\beta$ and $a\beta$ are not prime to one another, which is contrary to the hypothesis.

PROPOSITION 16.

If two numbers be prime to one another, the second will not be to any other number as the first is to the second.

For let the two numbers A, B be prime to one another;
I say that B is not to any other number as
A is to B.

For, if possible, as A is to B, so let B be
to C.

Now A, B are prime,

primes are also least, [VII. 21]

and the least numbers measure those which have the same ratio the same number of times, the antecedent the antecedent and the consequent the consequent; [VII. 20]

therefore A measures B as antecedent antecedent.

But it also measures itself;

therefore A measures A, B which are prime to one another: which is absurd.

Therefore B will not be to C, as A is to B.

Q. E. D.

If a, b are prime to one another, they can have no integral third proportional.

If possible, let $a : b = b : x$.

Therefore [VII. 20, 21] a measures b; and a, b have the common measure a, which is contrary to the hypothesis.

PROPOSITION 17.

If there be as many numbers as we please in continued proportion, and the extremes of them be prime to one another, the last will not be to any other number as the first to the second.

For let there be as many numbers as we please, A, B, C, D, in continued proportion,

and let the extremes of them, A, D, be prime to one another;

I say that D is not to any other number as A is to B.

For, if possible, as A is to B, so let D be to E;

therefore, alternately, as A is to D, so is B to E. [VII. 13]

But A, D are prime,

primes are also least,   [VII. 21]

and the least numbers measure those which have the same ratio the same number of times, the antecedent the antecedent and the consequent the consequent.   [VII. 20]

Therefore A measures B.

And, as A is to B, so is B to C.

Therefore B also measures C;

so that A also measures C.

And since, as B is to C, so is C to D,

and B measures C,

therefore C also measures D.

But A measured C;

so that A also measures D.

But it also measures itself;

therefore A measures A, D which are prime to one another: which is impossible.

Therefore D will not be to any other number as A is to B.

Q. E. D.

If a, a_2, a_3, ... a_n be a geometrical progression, and a, a_n are prime to one another, then a, a_2, a_n can have no integral fourth proportional.

For, if possible, let $a : a_2 = a_n : x$.

Therefore $a : a_n = a_2 : x$,

and hence [VII. 20, 21] a measures a_2.

Therefore a_2 measures a_3,   [VII. Def. 20]

and hence a measures a_3, and therefore also ultimately a_n.

Thus a, a_n are both measured by a: which is contrary to the hypothesis.

PROPOSITION 18.

Given two numbers, to investigate whether it is possible to find a third proportional to them.

Let A, B be the given two numbers, and let it be required to investigate whether it is possible to find a third proportional to them.

Now A, B are either prime to one another or not.

And, if they are prime to one another, it has been proved that it is impossible to find a third proportional to them.

[IX. 16]

Next, let A, B not be prime to one another,
and let B by multiplying itself make C.

Then A either measures C or does not measure it.

```
A ——          D ————————
B ——      C ————————————————————————
```

First, let it measure it according to D;
therefore A by multiplying D has made C.

But, further, B has also by multiplying itself made C;
therefore the product of A, D is equal to the square on B.

Therefore, as A is to B, so is B to D;   [VII. 19]
therefore a third proportional number D has been found to
A, B.

Next, let A not measure C;
I say that it is impossible to find a third proportional number
to A, B.

For, if possible, let D, such third proportional, have been
found.

Therefore the product of A, D is equal to the square on B.

But the square on B is C;
therefore the product of A, D is equal to C.

Hence A by multiplying D has made C;
therefore A measures C according to D.

But, by hypothesis, it also does not measure it:
which is absurd.

Therefore it is not possible to find a third proportional
number to A, B when A does not measure C.   Q. E. D.

Given two numbers a, b, to find the condition that they may have an
integral third proportional.

(1) a, b must not be prime to one another.   [IX. 16]
(2) a must measure b^2.

For, if a, b, c be in continued proportion,
$$ac = b^2.$$
Therefore a measures b^2.

Condition (1) is included in condition (2) since, if $b^2 = ma$, a and b cannot
be prime to one another.

The result is of course easily seen if the three terms in continued
proportion be written
$$a, \quad a\frac{b}{a}, \quad a\left(\frac{b}{a}\right)^2.$$

PROPOSITION 19.

Given three numbers, to investigate when it is possible to find a fourth proportional to them.

Let A, B, C be the given three numbers, and let it be required to investigate when it is possible to find a fourth proportional to them.

Now either they are not in continued proportion, and the extremes of them are prime to one another;

or they are in continued proportion, and the extremes of them are not prime to one another;

or they are not in continued proportion, nor are the extremes of them prime to one another;

or they are in continued proportion, and the extremes of them are prime to one another.

If then A, B, C are in continued proportion, and the extremes of them A, C are prime to one another,

it has been proved that it is impossible to find a fourth proportional number to them.   [IX. 17]

†Next, let A, B, C not be in continued proportion, the extremes being again prime to one another;

I say that in this case also it is impossible to find a fourth proportional to them.

For, if possible, let D have been found, so that,

as A is to B, so is C to D,

and let it be contrived that, as B is to C, so is D to E.

Now, since, as A is to B, so is C to D,

and, as B is to C, so is D to E,

therefore, *ex aequali*, as A is to C, so is C to E.   [VII. 14]

But A, C are prime,

primes are also least,   [VII. 21]

and the least numbers measure those which have the same ratio, the antecedent the antecedent and the consequent the consequent.   [VII. 20]

Therefore A measures C as antecedent antecedent.

But it also measures itself;
therefore A measures A, C which are prime to one another:
which is impossible.

Therefore it is not possible to find a fourth proportional
to A, B, C.†

Next, let A, B, C be again in continued proportion,
but let A, C not be prime to one another.

I say that it is possible to find a fourth proportional to
them.

For let B by multiplying C make D;
therefore A either measures D or does not measure it.

First, let it measure it according to E;
therefore A by multiplying E has made D.

But, further, B has also by multiplying C made D;
therefore the product of A, E is equal to the product of
B, C;

therefore, proportionally, as A is to B, so is C to E;   [VII. 19]
therefore E has been found a fourth proportional to A, B, C.

Next, let A not measure D;
I say that it is impossible to find a fourth proportional number
to A, B, C.

For, if possible, let E have been found;
therefore the product of A, E is equal to the product of B, C.
[VII. 19]

But the product of B, C is D;
therefore the product of A, E is also equal to D.

Therefore A by multiplying E has made D;
therefore A measures D according to E,
so that A measures D.

But it also does not measure it:
which is absurd.

Therefore it is not possible to find a fourth proportional
number to A, B, C when A does not measure D.

Next, let A, B, C not be in continued proportion, nor the
extremes prime to one another.

And let B by multiplying C make D.

Similarly then it can be proved that, if A measures D,
it is possible to find a fourth proportional to them, but, if it
does not measure it, impossible.   Q. E. D.

Given three numbers a, b, c, to find the condition that they may have an integral fourth proportional.

The Greek text of part of this proposition is hopelessly corrupt. According to it Euclid takes four cases.

(1) a, b, c not in continued proportion, and a, c prime to one another.

(2) a, b, c in continued proportion, and a, c not prime to one another.

(3) a, b, c not in continued proportion, and a, c not prime to one another.

(4) a, b, c in continued proportion, and a, c prime to one another.

(4) is the case dealt with in IX. 17, where it is shown that on hypothesis (4) a fourth proportional cannot be found.

The text now takes case (1) and asserts that a fourth proportional cannot be found in this case either. We have only to think of 4, 6, 9 in order to see that there is something wrong here. The supposed proof is also wrong. If possible, says the text, let d be a fourth proportional to a, b, c, *and let* e *be taken such that*

$$b : c = d : e.$$

Then, *ex aequali*, $a : c = c : e,$

whence a measures c :   [VII. 20, 21]

which is impossible, since a, c are prime to one another.

But this does not prove that a fourth proportional d cannot be found; it only proves that, if d is a fourth proportional, no integer e can be found to satisfy the equation

$$b : c = d : e.$$

Indeed it is obvious from IX. 16 that in the equation

$$a : c = c : e$$

e cannot be integral.

The cases (2) and (3) are correctly given, the first in full, and the other as a case to be proved "similarly" to it.

These two cases really give all that is necessary.

Let the product bc be taken.

Then, if a measures bc, suppose $bc = ad$;

therefore $a : b = c : d,$

and d is a fourth proportional.

But, if a does *not* measure bc, no fourth proportional can be found. For, if x were a fourth proportional, ax would be equal to bc, and a would measure bc.

The sufficient condition in any case for the possibility of finding a fourth proportional to a, b, c is that a should measure bc.

Theon appears to have corrected the proof by leaving out the incorrect portion which I have included between daggers and the last case (3) dealt with in the last lines. Also, in accordance with this arrangement, he does not distinguish four cases at the beginning but only two. "Either A, B, C are in continued proportion and the extremes of them A, C are prime to one another; or not." Then, instead of introducing case (2) by the words "Next let A, B, C...to find a fourth proportional to them," immediately following the second dagger above, Theon merely says "*But, if not,*" [i.e. if it is not the case that a, b, c are in G.P. and a, c prime to one another] "let B by multiplying C make D," and so on.

August adopts Theon's form of the proof. Heiberg does not feel able to do this, in view of the superiority of the authority for the text as given above (P); he therefore retains the latter without any attempt to emend it.

PROPOSITION 20.

Prime numbers are more than any assigned multitude of prime numbers.

Let A, B, C be the assigned prime numbers;
I say that there are more
prime numbers than A, B, C.

For let the least number
measured by A, B, C be
taken,
and let it be DE;

let the unit DF be added to DE.

Then EF is either prime or not.
First, let it be prime;

then the prime numbers A, B, C, EF have been found which are more than A, B, C.

Next, let EF not be prime;

therefore it is measured by some prime number.   [VII. 31]

Let it be measured by the prime number G.
I say that G is not the same with any of the numbers A, B, C.

For, if possible, let it be so.
Now A, B, C measure DE;

therefore G also will measure DE.

But it also measures EF.

Therefore G, being a number, will measure the remainder, the unit DF:

which is absurd.

Therefore G is not the same with any one of the numbers A, B, C.

And by hypothesis it is prime.

Therefore the prime numbers A, B, C, G have been found which are more than the assigned multitude of A, B, C.

Q. E. D.

We have here the important proposition that *the number of prime numbers is infinite*.

The proof will be seen to be the same as that given in our algebraical text-books.   Let a, b, c, ... k be any prime numbers.

Take the product $abc ... k$ and add unity.

Then $(abc ... k + 1)$ is either a prime number or not a prime number.

(1) If it *is*, we have added another prime number to those given.

(2) If it is *not*, it must be measured by some prime number [VII. 31], say p.

Now p cannot be identical with any of the prime numbers a, b, c, ... k.

For, if it is, it will divide $abc ... k$.

Therefore, since it divides $(abc ... k + 1)$ also, it will measure the difference, or unity:

which is impossible.

Therefore in any case we have obtained one fresh prime number.

And the process can be carried on to any extent.

PROPOSITION 21.

If as many even numbers as we please be added together, the whole is even.

For let as many even numbers as we please, AB, BC, CD, DE, be added together;
I say that the whole AE is even.

For, since each of the numbers AB, BC, CD, DE is even, it has a half part;   [VII. Def. 6]
so that the whole AE also has a half part.

But an even number is that which is divisible into two equal parts;   [*id.*]
therefore AE is even.

Q. E. D.

In this and the following propositions up to IX. 34 inclusive we have a number of theorems about odd, even, "even-times even" and "even-times odd" numbers respectively.   They are all simple and require no explanation in order to enable them to be followed easily.

PROPOSITION 22.

If as many odd numbers as we please be added together, and their multitude be even, the whole will be even.

For let as many odd numbers as we please, AB, BC, CD, DE, even in multitude, be added together;
I say that the whole AE is even.

For, since each of the numbers AB, BC, CD, DE is odd, if an unit be subtracted from each, each of the remainders will be even;   [VII. Def. 7]
so that the sum of them will be even.   [IX. 21]

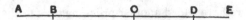

But the multitude of the units is also even.
Therefore the whole AE is also even.   [IX. 21]

Q. E. D.

PROPOSITION 23.

If as many odd numbers as we please be added together, and their multitude be odd, the whole will also be odd.

For let as many odd numbers as we please, AB, BC, CD, the multitude of which is odd,
be added together;
I say that the whole AD is
also odd.

Let the unit DE be subtracted from CD;
therefore the remainder CE is even.   [VII. Def. 7]
But CA is also even;   [IX. 22]
therefore the whole AE is also even.   [IX. 21]
And DE is an unit.
Therefore AD is odd.   [VII. Def. 7]

Q. E. D.

3.   Literally "let there be as many numbers as we please, of which *let* the multitude *be* odd." This form, natural in Greek, is awkward in English.

PROPOSITION 24.

If from an even number an even number be subtracted, the remainder will be even.

For from the even number AB let the even number BC be subtracted:
I say that the remainder CA is even.

For, since AB is even, it has a half part.   [VII. Def. 6]

For the same reason BC also has a half part ;

so that the remainder [CA also has a half part, and] AC is therefore even.

Q. E. D.

PROPOSITION 25.

If from an even number an odd number be subtracted, the remainder will be odd.

For from the even number AB let the odd number BC be subtracted ;

I say that the remainder CA is odd.

For let the unit CD be subtracted from BC ;

therefore DB is even.   [VII. Def. 7]

But AB is also even ;

therefore the remainder AD is also even.   [IX. 24]

And CD is an unit ;

therefore CA is odd.   [VII. Def. 7]

Q. E. D.

PROPOSITION 26.

If from an odd number an odd number be subtracted, the remainder will be even.

For from the odd number AB let the odd number BC be subtracted ;

I say that the remainder CA is even.

For, since AB is odd, let the unit
BD be subtracted ;

therefore the remainder AD is even.   [VII. Def. 7]

For the same reason CD is also even ;   [VII. Def. 7]

so that the remainder CA is also even.   [IX. 24]

Q. E. D.

PROPOSITION 27.

If from an odd number an even number be subtracted, the remainder will be odd.

For from the odd number AB let the even number BC be subtracted;

I say that the remainder CA is odd.

Let the unit AD be subtracted;

therefore DB is even.   [VII. Def. 7]

But BC is also even;

therefore the remainder CD is even.   [IX. 24]

Therefore CA is odd.   [VII. Def. 7]

Q. E. D.

PROPOSITION 28.

If an odd number by multiplying an even number make some number, the product will be even.

For let the odd number A by multiplying the even number B make C;

I say that C is even.

For, since A by multiplying B has made C,

therefore C is made up of as many numbers equal to B as there are units in A.   [VII. Def. 15]

And B is even;

therefore C is made up of even numbers.

But, if as many even numbers as we please be added together, the whole is even.   [IX. 21]

Therefore C is even.

Q. E. D.

PROPOSITION 29.

If an odd number by multiplying an odd number make some number, the product will be odd.

For let the odd number A by multiplying the odd number B make C;

I say that C is odd.

For, since A by multiplying B has made C,

therefore C is made up of as many numbers equal to B as there are units in A.   [VII. Def. 15]

And each of the numbers A, B is odd ;
therefore C is made up of odd numbers the multitude of which is odd.

Thus C is odd.   [IX. 23]

Q. E. D.

PROPOSITION 30.

If an odd number measure an even number, it will also measure the half of it.

For let the odd number A measure the even number B ;
I say that it will also measure the half of it.

For, since A measures B,
let it measure it according to C ;
I say that C is not odd.

For, if possible, let it be so.

Then, since A measures B according to C,
therefore A by multiplying C has made B.

Therefore B is made up of odd numbers the multitude of which is odd.

Therefore B is odd :   [IX. 23]
which is absurd, for by hypothesis it is even.

Therefore C is not odd ;
therefore C is even.

Thus A measures B an even number of times.
For this reason then it also measures the half of it.

Q. E. D.

PROPOSITION 31.

If an odd number be prime to any number, it will also be prime to the double of it.

For let the odd number A be prime to any number B,
and let C be double of B ;
I say that A is prime to C.

For, if they are not prime
to one another, some number
will measure them.

Let a number measure them, and let it be D.
Now A is odd;
therefore D is also odd.

And since D which is odd measures C,
and C is even,
therefore [D] will measure the half of C also.   [IX. 30]

But B is half of C;
therefore D measures B.

But it also measures A;
therefore D measures A, B which are prime to one another:
which is impossible.

Therefore A cannot but be prime to C.
Therefore A, C are prime to one another.

Q. E. D.

PROPOSITION 32.

Each of the numbers which are continually doubled beginning from a dyad is even-times even only.

For let as many numbers as we please, B, C, D, have been continually doubled beginning from the dyad A;
I say that B, C, D are even-times even only.

Now that each of the numbers B, C, D is even-times even is manifest; for it is doubled from a dyad.

I say that it is also even-times even only.

For let an unit be set out.

Since then as many numbers as we please beginning from an unit are in continued proportion,
and the number A after the unit is prime,
therefore D, the greatest of the numbers A, B, C, D, will not be measured by any other number except A, B, C.   [IX. 13]

And each of the numbers A, B, C is even;
therefore D is even-times even only.   [VII. Def. 8]

Similarly we can prove that each of the numbers B, C is even-times even only.

Q. E. D.

See the notes on VII. Deff. 8 to 11 for a discussion of the difficulties shown by Iamblichus to be involved by the Euclidean definitions of "even-times even," "even-times odd" and "odd-times even."

PROPOSITION 33.

If a number have its half odd, it is even-times odd only.

For let the number A have its half odd;
I say that A is even-times odd only.

Now that it is even-times odd is manifest; for the half of it, being odd, measures it an even number of times.   [VII. Def. 9]

I say next that it is also even-times odd only.

For, if A is even-times even also,
it will be measured by an even number according to an even number;   [VII. Def. 8]
so that the half of it will also be measured by an even number though it is odd:
which is absurd.

Therefore A is even-times odd only.   Q. E. D.

PROPOSITION 34.

If a number neither be one of those which are continually doubled from a dyad, nor have its half odd, it is both even-times even and even-times odd.

For let the number A neither be one of those doubled from a dyad, nor have its half odd;
I say that A is both even-times even and even-times odd.

Now that A is even-times even is manifest;
for it has not its half odd.   [VII. Def. 8]

I say next that it is also even-times odd.

For, if we bisect A, then bisect its half, and do this continually, we shall come upon some odd number which will measure A according to an even number.

For, if not, we shall come upon a dyad,
and A will be among those which are doubled from a dyad:
which is contrary to the hypothesis.

Thus A is even-times odd.

But it was also proved even-times even.

Therefore A is both even-times even and even-times odd.

Q. E. D.

PROPOSITION 35.

If as many numbers as we please be in continued proportion, and there be subtracted from the second and the last numbers equal to the first, then, as the excess of the second is to the first, so will the excess of the last be to all those before it.

Let there be as many numbers as we please in continued proportion, A, BC, D, EF, beginning from A as least, and let there be subtracted from BC and EF the numbers BG, FH, each equal to A ; I say that, as GC is to A, so is EH to A, BC, D.

For let FK be made equal to BC, and FL equal to D.

Then, since FK is equal to BC, and of these the part FH is equal to the part BG, therefore the remainder HK is equal to the remainder GC.

And since, as EF is to D, so is D to BC, and BC to A, while D is equal to FL, BC to FK, and A to FH, therefore, as EF is to FL, so is LF to FK, and FK to FH.

Separando, as EL is to LF, so is LK to FK, and KH to FH.   [VII. 11, 13]

Therefore also, as one of the antecedents is to one of the consequents, so are all the antecedents to all the consequents;   [VII. 12]

therefore, as KH is to FH, so are EL, LK, KH to LF, FK, HF.

But KH is equal to CG, FH to A, and LF, FK, HF to D, BC, A ;

therefore, as CG is to A, so is EH to D, BC, A.

Therefore, as the excess of the second is to the first, so is the excess of the last to all those before it.

Q. E. D.

This proposition is perhaps the most interesting in the arithmetical Books, since it gives a method, and a very elegant one, of *summing any series of terms in geometrical progression*.

Let $a_1, a_2, a_3, \ldots a_n, a_{n+1}$ be a series of terms in geometrical progression. Then Euclid's proposition proves that

$$(a_{n+1} - a_1) : (a_1 + a_2 + \ldots + a_n) = (a_2 - a_1) : a_1.$$

For clearness' sake we will on this occasion use the fractional notation of algebra to represent proportions.

Euclid's method then comes to this.

Since
$$\frac{a_{n+1}}{a_n} = \frac{a_n}{a_{n-1}} = \ldots = \frac{a_2}{a_1},$$

we have, *separando*,

$$\frac{a_{n+1} - a_n}{a_n} = \frac{a_n - a_{n-1}}{a_{n-1}} = \ldots = \frac{a_3 - a_2}{a_2} = \frac{a_2 - a_1}{a_1},$$

whence, since, as one of the antecedents is to one of the consequents, so is the sum of all the antecedents to the sum of all the consequents,   [VII. 12]

$$\frac{a_{n+1} - a_1}{a_n + a_{n-1} + \ldots + a_1} = \frac{a_2 - a_1}{a_1},$$

which gives $a_1 + a_2 + \ldots + a_n$, or S_n.

If, to compare the result with that arrived at in algebraical text-books, we write the series in the form

$$a, \quad ar, \quad ar^2, \ldots ar^{n-1} \quad (n \text{ terms}),$$

we have
$$\frac{ar^n - a}{S_n} = \frac{ar - a}{a},$$

or
$$S_n = \frac{a(r^n - 1)}{r - 1}.$$

PROPOSITION 36.

If as many numbers as we please beginning from an unit be set out continuously in double proportion, until the sum of all becomes prime, and if the sum multiplied into the last make some number, the product will be perfect.

For let as many numbers as we please, A, B, C, D, beginning from an unit be set out in double proportion, until the sum of all becomes prime,

let E be equal to the sum, and let E by multiplying D make FG;

I say that FG is perfect.

For, however many A, B, C, D are in multitude, let so many E, HK, L, M be taken in double proportion beginning from E;

therefore, *ex aequali*, as A is to D, so is E to M.   [VII. 14]

Therefore the product of E, D is equal to the product of A, M.   [VII. 19]

And the product of E, D is FG;
therefore the product of A, M is also FG.

Therefore A by multiplying M has made FG;
therefore M measures FG according to the units in A.

And A is a dyad;
therefore FG is double of M.

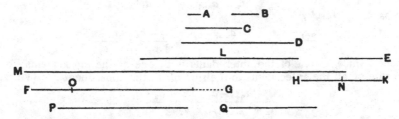

But M, L, HK, E are continuously double of each other;
therefore E, HK, L, M, FG are continuously proportional in double proportion.

Now let there be subtracted from the second HK and the last FG the numbers HN, FO, each equal to the first E;
therefore, as the excess of the second is to the first, so is the excess of the last to all those before it.   [IX. 35]

Therefore, as NK is to E, so is OG to M, L, KH, E.

And NK is equal to E;
therefore OG is also equal to M, L, HK, E.

But FO is also equal to E,
and E is equal to A, B, C, D and the unit.

Therefore the whole FG is equal to E, HK, L, M and A, B, C, D and the unit;
and it is measured by them.

I say also that FG will not be measured by any other number except A, B, C, D, E, HK, L, M and the unit.

For, if possible, let some number P measure FG,
and let P not be the same with any of the numbers A, B, C, D, E, HK, L, M.

And, as many times as P measures FG, so many units let there be in Q;
therefore Q by multiplying P has made FG.

But, further, E has also by multiplying D made FG;
therefore, as E is to Q, so is P to D.   [VII. 19]

And, since A, B, C, D are continuously proportional
beginning from an unit,
therefore D will not be measured by any other number except
A, B, C.   [IX. 13]

And, by hypothesis, P is not the same with any of the
numbers A, B, C;
therefore P will not measure D.

But, as P is to D, so is E to Q;
therefore neither does E measure Q.   [VII. Def. 20]

And E is prime;
and any prime number is prime to any number which it does
not measure.   [VII. 29]

Therefore E, Q are prime to one another.

But primes are also least,   [VII. 21]
and the least numbers measure those which have the same
ratio the same number of times, the antecedent the antecedent
and the consequent the consequent;   [VII. 20]
and, as E is to Q, so is P to D;
therefore E measures P the same number of times that Q
measures D.

But D is not measured by any other number except
A, B, C;
therefore Q is the same with one of the numbers A, B, C.

Let it be the same with B.

And, however many B, C, D are in multitude, let so many
E, HK, L be taken beginning from E.

Now E, HK, L are in the same ratio with B, C, D;
therefore, *ex aequali*, as B is to D, so is E to L.   [VII. 14]

Therefore the product of B, L is equal to the product of
D, E.   [VII. 19]

But the product of D, E is equal to the product of Q, P;
therefore the product of Q, P is also equal to the product of
B, L.

Therefore, as Q is to B, so is L to P.   [VII. 19]

And Q is the same with B;
therefore L is also the same with P:

which is impossible, for by hypothesis P is not the same with any of the numbers set out.

Therefore no number will measure FG except A, B, C, D, E, HK, L, M and the unit.

And FG was proved equal to A, B, C, D, E, HK, L, M and the unit;

and a perfect number is that which is equal to its own parts;

[VII. Def. 22]

therefore FG is perfect.

Q. E. D.

If the sum of any number of terms of the series

$$1, 2, 2^2, \ldots 2^{n-1}$$

be prime, and the said sum be multiplied by the last term, the product will be a "perfect" number, i.e. equal to the sum of all its factors.

Let $1 + 2 + 2^2 + \ldots + 2^{n-1} (= S_n)$ be prime;

then shall $S_n \cdot 2^{n-1}$ be "perfect."

Take $(n-1)$ terms of the series

$$S_n, 2S_n, 2^2 S_n, \ldots 2^{n-2} S_n.$$

These are then terms proportional to the terms

$$2, 2^2, 2^3, \ldots 2^{n-1}.$$

Therefore, *ex aequali*,

$$2 : 2^{n-1} = S_n : 2^{n-2} S_n,$$ [VII. 14]

or

$$2 \cdot 2^{n-2} S_n = 2^{n-1} \cdot S_n.$$ [VII. 19]

(This is of course obvious algebraically, but Euclid's notation requires him to prove it.)

Now, by IX. 35, we can sum the series $S_n + 2S_n + \ldots + 2^{n-2} S_n$,

and $\quad (2S_n - S_n) : S_n = (2^{n-1} S_n - S_n) : (S_n + 2S_n + \ldots + 2^{n-2} S_n).$

Therefore $\quad S_n + 2S_n + 2^2 S_n + \ldots + 2^{n-2} S_n = 2^{n-1} S_n - S_n,$

or $\quad 2^{n-1} S_n = S_n + 2S_n + 2^2 S_n + \ldots + 2^{n-2} S_n + S_n$

$$= S_n + 2S_n + \ldots + 2^{n-2} S_n + (1 + 2 + 2^2 + \ldots + 2^{n-1}),$$

and $2^{n-1} S_n$ is measured by every term of the right hand expression.

It is now necessary to prove that $2^{n-1} S_n$ cannot have any factor except those terms.

Suppose, if possible, that it has a factor x different from all of them,

and let $\quad 2^{n-1} S_n = x \cdot m.$

Therefore $\quad S_n : m = x : 2^{n-1}.$ [VII. 19]

Now 2^{n-1} can only be measured by the preceding terms of the series

$1, 2, 2^2, \ldots 2^{n-1}$, [IX. 13]

and x is different from all of these;

therefore x does not measure 2^{n-1},

so that S_n does not measure m. [VII. Def. 20]

And S_n is prime; therefore it is prime to m. [VII. 29]

It follows [VII. 20, 21] that

$$m \text{ measures } 2^{n-1}.$$

Suppose that $$m = 2^r.$$
Now, *ex aequali*, $$2^r : 2^{n-1} = S_n : 2^{n-r-1} S_n.$$
Therefore $$2^r . 2^{n-r-1} S_n = 2^{n-1} S_n \qquad \text{[VII. 19]}$$
$$= x . m, \text{ from above.}$$

And $m = 2^r$;

therefore $x = 2^{n-r-1} S_n$, one of the terms of the series $S_n, 2S_n, 2^2 S_n, \dots 2^{n-2} S_n$: which contradicts the hypothesis.

There $2^{n-1} S_n$ has no factors except

$$S_n, 2S_n, 2^2 S_n, \dots 2^{n-2} S_n, 1, 2, 2^2, \dots 2^{n-1}.$$

Theon of Smyrna and Nicomachus both define a "perfect" number and give the law of its formation. Nicomachus gives four perfect numbers and no more, namely 6, 28, 496, 8128. He says they are formed in "ordered" fashion, there being one among the units (i.e. less than 10), one among the tens (less than 100), one among the hundreds (less than 1000) and one among the thousands (less than 10000); he adds that they terminate in 6 or 8 alternately. They do all terminate in 6 or 8, as can easily be proved by means of the formula $(2^n - 1) 2^{n-1}$ (cf. Loria, *Le scienze esatte nell' antica Grecia*, pp. 840—1), but not alternately, for the fifth and sixth perfect numbers both end in 6, and the seventh and eighth both end in 8. Iamblichus adds a tentative suggestion that perhaps there may be, in like manner, one perfect number among the "first myriads" (less than 10000^2), one among the "second myriads" (less than 10000^3), and so on. This is, as we shall see, incorrect.

It is natural that the subject of perfect numbers should, ever since Euclid's time, have had a fascination for mathematicians. Fermat (1601—1655), in a letter to Mersenne (*Œuvres de Fermat*, ed. Tannery and Henry, Vol. II., 1894, pp. 197—9), enunciated three propositions which much facilitate the investigation whether a given number of the form $2^n - 1$ is prime or not. If we write in one line the exponents 1, 2, 3, 4, etc. of the successive powers of 2 and underneath them respectively the numbers representing the corresponding powers of 2 diminished by 1, thus,

$$1 \quad 2 \quad 3 \quad 4 \quad 5 \quad 6 \quad 7 \quad 8 \quad 9 \quad 10 \quad 11 \dots n$$
$$1 \quad 3 \quad 7 \quad 15 \quad 31 \quad 63 \quad 127 \quad 255 \quad 511 \quad 1023 \quad 2047 \dots 2^n - 1,$$

the following relations are found to subsist between the numbers in the first line and those directly below them in the second line.

1. If the exponent is not a prime number, the corresponding number is not a prime number either (since $a^{pq} - 1$ is always divisible by $a^p - 1$ as well as by $a^q - 1$).

2. If the exponent is a prime number, the corresponding number diminished by 1 is divisible by twice the exponent. $[(2^n - 2)/2n = (2^{n-1} - 1)/n$; so that this is a special case of "Fermat's theorem" that, if p is a prime number and a is prime to p, then a^{p-1} is divisible by p.]

3. If the exponent n is a prime number, the corresponding number is only divisible by numbers of the form $(2mn + 1)$. If therefore the corresponding number in the second line has no factors of this form, it has no integral factor.

The first and third of these propositions are those which are specially useful for the purpose in question. As usual, Fermat does not give his proofs but merely adds: "Voilà trois fort belles propositions que j'ay trouvées et prouvées non sans peine. Je les puis appeller les fondements de l'invention des nombres parfaits."

I append a few details of discoveries of further perfect numbers after the first four. The next are as follows :

fifth, $2^{12}(2^{13}-1) = 33\ 550\ 336$

sixth, $2^{16}(2^{17}-1) = 8\ 589\ 869\ 056$

seventh, $2^{18}(2^{19}-1) = 137\ 438\ 691\ 328$

eighth, $2^{30}(2^{31}-1) = 2\ 305\ 843\ 008\ 139\ 952\ 128$

ninth, $2^{60}(2^{61}-1) = 2\ 658\ 455\ 991\ 569\ 831\ 744\ 654\ 692\ 615\ 953\ 842\ 176$

tenth, $2^{88}(2^{89}-1)$.

It has further been proved that $2^{107}-1$ is prime, and so is $2^{127}-1$. Hence $2^{106}(2^{107}-1)$ and $2^{126}(2^{127}-1)$ are two more perfect numbers.

The fifth perfect number may have been known to Iamblichus, though he does not give it; it was however known, with all its factors, in the fifteenth century, as appears from a tract written in German which was discovered by Curtze (Cod. lat. Monac. 14908). The first eight perfect numbers were calculated by Jean Prestet (d. 1670). Fermat had stated, and Euler proved, that $2^{31}-1$ is prime. The ninth perfect number was found by P. Seelhoff (*Zeitschrift für Math. u. Physik*, XXXI., 1886, pp. 174—8) and verified by E. Lucas (*Mathésis*, VII., 1887, pp. 45—6). The tenth was discovered by R. E. Powers (see *Bulletin of the American Mathematical Society*, XVIII., 1912, p. 162). $2^{107}-1$ was proved to be prime by E. Fauquembergue and R. E. Powers (1914), while Fauquembergue proved that $2^{127}-1$ is prime .

There have been attempts, so far unsuccessful, to solve the question whether there exist other "perfect numbers" than those of Euclid, and, in particular, perfect numbers which are *odd*. (Cf. several notes by Sylvester in *Comptes rendus*, CVI., 1888 ; Catalan, "Mélanges mathématiques" in *Mém. de la Soc. de Liége*, 2e Série, xv., 1888, pp. 205—7 ; C. Servais in *Mathésis*, VII., pp. 228—30 and VIII., pp. 92—93, 135 ; E. Cesàro in *Mathésis*, VII., pp. 245—6 ; E. Lucas in *Mathésis*, x., pp. 74—6).

For the detailed history of the whole subject see L. E. Dickson, *History of the Theory of Numbers*, Vol. I., 1919, pp. iii—iv, 3—33.

INDEX OF GREEK WORDS AND FORMS.

ἄκρος, extreme (of numbers in a series) 328, 367: ἄκρον καὶ μέσον λόγον τετμῆσθαι, "to be cut in extreme and mean ratio" 189

ἄλογος, irrational 117-8

ἀναλογία, proportion: definitions of, interpolated 119

ἀνάλογον = ἀνὰ λόγον, proportional or in proportion: used as indeclinable adj. and as adv. 129, 165: μέση ἀνάλογον, mean proportional (of straight line) 129, similarly μέσος ἀνάλογον of numbers 295, 363 etc.: τρίτη (τρίτος) ἀνάλογον, third proportional 214, 407-8: τετάρτη (τέταρτος) ἀνάλογον, fourth proportional 215, 409: ἑξῆς ἀνάλογον in continued proportion 346

ἀνάπαλιν (λόγος), inverse (ratio), inversely 134

ἀναστρέψαντι, convertendo 135

ἀναστροφὴ λόγου, conversion of a ratio 135

ἀνισάκις ἀνισάκις ἴσος, unequal by unequal by equal (of solid numbers) = scalene, σφηνίσκος, σφηκίσκος or βωμίσκος 290

ἀνομοίως τεταγμένων τῶν λόγων (of perturbed proportion) in Archimedes 136

ἀνταναίρεσις, ἡ αὐτή, definition of same ratio in Aristotle (ἀνθυφαίρεσις Alexander) 120: terms explained 121

ἀντιπεπονθότα σχήματα, reciprocal (= reciprocally related) figures, interpolated def. of, 189

ἅπτεσθαι, to meet, occasionally to touch (instead of ἐφάπτεσθαι) 2: also = to pass through, to lie on 79

ἀριθμός, number, definitions of, 280

ἀρτιάκις ἀρτιοδύναμον (Nicomachus) 282

ἀρτιάκις ἄρτιος, even-times even 281-2

ἀρτιάκις περισσός, even-times odd 282-4

ἀρτιοπέριττος, even-odd (Nicomachus etc.) 282

ἄρτιος (ἀριθμός), even (number) 281

ἀσύνθετος, (prime and) incomposite (of numbers) 284

βεβηκέναι, to stand (of angle standing on circumference) 4

βωμίσκος, altar-shaped (of "scalene" solid numbers) 290

γεγονέτω (in constructions), "let it be made" 248

γεγονὸς ἂν εἴη τὸ ἐπιταχθέν, "what was enjoined will have been done" 80, 261

γενόμενος, ὁ ἐξ αὐτῶν, "their product" 316, 326 etc.: ὁ ἐκ τοῦ ἑνὸς γενόμενος = "the square of the one" 327

γνώμων, gnomon: Democritus περὶ διαφορῆς γνώμονος (γνώμης or γωνίης?) ἢ περὶ ψαύσιος κύκλου καὶ σφαίρης 40: (of numbers) 289

γραμμικός, linear (of numbers in one dimension) 287: (of prime numbers) 285

γράφεσθαι, "to be proved" (Aristotle) 120

δεύτερος, secondary (of numbers): in Nicomachus and Iamblichus a subdivision of odd 286, 287

δεχόμενον, "admitting" (of segment of circle admitting or containing an angle) 5

διαιρεῖσθαι (used of "separation" of ratios): διαιρεθέντα, separando, opp. to συγκείμενα, componendo 168

διαίρεσις λόγου, separation, literally division, of ratio 135

διεζευγμένη (ἀναλογία), disjoined, = discrete (proportion) 293

διελόντι, separando, literally dividendo (of proportions) 135

διῃρημένη (ἀναλογία), discrete (proportion), i.e. in four terms, as distinct from continuous (συνεχής, συνημμένη) in three terms 131, 293

διήχθω (διάγειν), "let it be drawn through" or "across" 7

δι' ἴσου, ex aequali (of ratios) 136: δι' ἴσου ἐν τεταραγμένῃ ἀναλογίᾳ, "ex aequali in perturbed proportion" 136

δικόλουρος, twice-truncated (of pyramidal numbers) 291

διπλάσιος λόγος, double ratio: διπλασίων λόγος, duplicate ratio, contrasted with, 133

δύναμις, power: = actual value of a submultiple in units (Nicomachus) 282: = side of number not a complete square (i.e. root or surd) in Plato 288, 290: = square in Plato 294-5

εἶδος, figure 234: = form 254

ἕκαστος, each: curious use of, 79

ἔλλειμμα, defect (in application of areas) 262

ἐλλείπειν, "fall short" (in application of areas) 262

ἐμπίπτειν, *fall in* (= be interpolated) 358

ἕνα πλείω, "several *ones*" (def. of number) 280

ἐναλλάξ (λόγος), alternate (ratio): alternately, *alternando* 134

ἐναρμόζειν, to *fit in* (active), IV. Def. 7 and IV. I, 79, 80, 81

ἐντός, within (of internal contact of circles) 13

ἑξῆς ἀνάλογον, in continued proportion (of terms in geometrical progression) 346

ἐπιμόριος λόγος, *superparticularis ratio*, = the ratio (*n* + 1) : *n*, 295

ἐπίπεδος (ἀριθμός), plane (number) 287-8

ἑπόμενα, consequents (= "following" terms) in a proportion 134, 238

ἑτερομήκης, oblong (of numbers): in Plato = προμήκης, which however is distinguished from ἑτερομήκης by Nicomachus etc. 289-90, 293

εὐθυγραμμικός, rectilinear (term for prime numbers) 285

εὐθυμετρικός, euthymetric (of primes) 285

ἡγούμενα, antecedents ("leading" terms) in a proportion 134

ᾗπερ, than: construction after διπλασίων etc. 133

ἰδιομήκης, of square number (Iamblichus) 293

ἰσάκις ἰσάκις ἴσος, equal multiplied by equal and again by equal (of a cube number) 290, 291

ἰσάκις ἴσος, equal multiplied by equal (of a square number) 291

ἰσάκις ἴσος ἐλαττονάκις (μειζονάκις), species of solid numbers, = πλινθίς (δοκίς or στηλίς) 291

καλείσθω, "let it be called," indicating originality of a definition 129

καταμετρεῖν, measure 115: without remainder, completely (πληρούντως) 280

κατασκευάζω, construct: τῶν αὐτῶν κατασκευασθέντων, "with the same construction" 11

κατατομή κανόνος, *Sectio canonis* of Euclid 295

κέντρον, centre: ἡ ἐκ τοῦ κ. = radius 2

κερατοειδὴς γωνία, *hornlike* angle 4, 39, 40

κλᾶν, to *break off*, *inflect*: κεκλάσθω δὴ πάλιν 47: κεκλάσθαι, def. of, alluded to by Aristotle 47

κόλουρος, *truncated* (of pyramidal number *minus* vertex) 291

κυκλικός, *cyclic*, a particular species of square number 291

λόγος, ratio: meaning 117: definition of, 116-9: original meaning (of something *expressed*) accounts for use of ἄλογος, *having no ratio*, irrational 117

μεμονῶσθαι, to be *isolated*, of μονάς (Theon of Smyrna) 279

μέρος, part: two meanings 115: generally = submultiple 280: μέρη, *parts* (= proper fraction) 115, 280

μέση ἀνάλογον (εὐθεῖα), μέσος ἀνάλογον (ἀριθμός), mean proportional (straight line or number) 129, 295, 363 etc.

μὴ γάρ, "suppose it is not" 7

μῆκος, length (of number in one dimension): = side of complete square in Plato 288

μονάς, unit, monad: supposed etymological connexion with μόνος, solitary, μονή, rest 279

ὅμοιος, similar: (of rectilineal figures) 188: (of plane and solid numbers) 293

ὁμοιότης λόγων, "similarity of ratios" (interpolated def. of proportion) 119

ὁμόλογος, homologous, corresponding 134: exceptionally "in the same ratio with" 238

ὅρος, *term*, in a proportion 131

παραβάλλειν ἀπό, used, exceptionally, instead of παραβάλλειν παρά or ἀναγράφειν ἀπό 262

παραλλάττω, "fall sideways" or "awry" 54

πεντάγραμμον 99

περαίνουσα ποσότης, "limiting quantity" (Thymaridas' definition of unit) 279

περισσάκις ἄρτιος, *odd-times even* 282-4

περισσάκις περισσός, *odd-times odd* 284

περισσάρτιος, *odd-even* (Nicomachus etc.) 283

περισσός, odd 281

πηλίκος, how great: refers to *continuous* (geometrical) magnitude as ποσός to *discrete* (multitude) 116-7

πηλικότης, used in V. Def. 3, and spurious Def. 5 of VI.: = *size* (not *quantuplicity* as it is translated by De Morgan) 116-7, 189-90: supposed multiplication of πηλικότητες (VI. Def. 5) 132: distinction between πηλικότης and μέγεθος 117

πλάτος, breadth: (of numbers) 288

πλευρά, side: (of factors of "plane" and "solid," numbers) 288

πλῆθος ὡρισμένον or πεπερασμένον, defined or finite multitude (definition of number) 280: ἐκ μονάδων συγκείμενον πλῆθος (Euclid's def.) 280

πολλαπλασιάζειν, multiply: defined 287

πολλαπλασιασμός, multiplication: καθ' ὁποιονοῦν πολλαπλασιασμόν "(arising) from any multiplication whatever" 120

πολλαπλάσιος, multiple: ἰσάκις πολλαπλάσια, equimultiples 120 etc.

πολύπλευρον, multilateral: excludes τετράπλευρον, quadrilateral 239

πορίσασθαι, to *find* 248

ποσάκις ποσάκις ποσοί, "so many times so many times so many" (of solid numbers, in Aristotle) 286, 290

ποσάκις ποσοί, "so many times so many" (of plane numbers, in Aristotle) 286

ποσόν, *quantity*, in Aristotle 115: refers to multitude as πηλίκον to magnitude 116-7

προμήκης, *oblong* (of numbers): in Plato=
ἑτερομήκης, but distinguished by Nicomachus etc. 289–90, 293
προσαναγράψαι, to *draw on to*: (of a circle) to
complete, when segment is given 56
προσευρεῖν, to find in addition (of finding
third and fourth proportionals) 214
πρῶτος, prime 284–5
πρῶτοι πρὸς ἀλλήλους, (numbers) prime to
one another 285–6

ῥητός, rational (literally " expressible ") 117

συνεχής, continuous: συνεχὴς ἀναλογία, "continuous proportion" (in three terms) 131
συνημμένη ἀναλογία, *connected* (i.e. continuous) proportion 131, 293: συνημμένος,
of *compound* ratio in Archimedes 133
συνθέντι, *componendo* 134–5
σύνθεσις λόγου, composition of a ratio, distinct from *compounding* of ratios 134–5
σύνθετος, composite (of numbers): in Nicomachus and Iamblichus a subdivision of
odd 286
συνίστασθαι, *construct*: οὐ συσταθήσεται 53
συντίθημι, σύγκειμαι (of ratios) 135, 189–90:
συγκείμενα and διαιρεθέντα (*componendo* and
separando) used relatively to one another
168, 170
σύστημα μονάδων, "collection of units" (def.
of number) 280
συστηματικός, collective 279
σφαιρικός, *spherical* (of a particular species of
cube number) 291
σφηκίσκος, or σφηνίσκος, of solid number
with all three sides unequal (= " scalene ")
290

σχέσις, "relation": ποιὰ σχέσις, "a sort of
relation" (in def. of ratio) 116–7

ταὐτομήκης, of square number (Nicom.) 293
ταὐτότης λόγων, "sameness of ratios" 119
τέλειος, *perfect* (of a class of numbers) 293–4
τεταγμένη (ἀναλογία), "ordered (proportion)"
137
τεταραγμένη ἀναλογία, *perturbed proportion*
136
τετράπλευρον, quadrilateral, not a "polygon"
239
τμῆμα (κύκλου), segment (of circle): τμήματος
γωνία, angle *of* a segment 4: ἐν τμήματι
γωνία, angle *in* a segment 4
τομεύς (κύκλου), sector (of circle): σκυτοτομικὸς τομεύς, "shoemaker's knife" 5
τομοειδής (of figure), *sector-like* 5
τοσαυταπλάσιον, "the same multiple" 146
τρίγωνον: τὸ τριπλοῦν, τὸ δι' ἀλλήλων, triple,
interwoven triangle, = pentagram 99
τριπλάσιος, triple, τριπλασίων, triplicate (of
ratios) 133
τυγχάνειν, happen: ἄλλα, ἃ ἔτυχεν, ἰσάκις
πολλαπλάσια, "other, chance, equimultiples" 143–4: τυχοῦσα γωνία, "*any* angle"
212

ὑπερτελής or ὑπερτέλειος, "over-perfect" (of
a class of numbers) 293–4
ὑποδιπλάσιος, *sub-duplicate*, = half (Nicomachus) 280
ὑποπολλαπλάσιος, submultiple (Nicomachus)
280
ὕψος, height 189

χωρίον, area 254

ENGLISH INDEX.

and even by one another unscientific
(Aristotle) 281 : Nicom. divides even into
three classes (1) *even-times even* and (2) *even-times odd* as extremes, and (3) *odd-times even* as intermediate 282-3

Even-times even: Euclid's use differs from
use by Nicomachus, Theon of Smyrna and
Iamblichus 281-2

Even-times odd in Euclid different from *even-odd* of Nicomachus and the rest 282-4

Ex aequali, of ratios, 136: *ex aequali* propositions (v. 20, 22), and *ex aequali* "in
perturbed proportion" (v. 21, 23) 176-8

Faifofer 126
Fauquembergue, E. 426
Fermat, 425, 426
Fourth proportional: assumption of existence
of, in v. 18, and alternative methods for
avoiding (Saccheri, De Morgan, Simson,
Smith and Bryant) 170-4: Clavius made
the assumption an axiom 170: sketch of
proof of assumption by De Morgan 171:
condition for existence of number which
is fourth proportional to three numbers
409-11

Galileo Galilei: on *angle of contact* 42
Geometric means 357 sqq.: one mean between
square numbers 294, 363, or between
similar plane numbers 371-2: two means
between cube numbers 294, 364-5, or
between similar solid numbers 373-5
Geometrical progression 346 sqq.: summation
of *n* terms of (IX. 35) 420-1
Gherard of Cremona 47
Gnomon (of numbers) 289
Golden section (section in extreme and mean
ratio), discovered by Pythagoreans 99:
theory carried further by Plato and Eudoxus 99
Greater ratio: Euclid's criterion not the only
·one 130: arguments from greater to less
ratios etc. unsafe unless they go back to
original definitions (Simson on v. 10) 156-7:
test for, cannot coexist with test for equal
or less ratio 130-1
Greatest common measure: Euclid's method
of finding corresponds exactly to ours 118,
299: Nicomachus gives the same method 300
Gregory, D. 116, 143

Häbler, Th. 294 *n*.
Hankel, H. 116, 117
Hauber, C. F. 244
Heiberg, J. L. *passim*
Henrici and Treutlein 30
Heron of Alexandria: Eucl. III. 12 interpolated from, 28: extends III. 20; 21 to angles
in segments less than semicircles 47-8: does
not recognise angles equal to or greater than
two right angles 47-8: proof of formula for
area of triangle, $\Delta = \sqrt{s(s-a)(s-b)(s-c)}$
87-8: 5, 16-17, 24, 28, 34, 36, 44, 116,
189, 302, 320, 383, 395

Hippasus 97
Hippocrates of Chios 133
Hornlike angle (κερατοειδὴς γωνία) 4, 39, 40:
hornlike angle and angle *of* semicircle, controversies on, 39-42: Proclus on, 39-40:
Democritus may have written on *hornlike*
angle 40: Campanus ("not angles in same
sense") 41: Cardano (*quantities* of different
orders or kinds): Peletier (*hornlike* angle
no angle, no quantity, nothing; angles of
all semicircles *right angles* and equal) 41:
Clavius 42: Vieta and Galileo ("angle of
contact no angle") 42: Wallis (angle of
contact not *inclination* at all but *degree of
curvature*) 42
Hultsch, F. 133, 190

Iamblichus 97, 116, 279, 280, 281, 283, 284,
285, 286, 287, 288, 289, 290, 291, 292, 293,
419, 425, 426
Icosahedron 98
Incommensurables: method of testing incommensurability (process of finding G.C.M.)
118: means of expression consist in power
of approximation without limit (De Morgan)
119: approximations to $\sqrt{2}$ (by means of
side- and *diagonal*-numbers) 119, to $\sqrt{3}$
and to π, 119: to $\sqrt{4500}$ by means of
sexagesimal fractions 119
Incomposite (of number) = prime 284
Ingrami, G. 30, 126
Inverse (ratio), inversely (ἀνάπαλιν) 134: inversion is subject of v. 4, Por. (Theon)
144, and of v. 7, Por. 149, but is not
properly put in either place 149: Simson's
Prop. B on, directly deducible from v.
Def. 5, 144
Isosceles triangle of IV. 10: construction of,
by Pythagoreans 97-9

Jacobi, C. F. A. 188

Lachlan, R. 226, 227, 245-6, 247, 256, 272
Lardner, D. 58, 259, 271
Least common multiple 336-41
Legendre 30: proves VI. 1 and similar pro-
positions in two parts (1) for commen-
surables, (2) for incommensurables 193-4
Lemma assumed in VI. 22, 242-3: alternative
propositions on duplicate ratios and ratios
of which they are duplicate (De Morgan
and others) 242-7
Length, μῆκος (of numbers in one dimension)
287: Plato restricts term to side of inte-
gral square number 287
Leotaud, Vincent 42
Linear (of numbers) = (1) in one dimension
287, (2) prime 285
Logical inferences, not made by Euclid 22, 29
Loria, G. 425
Lucas, E. 426
Lucian 99

Means: three kinds, arithmetic, geometric
and harmonic 292-3: geometric mean is

"proportion *par excellence*" (κυρίως) 292–3: one geometric mean between two square numbers, two between two cube numbers (Plato) 294, 363–5: one geometric mean between similar plane numbers, two between similar solid numbers 371–5: no numerical geometric mean between *n* and *n*+1 (Archytas and Euclid) 295

Moderatus, a Pythagorean 280

Multiplication, definition of 287

an-Nairīzī 5, 16, 28, 34, 36, 44, 47, 302, 320, 383

Naṣīraddīn aṭ-Ṭūsī 28

Nesselmann, G. H. F. 287, 293

Nicomachus 116, 119, 131, 279, 280, 281, 282, 283, 284, 285, 286, 287, 288, 289, 290, 291, 292, 293, 294, 300, 363, 425

Nixon, R. C. J. 16

Number: defined by Thales, Eudoxus, Moderatus, Aristotle, Euclid 280: Nicomachus and Iamblichus on, 280: represented by lines 288, and by points or dots 288–9

Oblong (of number): in Plato either προμήκης or ἑτερομήκης 288: but these terms denote two distinct divisions of plane numbers in Nicomachus, Theon of Smyrna and Iamblichus 289–90

Octahedron 98

Odd (number): defs. of in Nicomachus 281: Pythagorean definition 281: def. of odd and even by one another unscientific (Aristotle) 281: Nicom. and Iambl. distinguish three classes of odd numbers (1) prime and incomposite, (2) secondary and composite, as extremes, (3) secondary and composite in itself but prime and incomposite to one another, which is intermediate 287

Odd-times even (number): definition in Eucl. spurious 283–4, and differs from definitions by Nicomachus etc. *ibid.*

Odd-times odd (number): defined in Eucl. but not in Nicom. and Iambl. 284: Theon of Smyrna applies term to prime numbers 284

Oenopides of Chios 111

"Ordered" proportion (τεταγμένη ἀναλογία), interpolated definition of, 137

Pappus: lemma on Apollonius' *Plane νεύσεις* 64–5: problem from same work 81: assumes case of VI. 3 where external angle bisected (Simson's VI. Prop. A) 197: theorem from Apollonius' *Plane Loci* 198: theorem that ratio compounded of ratios of sides is equal to ratio of rectangles contained by sides 250: 4, 27, 29, 67, 79, 81, 113, 133, 211, 250, 251, 292

"Parallelepipedal" (solid) numbers: two of the three factors differ by unity (Nicomachus) 290

Peletarius (Peletier): on *angle of contact* and

angle *of* semicircle 41: 47, 56, 84, 146, 190

Pentagon: decomposition of regular pentagon into 30 elementary triangles 98: relation to pentagram 99

Pentagonal numbers 289

"Perfect" (of a class of numbers) 293–4, 421–6: Pythagoreans applied term to 10, 294: 3 also called "perfect" 294

Perturbed proportion (τεταραγμένη ἀναλογία) 136, 176–7

Pfleiderer, C. F. 2

Philoponus 234, 282

Plane numbers, product of two factors ("sides" or "length" and "breadth") 287–8: in Plato either square or oblong 287–8: similar plane numbers 293: one mean proportional between similar plane numbers 371–2

Plato: construction of regular solids from triangles 97–8: on *golden section* 99: 7/5 as approximation to √2, 119: on square and oblong numbers 288, 293: on δυνάμεις (square roots or *surds*) 288, 290: theorem that between square numbers one mean suffices, between cube numbers two means necessary 294, 364

Playfair, John 2

Plutarch 98, 254

Porism (corollary) to proposition precedes "Q.E.D." or "Q.E.F." 8, 64: Porism to IV. 15 mentioned by Proclus 109: Porism to VI. 19, 234

Polygonal numbers 289

Powers, R. E. 426

Prestet, Jean 426

Prime (number): definitions of, 284–5: Aristotle on two senses of "prime" 285: 2 admitted as prime by Eucl. and Aristotle, but excluded by Nicomachus, Theon of Smyrna and Iamblichus, who make prime a subdivision of *odd* 284–5: "prime and incomposite (ἀσύνθετος)" 284: different names for prime, "odd-times odd" (Theon), "linear" (Theon), "rectilinear" (Thymaridas), "euthymetric" (Iamblichus) 285: prime absolutely or in themselves as distinct from prime to one another (Theon) 285: definitions of "prime to one another" 285–6

Proclus: on absence of formal divisions of proposition in certain cases, e.g. IV. 10, 100: on use of "quindecagon" for astronomy 111: 4, 39, 40, 193, 247, 269

Proportion: complete theory applicable to incommensurables as well as commensurables is due to Eudoxus 112: old (Pythagorean) theory practically represented by arithmetical theory of Eucl. VII. 113: in giving older theory as well Euclid simply followed tradition 113: Aristotle on general proof (new in his time) of theorem (*alternando*) in proportion 113: x. 5 as connecting two theories 113: De Morgan on extension of meaning of *ratio* to cover incommensurables 118: power of *expressing* incommensurable ratio is power of approximation without limit 119: interpolated definitions of proportion as "sameness" or "similarity of ratios" 119: definition in v. Def. 5 substituted for that of VII. Def. 20 because latter found inadequate, not *vice versa* 121: De Morgan's defence of v. Def. 5 as necessary and sufficient 122–4: v. Def. 5 corresponds to Weierstrass' conception of number in general and to Dedekind's theory of irrationals 124–6: alternatives for v. Def. 5 by a geometer-friend of Saccheri, by Faifofer, Ingrami, Veronese, Enriques and Amaldi 126: proportionals of VII. Def. 20 (numbers) a particular case of those of v. Def. 5 (Simson's Props. C, D and notes) 126–9: proportion in three terms (Aristotle makes it four) the "least" 131: "continuous" proportion (συνεχὴς or συνημμένη ἀναλογία, in Euclid ἑξῆς ἀνάλογον) 131, 293: three "proportions" 292, but proportion *par excellence* or primary is continuous or geometric 292–3: "discrete" or "disjoined" (διῃρημένη, διεζευγμένη) 131, 293: "ordered" proportion (τεταγμένη), interpolated definition of, 137: "perturbed" proportion (τεταραγμένη) 136, 176–7: extensive use of proportions in Greek geometry 187: proportions enable any quadratic equation with real roots to be solved 187: supposed use of propositions of Book V. in arithmetical Books 314, 320

Psellus 234

Ptolemy, Claudius: lemma about quadrilateral in circle (Simson's VI. Prop. D) 225–7: 111, 117, 119

Pyramidal numbers 290: pyramids truncated, twice-truncated etc. 291

Pythagoras: reputed discoverer of construction of five regular solids 97: introduced "the most perfect proportion in four terms and specially called 'harmonic'" into Greece 112: construction of figure equal to one and similar to another rectilineal figure 254

Pythagoreans: construction of dodecahedron in sphere 97: construction of isosceles triangle of IV. 10 and of regular pentagon due to, 97–8: possible method of discovery of latter 97–9: theorem about only three regular polygons filling space round a point 98: distinguished three sorts of *means*, arithmetic, geometric, harmonic 112: had theory of proportion applicable to commensurables only 112: 7/5 as approximation to √2, 119: definitions of unit 279: of even and odd 281: called 10 "perfect" 294

Quadratic equations: solution by means of proportions 187, 263–5, 266–7: διορισμός or condition of possibility of solving equation of Eucl. VI. 28, 259: one solution

only given, for obvious reasons 260, 264, 267: but method gives both roots if real 258: exact correspondence of geometrical to algebraical solution 263–4, 266–7

Quadrilateral: inscribing in circle of quadrilateral equiangular to another 91–2: condition for inscribing circle in, 93, 95: quadrilateral in circle, Ptolemy's lemma on (Simson's VI. Prop. D), 225–7: quadrilateral not a "polygon" 239

"Quindecagon" (fifteen-angled figure): useful for astronomy 111

Radius: no Greek word for, 2

Ramus, P. 121

Ratio: definition of, 116–9, no sufficient ground for regarding it as spurious 117, Barrow's defence of it 117: method of transition from arithmetical to more general sense covering incommensurables 118: means of *expressing* ratio of incommensurables is by approximation to any degree of accuracy 119: def. of greater ratio only *one* criterion (there are others) 130': tests for greater equal and less ratios mutually exclusive 130–1: test for greater ratio easier to apply than that for equal ratio 129–30: arguments about greater and less ratios unsafe unless they go back to original definitions (Simson on V. 10) 156–7: *compound* ratio 132–3, 189–90, 234: operation of compounding ratios 234: "ratio compounded of their sides" (careless expression) 248: *duplicate, triplicate* etc. ratio as distinct from *double, triple* etc. 133: *alternate* ratio, *alternando* 134: *inverse* ratio, *inversely* 134: *composition* of ratio, *componendo*, different from *compounding* ratios 134–5: *separation* of ratio, *separando* (commonly *dividendo*) 135: *conversion* of ratio, *convertendo* 135: ratio *ex aequali* 136, *ex aequali* in *perturbed proportion* 136: *division* of ratios used in *Data* as general method alternative to compounding 249–50: names for particular arithmetical ratios 292

Reciprocal or *reciprocally related* figures: definition spurious 189

Reductio ad absurdum, the only possible method of proving III. 1, 8

"Rule of three": VI. 12 equivalent to, 215

Saccheri, Gerolamo 126, 130: proof of existence of fourth proportional by VI. 1, 2, 12, 170

Savile, H. 190

Scalene, a class of solid numbers 290

Scholia: IV. No. 2 ascribes Book IV. to Pythagoreans 97: V. No. 1 attributes Book V. to Eudoxus 112

Scholiast to *Clouds* of Aristophanes 99

Sectio canonis attributed to Euclid 295

Sector (of circle): explanation of name: two kinds (1) with vertex at centre, (2) with vertex at circumference 5

Printed in the United States
By Bookmasters